TRIBOPHYSICS

TRIBOPHYSICS

NAM P. SUH
Massachusetts Institute of Technology

PRENTICE-HALL, INC., Englewood Cliffs, New Jersey 07632

Library of Congress Cataloging-in-Publication Data

Suh, Nam P., (date)
  Tribophysics.

Includes bibliographies and index.
1. Tribology. 2. Surfaces (Physics)
I. Title.
TJ1075.S84 1986   621.8'9   85-12149
ISBN 0-13-930983-7

*To Dr. & Mrs. D. S. Suh
and Mr. & Mrs. K. H. Surh*

Editorial/production supervision and
  interior design: David Ershun/Nancy Menges
Cover design: Ben Santora
Manufacturing buyer: Rhett Conklin

© 1986 by Prentice-Hall, Inc.
A Division of Simon & Schuster
Englewood Cliffs, New Jersey 07632

All rights reserved. No part of this book may be
reproduced, in any form or by any means,
without permission in writing from the publisher.

Printed in the United States of America

10 9 8 7 6 5 4 3 2 1

ISBN 0-13-930983-7 025

PRENTICE-HALL INTERNATIONAL (UK) LIMITED, *London*
PRENTICE-HALL OF AUSTRALIA PTY. LIMITED, *Sydney*
PRENTICE-HALL CANADA INC., *Toronto*
PRENTICE-HALL HISPANOAMERICANA, S.A., *Mexico*
PRENTICE-HALL OF INDIA PRIVATE LIMITED, *New Delhi*
PRENTICE-HALL OF JAPAN, INC., *Tokyo*
PRENTICE-HALL OF SOUTHEAST ASIA PTE. LTD., *Singapore*
EDITORA PRENTICE-HALL DO BRASIL, LTDA., *Rio de Janeiro*
WHITEHALL BOOKS LIMITED, *Wellington, New Zealand*

CONTENTS

PREFACE xi
ACKNOWLEDGMENTS xiii

1 INTRODUCTION 1

    1.1   Three Aspects of Tribological Problems 1
    1.2   Phenomenological Observations Related to Friction and Early Theories 3
    1.3   Phenomenological Observations Related to Wear and Early Theories 11
    1.4   Comments on Surface Topography 19
    1.5   Introduction to the Book 21
    References 22
    Appendix 1.A: Classical Model for Friction and Wear by Abrasive Particles 23

2 CHEMICAL AND PHYSICAL STATE OF THE SOLID SURFACE 26

    2.1   Introduction 26
    2.2   Brief Introduction to Metals, Polymers, and Ceramics 27

2.3 General Characteristics of a Solid Surface and Tribology 30
2.4 Chemical Interaction of Metal Surfaces with Lubricants 34
2.5 Mechanical Properties of Solid Surfaces 35
2.6 Thermodynamic Analysis of an Interface 41
2.7 Concluding Remarks 45
References 46

Appendix 2.A: Elements of Continuum Mechanics 48

Appendix 2.B: Determination of the Equivalent Strain of Deformed Grains 52

Appendix 2.C: Analysis of Plastic Deformation by the Slip-Line Field Method 57

3 GENERATION AND TRANSMISSION OF FORCE AT THE INTERFACE: THE GENESIS OF FRICTION 63

3.1 Introduction 63
3.2 Typical Friction Tests and Experimental Observations of Frictional Behavior of Metals 65
3.3 Genesis of Friction 73
3.4 Analysis of the Friction—Generating Mechanisms 75
3.5 Other Friction Components 82
3.6 Relative Contributions of μ_d, μ_a, and μ_p to the Overall Friction Force 83
3.7 Concluding Remarks 90
References 90

Appendix 3.A: Slip-Line Field Solution for Deformation of Asperity Contacts at the Sliding Contact 92

Appendix 3.B: Adhesion Component of Friction 95

Appendix 3.C: Plowing Component of Friction 96

Appendix 3.D: Measurement of the Coefficient of Friction and Wear Rates 100

4 RESPONSE OF MATERIALS TO SURFACE TRACTION 103

4.1 Introduction 103
4.2 Response of Elastic Solids to the Cyclic Load Applied at Sliding Contacts 104

4.3 Plastic Deformation of the Surface Layer 128
4.4 Void and Crack Nucleation at the Subsurface 141
4.5 Crack Propagation Due to Surface Traction 153
4.6 Concluding Remarks 171
References 172
Appendix 4.A: Fracture and Fatigue 176
Appendix 4.B: The Boussinesq Solution and Hertzian Stress 186
Appendix 4.C: Calculation of Elastic–Plastic Subsurface Stresses in Sliding Contacts by the Merwin–Johnson Method 192

5 SLIDING WEAR OF METALS 195

5.1 Introduction 195
5.2 Asperity Removal during Sliding 196
5.3 Delamination Theory of Wear 199
5.4 Microstructural Effects in Delamination Wear 209
5.5 Concluding Remarks 221
References 221

6 FRICTION AND WEAR OF POLYMERS AND COMPOSITES 223

6.1 Introduction 223
6.2 Phenomenological Observations on the Friction and Wear Behavior of Highly Linear Semicrystalline Polymers: Polytetrafluoroethylene (PTFE), High-Density Polyethylene (HDPE), and Polyoxymethylene (POM) 227
6.3 Phenomenological Observations on the Tribological Behavior of Other Polymers 231
6.4 Wear of Polymeric Composites 232
6.5 Basic Mechanisms for Friction in Polymers 237
6.6 Model for Wear of Fiber-Reinforced Composites 248
6.7 Wear Mechanisms in Single-Phase Polymers 255
6.8 Concluding Remarks 257
References 257
Appendix 6.A: Time–Temperature Superposition and the Shift Factor $a(T)$ 260

7 FRICTION AND WEAR DUE TO HARD PARTICLES AND HARD ROUGH SURFACES: ABRASION AND EROSION   264

7.1 Introduction   264
7.2 Friction and Wear Due to Abrasive Action   266
7.3 Abrasive Friction and Wear Mechanisms   282
7.4 Erosive Wear by Solid Particle Impingement   305
7.5 Concluding Remarks   342
     References   343
Appendix 7.A: Slip-Line Analysis for Plowing Mechanisms   346
Appendix 7.B: Proof of Theorems of Limit Analysis   351

8 WEAR DUE TO CHEMICAL INSTABILITY   356

8.1 Introduction   356
8.2 Brief Introduction to Metal Cutting   357
8.3 Cutting Tool Materials and Phenomenological Aspects of Tool Wear   364
8.4 Wear Mechanisms of Cutting Tools   366
8.5 Solution Wear   367
8.6 Diffusion-Controlled Wear   376
8.7 Concluding Remarks   379
     References   379
Appendix 8.A: Brief Introduction to the Thermodynamics of Solids   382
Appendix 8.B: Estimation of Thermodynamic Properties   388
Appendix 8.C: Temperature Distribution at the Sliding Interface   395

9 NOVEL METHODS OF IMPROVING TRIBOLOGICAL BEHAVIOR OF SLIDING SURFACES   413

9.1 Introduction   413
9.2 Prevention of Wear of Composites by the Creation of Microvoids to Trap Wear Particles   416
9.3 Control of Electrical Contact Resistance   424

- 9.4 Soft-Metal Coatings on a Hard Substrate to Lower the Wear Rate 443
- 9.5 Hard Coating of Single-Phase Materials on the Surface 454
- 9.6 Ion Implantation 459
- 9.7 Minimization of Wear of Polymers by Plasma Treatment 479
- 9.8 Concluding Remarks 486

  References 487

INDEX 491

ACKNOWLEDGMENTS

There are many who contributed to the making of this book. The author cannot thank them all adequately, but let him try.

First, his thanks go to Ms. Marjorie Wetzel, Ms. Fran Justice, Ms. Theresa Harrison, Mrs. Marge McDonald, and Mrs. Marcia Weir. They typed many different versions of the text with understanding and patience. His verbal appreciation can hardly express his sincere gratitude to them.

Then there are those who not only contributed as scholars but also editorially and otherwise. They are Dr. Hyo-Chol Sin, Dr. Nannaji Saka, and Mr. Kyriakos Komvopoulos. They read the entire manuscript and provided many important comments. The author is greatly indebted to them for their many contributions. Dr. Saka and Dr. Sin have collaborated with the author for many years and have made many significant contributions to the field of tribology.

Many of the author's former students made important contributions through their collaboration with the author in various phases of the research reported in this book. They are Dr. Said Jahanmir, Professor Bruce M. Kramer, Mr. Ming J. Liou, Mrs. Sharon Shepard Rinderle, Mr. J. Fleming, Mr. Steve Burgess, Mr. J. Pamies-Teixeira, and Mr. N. Behbehani. Dr. Said Jahanmir was the first to conduct all the experimental work on the delamination theory of wear. Professor Bruce Kramer, now a colleague at MIT, took a rather qualitative concept and made a fine contribution in the form of *solution wear*.

The author was also fortunate to have had the opportunity to work with Dr. E. P. Abrahamson II, who provided many insights to metallurgical aspects of wear along with Dr. Nannaji Saka. He also derived many benefits from his association with many distinguished colleagues at MIT, Professors Frank McClintock, Nathan Cook, Ernest Rabinowicz, and Ali Argon.

This book would not have been possible without the support received from the Defense Advanced Research Projects Agency, the National Science Foundation, and especially the many years of continuing support from the Office of Naval Research. The author is particularly indebted to Dr. Edward van Reuth, Dr. Richard Miller, Mr. Harold Martin, and Mr. Vernon Westcott. The author is also grateful to the many industrial sponsors of his research at MIT, especially the member firms of the MIT-Industry Polymer Processing Program and the Digital Equipment Corporation. They provided the reason for a scientific inquiry and the bounds to technical solutions.

Finally, the author must acknowledge the prime source of his moral support and love: his wife Young and his children Mary, Helen, Grace and Caroline. They sacrificed much for what is in this book.

1

INTRODUCTION

1.1 THREE ASPECTS OF TRIBOLOGICAL PROBLEMS

Tribology is concerned with the science and the technology of the interface between two or more bodies in relative motion. The nature and the consequence of the interactions that take place at the interface control the friction and wear behavior of the materials involved. During these interactions, forces are transmitted; energy is consumed; physical and chemical natures of the materials are changed; the surface topography is altered; and sometimes loose wear particles are generated. All these seemingly random and complex surface phenomena follow a certain order and satisfy the laws of nature. Understanding the nature of these interactions and solving the technological problems associated with the interfacial phenomena constitute the essence of tribology and also the purpose of this book. To achieve these dual goals, this and the following chapters describe the basic mechanisms that govern interfacial behavior and illustrate how the basic theories can be applied to provide practical solutions to important friction and wear problems.

There are three fundamental aspects to the science of tribology that must be understood to deal with the technology of the field:

1. The effect of environment on surface characteristics through physicochemical interactions
2. The force generation and transmission between the surfaces in contact
3. The behavior of the material near the surface in response to the external forces acting at the contact points of the surface

These three aspects of tribology are clearly interrelated and therefore must be understood to deal with the totality of tribological problems. However, since an understanding of all three aspects of tribology requires a multidisciplinary background, most researchers and practitioners deal with only a limited aspect of tribology. Unfortunately, such an approach may not yield coherent solutions to some tribological problems.

To solve a boundary lubrication problem, one may deal only with the first two aspects of tribology (i.e., physicochemical interactions of the surface with the environment and the asperity interactions). However, the third aspect of tribology (i.e., the macroscopic deformation of the surface in response to the external force) cannot be ignored, since it affects the transmission of forces at the interface through the generation of new surfaces, loose wear particles, and new surface asperities. Conversely, the physical and chemical interactions of lubricants (and gaseous environment) with the sliding surface affect the material response by producing compounds which are actually present at the interface. These compounds affect the nature of force generation and transmission at the asperity contact and thus the frictional force.

The basic mechanisms of boundary lubrication are not yet completely understood, perhaps because the third aspect of the interaction was not fully integrated with the first two aspects. To date, lubricants and their additives have been developed through much trial and error. Much of the research done on boundary lubrication through the chemical analyses of the surface layers has not yielded any useful information on the basic mechanisms by which these additives impart beneficial effects.

To comprehend the frictional behavior of materials, all three aspects of tribology must also be understood, including the details of the chemical and physical interactions at the interface, the role of entrapped wear particles, and deformation of asperities and the surface layers. Contrary to the folklore, each one of these contributions to friction changes as a function of sliding distance (or time), the environment, and even the geometry of the sliding surfaces. Therefore, the coefficient of friction is not an invariant that depends only on material properties. Even for the same pair of materials it can assume different values depending on the specific application and the sliding distance.

Wear under normal sliding situations is primarily a consequence of the response of materials to given surface tractions. To understand sliding wear, it is therefore necessary to deal with the mechanics of deformation of the surface layer, crack nucleation and propagation mechanisms, and the mechanisms of force transmission at the surface. On the other hand, wear under extremely high speeds and/or high loads, which generate high interfacial temperatures, can be due to chemical dissolution of the contacting materials, requiring an understanding of chemical thermodynamics.

In a given tribological situation, wear may be caused by all of these different mechanisms, but in many cases, there is usually only one wear-rate-determining mechanism that dominates the wear process. Therefore, to deal with tribological problems, one must be able to identify the rate-controlling mechanism among various plausible mechanisms. Once the rate-determining wear process is identified in a given situation, it is relatively easy to devise solutions for it.

In dealing with all of these tribology problems, it is important to note that the surface properties, both chemical and physical, are different from the bulk properties. The chemical composition of the surface layer may be different from the bulk; dislocations at the surface may have different mobilities and experience different forces than those in the bulk; electronic configurations at the surface may be different from the bulk configurations; and even the atomic structure at the surface differs from that of the bulk. Therefore, the fundamental understanding of the surface chemistry and physics is a prerequisite in making progress in tribology.

1.2 PHENOMENOLOGICAL OBSERVATIONS RELATED TO FRICTION AND EARLY THEORIES

The fact that friction exists between any two sliding surfaces in relative motion has been known to humankind from prehistoric times. This knowledge has been used to our advantage, albeit intuitively and empirically in most cases. Even today, many of the macroscopic, phenomenological observations made on frictional behavior still form the basis for much of current engineering practice. In recent decades scientific efforts have been made to provide rational explanations to these ancient observations through scientific investigations. Therefore, it is most appropriate to discuss the phenomenological aspect of friction as part of this introductory chapter.

Frictional behavior is affected by the following factors:

1. Kinematics of the surfaces in contact (i.e., the direction and the magnitude of the relative motion between the surfaces in contact)
2. Externally applied load and/or displacements
3. Environmental conditions such as temperature and lubricants
4. Surface topography
5. Materials properties

This list of important factors that control friction clearly indicates that the coefficient of friction (i.e., the ratio of tangential force to the normal load) is not a simple material property.

Most metals behave in such a manner that the coefficient of friction under normal sliding conditions is, to a first approximation, independent of normal load and sliding speed. However, as the normal load is increased to a very high value, such as found in metal cutting, the coefficient of friction often decreases with the increasing normal load and the sliding speed. In the intermediate normal load and speed ranges, the frictional coefficient may reach a peak value, as shown in Fig. 1.1.

Polymeric materials behave differently from metals in that their coefficient of friction is a function of the normal load and sliding velocity. As the load is increased, most polymers, be they a thermoplastic or a thermoset, exhibit lower friction coefficients, even at a low sliding speed, as shown in Fig. 1.2. As the sliding speed is increased, the coefficient of friction first increases and then at a very high speed, the coefficient of friction decreases with an increase in sliding speed, as shown in Fig. 1.3.

In the past these phenomenological observations of the frictional behavior of all materials were explained in terms of the adhesion model. In essence, this theory assumes that the surface consists of asperities and the interface is made up of asperity contacts, as shown in Fig. 1.4. The real area of contact is much smaller than the apparent area of contact except in such applications as in metal cutting, where the real area of contact approaches the apparent area of contact.

According to the adhesion theory of friction, when a relative motion is imparted to the interface by applying a tangential force, each pair of contacting asperities weld together and shear to accommodate the relative motion. Then, to a very rough first approximation, the frictional force is given by

$$F = A_r \tau \tag{1.1}$$

where τ is the shear strength of each junction, which must be overcome to satisfy kinematics of the sliding motion, and A_r is the real area of contact. The real area of contact was related to the hardness of the material H, based on the assumption that the real area of contact must be large enough

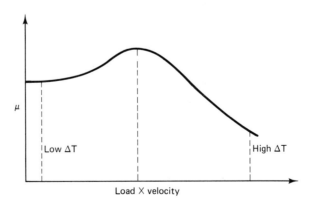

Figure 1.1 Qualitative representation of the coefficient of friction of metals versus load times speed. T is the interfacial temperature rise.

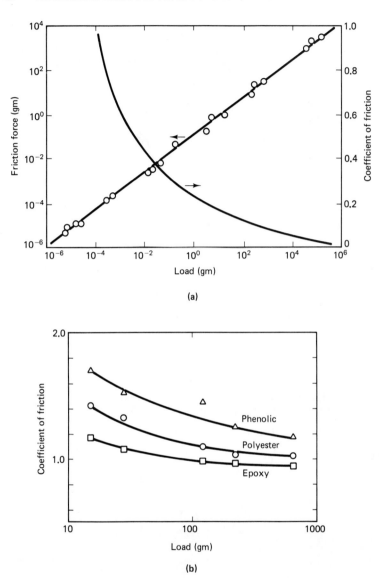

Figure 1.2 Coefficient of friction of thermoplastics and thermosetting plastics. [(a), From Allan, 1958; Reprinted by permission of the American Society of Lubrication Engineers. (b), from Pinchibeck, P. H., "A Review of Plastic Bearings," *Wear*, Vol. 5, 1962, pp. 85–113.

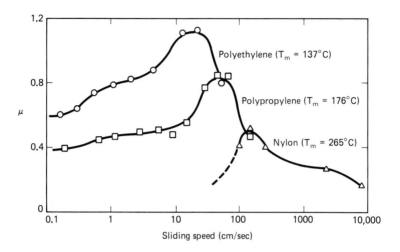

Figure 1.3 Coefficient of friction of thermoplastics as a function of sliding speed. (From McLaren and Tabor, 1963.)

to support the given normal load, L. That is,

$$A_r = \frac{L}{H} \tag{1.2}$$

Ever since the introduction of the adhesion theory of friction, a great deal of discussion has taken place as to how τ is related to H. Many believers of the adhesion theory assumed that τ must be greater than the critical shear strength of the bulk material, due to the work hardening of the surface layer during sliding. This type of argument was necessitated by the low predicted values of the friction coefficient compared to experimental results. If we assume that τ is equal to the critical shear strength of the bulk k, then

$$F = A_r \tau = \frac{L}{H}\tau = \frac{L}{6k}k = \frac{L}{6}$$
$$\mu = \frac{F}{L} = \frac{1}{6} \tag{1.3}$$

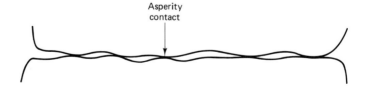

Figure 1.4 Asperity contact.

Sec. 1.2 Observations Related to Friction and Early Theories

This predicted value is much smaller than the typical values observed under steady-state sliding conditions, which range from 0.3 to 0.7. To improve the correlation between the experimental results and the adhesion theory, a number of theories have been advanced.

One of these theories is due to Rabinowicz (1965), who postulated that the actual area of contact is much larger than that given by Eq. (1.2) because of the surface energy of adhesion. His reasoning is as follows. If the overall surface energy change is denoted by W_{ab} (i.e., the surface energy of adhesion), then

$$W_{ab} = \gamma_a + \gamma_b - \gamma_{ab} \quad (1.4)$$

where γ_a and γ_b are the surface energies of the two contacting surfaces and γ_{ab} is the interface energy. The sum W_{ab} is always positive; that is, the overall energy is decreased by bonding. Idealizing the indentation of asperities as an indentation by a conical indenter of material b penetrating into a half-space of material a, as shown in Fig. 1.5, the work done by the normal load L during an infinitesimal indentation dx may be equated to the difference in the work done in deforming the material plastically and the surface energy change; that is,

$$L\,dx = \pi r^2 H\,dx - (2\pi r)W_{ab}\frac{dx}{\sin\theta} \quad (1.5)$$

Equation (1.5) may be rewritten for the area in contact as

$$\pi r^2 = \frac{L}{H} + \frac{2\pi r}{\sin\theta}\frac{W_{ab}}{H} \quad (1.6)$$

Equation (1.6) states that when the interfacial energy change is included in considerations, the projected area (πr^2) is larger than that given by Eq. (1.2) by an amount $(2\pi r/\sin\theta)(W_{ab}/H)$. Substituting Eq. (1.6) into Eq.

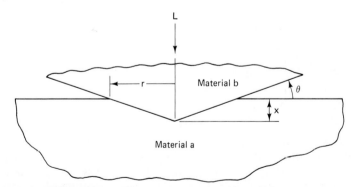

Figure 1.5 Indentation of conical indenter to represent the indentation of asperities. (From Rabinowicz, 1965.) Reprinted with permission from Rabinowicz, E., *Friction and Wear of Materials*, Wiley, New York, 1965. Copyright 1965 by John Wiley & Sons, Inc.

(1.3), the coefficient of friction may be expressed as

$$\mu = \frac{F}{L} = \frac{k}{H} \frac{1}{1 - 2W_{ab}/rH \sin \theta} \simeq \frac{k}{H}\left(1 + K\frac{W_{ab}}{H}\right) \quad (1.7)$$

where K is a geometric factor. According to Eq. (1.7), the coefficient of friction is high when the ratio of the surface energy of adhesion W_{ab} to hardness H is large and when the roughness angle θ is small. Figure 1.6 is a compatibility chart for various metal combinations of preferred antifriction surface presented in support of this theory.

In spite of the correlation, some basic considerations raise issues on the importance of surface energy in determining friction. The most obvious difficulty is the relatively small magnitude of the surface energy change

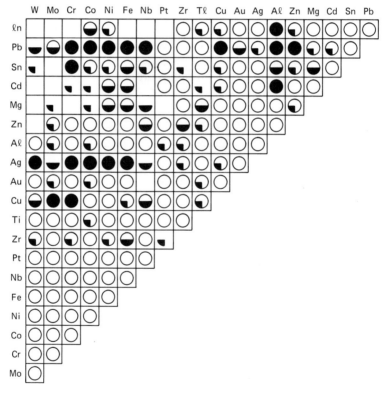

Figure 1.6 Rabinowicz's compatibility chart for various metal combinations derived from binary diagrams of the respective elements in terms of preferred antifriction surfaces. ●, Two liquid phases, solid solution less than 0.1% solubility (lowest adhesion); ◐, two liquid phases, solid solution greater than 0.1%, or one liquid phase, solid solution less than 0.1% solubility (next lowest adhesion); ◑, one liquid phase, solid solution between 0.1 and 1% solubility (higher adhesion); ○, one liquid phase, solid solution over 1% (higher adhesion). Blank boxes indicate insufficient information. (From Rabinowicz, 1971.)

compared to the total work done. An order-of-magnitude analysis of Eq. (1.6) shows that the first term on the right-hand side of Eq. (1.6) is much larger than the second term on the right-hand side. Another difficulty is that most surfaces are so highly contaminated by absorbants and impurity atoms in the metal (Buckley, 1980) that the validity of experimentally measured W_{ab} for various surfaces is in doubt. It will be shown in a later section that the correlation shown in Fig. 1.6 can also be explained in terms of the mechanical properties of the surface.

Another adhesion model that stresses the real area of contact has been advanced by Green (1955a, b). Green analyzed the deformation of the surface asperity contact using the slip-line field for a rigid–perfectly plastic material. The plasticity analysis of the junction shown in Fig. 1.7 showed that the coefficient of friction can be larger than 0.17, as shown in Fig. 1.8. For a typical surface, the angles δ and θ are less than 10°, yielding a coefficient of 0.6. Based on the analysis and experiments with plasticine, Green concluded that some of the strongly adhering junctions may also support tensile stress during the deformation process. Since the total compressive normal load acting on a surface must be supported by the asperity junctions, the equilibrium of forces requires that the real area of contact under the steady-state sliding situation be greater when some of the junctions are under tension than when all the junctions are under compression. Therefore, he reasoned that the coefficient of friction can be extremely large when a large number of junctions are under tension. This line of thought cannot, however, explain the existence of friction, even when adhesion is totally absent.

One of the early attempts to explain friction was to relate it to surface roughness, because surface is not generally smooth, consisting of asperities (i.e., short-range perturbations from the mean) and waviness (i.e., long-range perturbations from the mean). The roughness theory assumed that the frictional force is equal to the force required to climb up the asperity

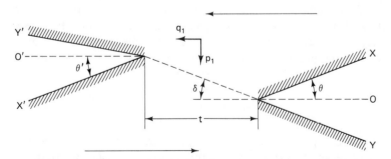

Figure 1.7 Strong junction during steady sliding. Reprinted with permission from *Journal of the Mechanics and Physics of Solids*, Vol. 2, Green, A. P., "The Plastic Yielding of Metal Junctions due to Combined Shear and Pressure," Copyright 1955, Pergamon Press, Ltd.

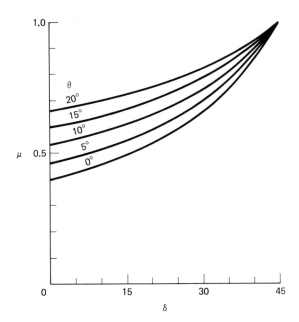

Figure 1.8 Theoretical results for strong junctions for μ versus slope of asperities. Reprinted with permission from *Journal of the Mechanics and Physics of Solids*, Vol. 2, Green, A. P., "The Plastic Yielding of Metal Junctions due to Combined Shear and Pressure," Copyright 1955, Pergamon Press, Ltd.

of slope θ. Then the coefficient of friction is given by

$$\mu = \tan \theta \tag{1.8}$$

It is, however, clear that asperities undergo deformation due to the sliding action rather than simply sliding over each other. Moreover, asperities cannot continue to climb up the asperities of the counterface throughout the sliding action.

Another school of thought attributes the frictional force to a combined effect of adhesion, plowing, and roughness. Shaw and Macks (1949) expressed the frictional force to be the sum of the adhesion component given by Eq. (1.1), the roughness contribution given by Eq. (1.8), and the plowing component, P. Then μ can be expressed as

$$\mu = \frac{\tau}{H} + \tan \theta + P \tag{1.9}$$

Kragelskii (1980) also believes that plowing contributes to the frictional force as well as adhesion. He expressed the coefficient of friction as

$$\mu = \frac{\tau_0}{p_r} + \beta + K\alpha_h \sqrt{\frac{h}{r}} \tag{1.10}$$

where τ_0 is the shear strength of the surface due to molecular bonds when the normal pressure is equal to zero; p_r is the contact pressure, which for a surface that has been run in, depends on elastic modulus and the surface roughness parameter; α_h the coefficient of hysteresis losses in friction; β the molecular bond strengthening coefficient; h the height of an asperity; r the asperity radius; and K the friction parameter. In many sliding situations the ideas embodied in developing Eqs. (1.9) and (1.10) may be closer to the real picture than theories that include only adhesional forces. However, as will be shown later, the frictional force is not a constant as is implied by these equations; plowing by wear particles is more important than that due to hard asperities in most cases; Eqs. (1.9) and (1.10) may not satisfy the equilibrium condition for a vertical component of forces; and contrary to Eq. (1.10), even after run-in, there can be plastic deformation due to adhesion and plowing. Another theory of friction, which overcomes these difficulties, is presented in Chapter 3.

The friction behavior is greatly affected by the interfacial temperature and the environment. The interfacial temperature and environment govern the degree of adhesion, the nature of chemical reaction at the interface, and the hardness of the surfaces. In normal sliding situations at low speeds, the interfacial temperature rise is small relative to its melting points. However, in high-speed cutting of metals the interfacial temperature rise can be quite significant. As a consequence, the frictional behavior can be very different in two situations. At low temperatures, the mechanical behavior of solids dominates the frictional behavior; whereas at high temperatures (i.e., when the homologous temperature, that is, the ratio of the actual temperature to the melting point, is greater than $\frac{1}{2}$), chemical interactions and mutual solubility may govern the frictional behavior as well as the mechanical behavior of solids.

1.3 PHENOMENOLOGICAL OBSERVATIONS RELATED TO WEAR AND EARLY THEORIES

Wear of materials occurs by many different mechanisms, depending on the materials, the environmental and the operating conditions, and the geometry of the wearing bodies. These wear mechanisms can be classified into two groups, as shown in Table 1.1: those dominated primarily by the mechanical behavior of solids and those dominated primarily by the chemical behavior of materials. In many wear situations there are many mechanisms operating simultaneously, but there is usually only one primary rate-determining mechanism which must be identified to deal with the wear problem. What determines the dominant wear behavior are mechanical properties, chemical stability of materials, temperature, and operating conditions. More commonly known phenomenological aspects of the wear behavior of metals and polymers under sliding conditions are described in this section.

When two materials slide against each other, the wear volume V is to a very rough approximation, linearly proportional to the distance slid S

TABLE 1.1 Classification of Wear

(a) Wear processes that are dominated primarily by the mechanical behavior of materials under a given loading condition

Type	Typical Characteristics and Definitions	Observed In:
Sliding wear (delamination wear)	Plastic deformation, crack nucleation, and propagation in the subsurface.	Sliders, bearings, gears, and cams, where surfaces undergo relative motion.
Fretting wear	The early stages of fretting wear are the same as sliding wear but depend on relative amplitude. The entrapped wear particles can have significant effect on wear. The relative displacement amplitude is important.	Press fit parts with a small relative sliding motion.
Abrasive wear	Hard particles or hard surface asperities plowing and cutting the surface in relative motion.	Sliding surfaces, earth-removing equipment.
Erosive wear (solid particle impingement)	Due to solid particle impingement, large subsurface deformation, crack nucleation, and propagation. Sometimes, the surface is cut by solid particles when the impingement angle is shallow.	Turbines, pipes for coal slurries, and helicopter blades.
Fatigue wear	Fatigue crack propagation takes place, normally perpendicular to the surface, without gross plastic deformation under cyclic loading conditions.	Ball bearings, roller bearings, and glassy solid sliders.

(b) Wear processes that are controlled primarily by chemical processes and thermally activated processes

Type	Typical Characteristics and Definitions	Observed In:
Solution wear	Formation of new compounds of a lower free energy of formations; high temperature; no gross plastic deformation; atomic-level wear process.	Carbide tools in cutting steel at high speeds.
Diffusive wear	Diffusion of elements across the interface.	High-speed-steel (HSS) tool in cutting steel at high speeds.
Oxidative wear	Formation of weak, mechanically incompatible oxide layer.	Sliding surfaces in highly oxidative environment (not common).
Corrosive wear	Corrosive of grain boundaries and formation of pits.	Lubricated and corrosive atmosphere.

and normal load L, but inversely proportional to the hardness H of the material. This may be expressed as

$$V \propto \frac{LS}{H} \qquad (1.11)$$

Equation (1.11) is normally written as

$$V = K\frac{LS}{3H} \qquad (1.12)$$

where K is a dimensionless proportionality constant commonly known as the wear coefficient. The factor 3 is a result of Archard's model for adhesion theory of wear (Archard, 1953). Although Eq. (1.12) implies that harder materials wear less, there are many exceptions to the foregoing statement. For example, a soft, commercially pure copper can be much more resistant to wear than AISI 1045 steel, which is much harder. In normal sliding situations the wear coefficient for most metals is in the range 10^{-4} to 10^{-3}.

In abrasive wear, the surface of a softer metal is plowed by hard abrasive particles or wear particles or hard asperities of the counterface. Abrasive wear follows the relationship given by Eq. (1.12) reasonably well; that is, the harder the material, the less the wear rate. The wear coefficient of typical abrasive wear is on the order of 10^{-2} to 10^{-1}. The wear particles generated by the abrasive mechanism resemble metal chips generated by cutting action.

In fretting wear the interface undergoes a small oscillatory motion, which results in wear of materials. The wear coefficient in this case depends on the amplitude of oscillation, as shown in Fig. 1.9, when the relative displacement at the interface is less than a critical value. At large amplitudes of oscillation the fretting wear coefficient approaches that of unidirectional sliding wear.

At high sliding speeds and loads, such as in metal cutting, the wear rate depends sensitively on the chemical nature of the material. This is shown in Fig. 1.10 for various carbides cutting steel. In this case, the wear rate correlates with temperature but does not increase linearly with normal load and the sliding distance. The hardness of the tool material has little bearing on the wear life of these tools, since they are much harder than the workpiece and do not deform under typical cutting conditions. On the other hand, an alumina tool will not wear, due to chemical instability but rather by mechanical deformation of the surface layer (Suh, 1980; Kramer and Suh, 1980).

All these wear phenomena, except abrasive wear, were explained in terms of the adhesion theory of wear until the advent of the delamination theory of wear (Suh, 1973a) for sliding wear and the solution wear theory (Suh, 1973b; Kramer and Suh, 1980). The adhesion theory of wear is very similar to the adhesion theory for friction. It states that the wear of materials

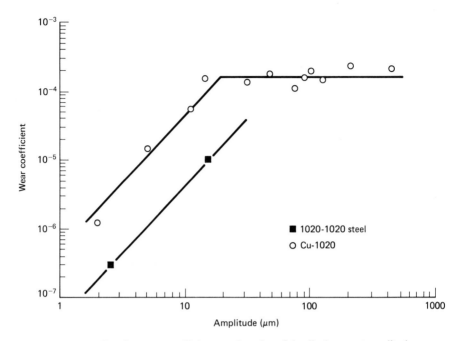

Figure 1.9 Fretting wear coefficient as a function of the displacement amplitude. (From Stowers, 1974.)

is due to the welding of asperity junctions, which create a hemispheric wear particle when the weaker material near the welded junction fractures. Archard (1953) developed a mathematical model to describe this process, which yielded Eq. (1.12). The difficulty with this model was that there were too many exceptions and that this theory could not form the basis for improvement and development of wear-resistant materials. It also violates the conservation law for energy because the actual work done is several orders of magnitude larger than the fracture energy.

The fact that this adhesion is not a complete description of the sliding wear behavior may be appreciated by examining the physical significance of the wear coefficient. The wear coefficient is a dimensionless quantity defined by Eq. (1.12), which may be rewritten as

$$K = \frac{3VH}{LS} \tag{1.13}$$

Since L/H is the real area of contact and since the cross-sectional area of the plastically deformed subsurface zone under the asperity contact, A_p, is of the order of A_r, Shaw (1977) showed that Eq. (1.13) may be rewritten as

$$K = \frac{3VH}{LS} = \frac{V}{A_p S} = \frac{\text{worn volume}}{\text{volume of the plastically deformed zone}} \tag{1.14}$$

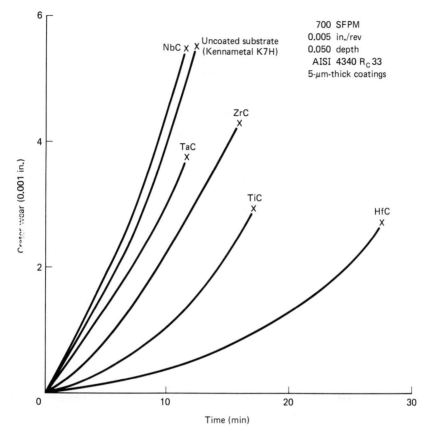

Figure 1.10 Wear rate of various carbides cutting AISI 4340 steel of $R_c 33$ at 700 surface feet per minute.

Therefore, the wear coefficient K for *sliding wear* may be interpreted as a dimensionless quantity that represents the ratio of the worn volume to the volume of the plastically deformed zone. Since K is on the order of 10^{-4} to 10^{-3}, the volume of the material removed by wear is a very small fraction of the material undergoing plastic deformation below the asperity contact. Therefore, it is clear that sliding wear cannot be understood properly without comprehending the plastic deformation process at the subsurface. This process is the primary mode of energy dissipation during sliding wear.

The abrasive wear was modeled in the past as a cutting process. This assumes that an abrasive particle leaves a wear track of the same cross-sectional shape. For example, if the abrasive grain can be idealized as a cone, as shown in Fig. 1.11, the wear volume, removed after the abrasive traverses a distance S, is the area shown by the shaded area. However, this type of oversimplified model misrepresents the true picture. To illustrate this point, we again examine the physical significance of wear coefficient,

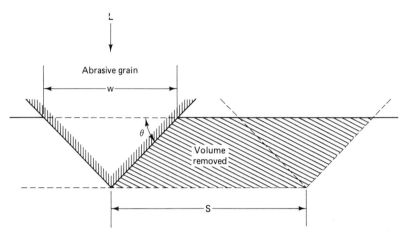

Figure 1.11 Abrasive wear model. (Reprinted with permission from Rabinowicz, E., *Friction and Wear of Materials*, Wiley, New York, 1965. Copyright 1965 by John Wiley & Sons, Inc.)

but this time for abrasive wear, and compare its predictions with experimental results. It can be shown that when wear particles in the form of chips are generated without plastic deformation, the specific energy u (i.e., the work done to remove a unit volume of material by a cutting mechanism) is equal to the hardness of the material for the idealized conical model shown in Fig. 1.11 (see Appendix 1.A for the derivation). Then Eq. (1.12) may be rewritten as

$$K = \frac{3\mu VH}{\mu LS} \simeq 3\mu \frac{Vu}{FS} = \frac{Vu}{FS} = \frac{\text{work done to create abrasive wear particles by cutting}}{\text{external work done}} \quad (1.15)$$

assuming that $3\mu \simeq 1$. It is then seen that the dimensionless quantity K is simply the ratio of the work done to generate wear particles, in the form of cut chips, to the total external work done. Therefore, when the entire work done is consumed to cut the surface, as the classical theories assumed, the wear coefficient should be equal to or greater than unity. However, the experimentally determined maximum coefficients are one or two orders of magnitude less than unity.

The fact that the abrasive wear theory based on the cutting mechanism predicts too large a wear coefficient can also be seen from the results of the cutting tests. Rabinowicz's theory, which is one of the more commonly cited works, states that the volume of material cut is equal to the volume displaced by an abrasive grain (Rabinowicz, 1965). Then it may be shown that the wear coefficient is related to geometry as (see Appendix 1.A for the derivation)

$$K = \frac{3 \tan \theta}{\pi} \quad (1.16)$$

The same model for friction by plowing yields

$$\mu = \frac{\tan \theta}{\pi} \tag{1.17}$$

Comparing Eqs. (1.16) and (1.17), the friction coefficient is related to the wear coefficient as

$$K = 3\mu$$

When a diamond stylus with a cutting angle of $\theta = 35°$ was used to cut AISI 1095 steel (Sin et al., 1979), the experimentally determined values of K and μ were 0.23 and 0.6, respectively. The theoretically predicted value for K was 0.67.

The discrepancy between the simple cutting model and the actual abrasive wear process is caused primarily by the deformation of the subsurface layer. A large plastic deformation occurs at and below the surface during abrasion. Plowing also forms ridges, which subsequently have to be removed. For this reason the wear rate depends on the ductility of the material; less ductile materials have generally higher wear coefficients than those of more ductile materials. Figure 1.12 shows the difference in wear rates between

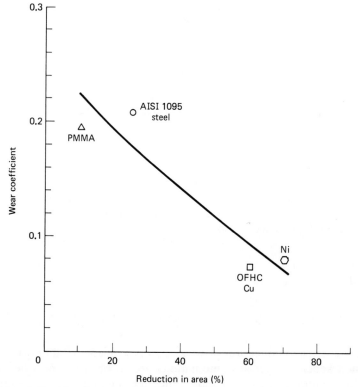

Figure 1.12 Abrasive wear rate versus ductility. From Sin, H.-C.; Saka, N.; and Suh, N. P., "Abrasive Wear Mechanisms and the Grit Size Effect," *Wear*, Vol. 55, 1979, pp. 163–190.

ductile materials, such as commercially pure nickel and copper, and more brittle materials, such as AISI 1095 steel and PMMA (polymethyl methacrylate).

The values of wear coefficients can give a misleading notion of the wear resistance of materials because of its overdependence on hardness. As shown in Table 1.2, the actual wear rate does not depend on the prior cold work, whereas the computed wear coefficient is very different between the annealed and cold-worked nickel. The large difference in numerical values of the wear coefficient, although the wear rate was the same, is due to the different values of the hardness used in computation. The cold-worked metal has the same wear rate as the annealed metal because the amount of cold work done during the wear process is much larger than the cold work done during metal processing.

TABLE 1.2   Effect of Cold Work on Abrasion[a]

Condition	Hardness (kg/mm^2)	Wear Rate (m^3/m)	Wear Coefficient	Hardness after Test (kg/mm^3)
Annealed	88.5	8.07 × 10	0.053	240
Fully cold worked	242.0	8.66 × 10	0.157	242

[a] Material, nickel; applied load, 4 kg; abrasive size, 60 grit.

A very useful piece of information to keep in mind in considering wear problems is that by knowing the wear coefficient, we can begin to speculate as to the cause of wear. With the caveat that there are dangers in using numerical values indiscriminately, the following "ballpark" figures are given for typical wear processes:

> During sliding wear (by delamination or sometimes referred to as adhesive wear): 10^{-4} to 10^{-3}
> Abrasive wear: 10^{-2} to 1
> Fretting wear (at large displacement amplitudes, $100 \mu m$): 10^{-4} to 10^{-2}

A rather extensive collection of wear data is available (Rabinowicz, 1980), but when special material combinations are to be used, or when unique operating conditions exist, the most prudent thing to do is to perform actual tests in addition to using the wear coefficients available in handbooks.

1.4 COMMENTS ON SURFACE TOPOGRAPHY

In specifying the surface finish of machine components, several factors must be considered based on the understanding of the causality relationship between the surface topography and the specific functional requirements of the parts. The first step in this process is the characterization of the surface finish, followed by the characterization of its effect on such functional requirements as wear resistance, frictional coefficient, fatigue life, and manufacturing cost.

A great deal of work has been done to characterize the surface finish mathematically (see, e.g., Whitehouse, 1980). However, the relationships between the surface topography and the functional requirements for friction and wear are not yet fully understood. Based on the available information, it appears that the original surface finish does not affect the *steady-state* friction and wear behavior significantly in low-speed dry sliding applications, whereas it is of importance when lubricants are used. In the former case, the surface geometry is drastically altered by sliding actions, whereas in the latter case, not only does the initial surface geometry change slowly, but also the pressure and temperature in the lubricant and the metal surface depend sensitively on the surface topography.

The initial surface finish does affect the initial wear rate of materials under dry sliding conditions. When the harder surface slides over the softer surface, the softer asperities either fracture immediately or deform. The rate at which these asperities are removed by the sliding process and the mechanism of removal depend on the initial surface roughness, the applied load, and the mechanical properties of the asperities. For example, the initial wear rate of AISI 1018 steel is higher for rougher surfaces than for smoother surfaces when the applied normal load is high, but the opposite is true under a lighter normal load, as shown in Fig. 1.13. The wear particles generated by the fracture of deforming asperities will be smaller than delaminated wear particles produced by the subsurface crack propagation process (see Chapter 5 for a further discussion). Figure 1.14 shows asperities of AISI 1018 steel (generated by a machining process) which are deformed, and some of which are fractured, under the sliding action of an AISI 52100 steel slider. High asperities are deformed first and these covered the lower ones. Upon further deformation the lower asperities also deform, eventually forming a layered structure of deformed asperities. Some of the asperities must fracture during the deformation process without undergoing much deformation, whereas the rest of the asperities eventually fracture after significant deformation upon repeated loading. Once the steady-state sliding condition is reached, new asperities will be generated by the wear process, and therefore the initial surface finish has little effect. However, under certain *special* conditions the initial surface roughness can have lasting

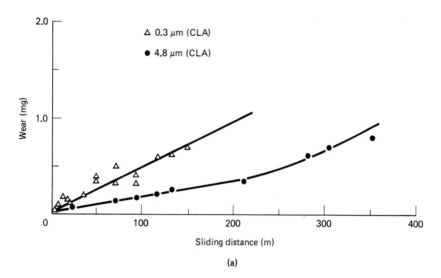

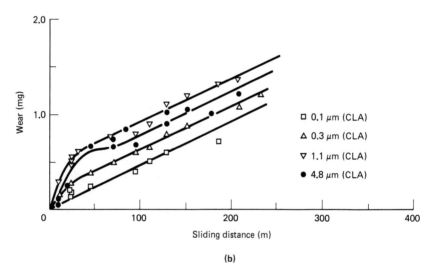

Figure 1.13 Weight loss of AISI 1018 steel as a function of sliding distance and the normal load under dry sliding in argon atmosphere: (a) normal load = 75 g; (b) normal load = 300 g. (From Abrahamson et al., 1975.)

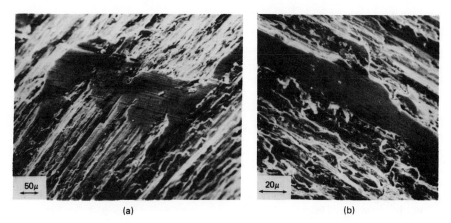

Figure 1.14 Plastic deformation of the original asperities on machined AISI 1018 steel during cylinder-on-cylinder wear tests: (a) sliding perpendicular to the machining marks, $2.0\mu m$ (CLA) surface finish, under a normal load of 0.91 kg after 0.2 m of sliding; (b) sliding parallel to the machining marks, 3.3 μm (CLA) surface finish, under a normal load of 0.35 kg after 0.25 m of sliding. From Suh, N. P., "An Overview of the Delamination Theory of Wear," *Wear*, Vol. 44, 1977, pp. 1–16.

effects on friction and wear if the initial wear particles generated by the fracture of the original asperities affect the number of steady-state wear particles entrapped between the sliding surfaces. (See Chapters 3 and 9.)

1.5 INTRODUCTION TO THE BOOK

There are two kinds of surface interactions: chemical and physical. Therefore, we first examine the chemical and physical state of solid surface, followed by discussions of the chemical and the physical behavior of solid surfaces in Chapter 2. Then, in Chapter 3, the generation and transmission of forces at the interface are examined in order to understand the genesis of friction. Chapter 4 deals with the response of materials when external forces are applied to the metallic and polymeric surface by the asperities and wear particles. Based on the basic analytical results of Chapter 4, various wear processes, including delamination wear, are covered in Chapter 5. Chapter 6 examines the wear of polymers and composites such as thermoplastic bearing materials and graphite-fiber-reinforced polyurethane. Abrasive and erosive wear are covered in greater detail in Chapter 7. Chapter 8 deals with the wear due to chemical instability of materials, which is in some ways in the realm of chemical physics. Chapter 9 illustrates how the basic theories given in this book can be applied in controlling the friction and wear phenomena using various coating techniques, surface topography, surface modification techniques, and lubrication.

Tribology is an interdisciplinary subject, requiring an understanding of many different subjects, such as materials science, mechanics, thermodynamics, and chemistry. Assuming that many readers might have forgotten some of these subjects, appendices are added at the end of many chapters to furnish the background information.

REFERENCES

ABRAHAMSON, E. P., II, JAHANMIR, S., and SUH, N. P., "The Effect of Surface Finish on the Wear of Sliding Surfaces," *CIRP Annals,* International Institution for Production Engineering Research, Vol. 24, 1975, pp. 513–514.

ALLAN, A. J. G., "Plastics as Solid Lubricants and Bearings," *Lubrication Engineering,* Vol. 14, 1958, pp. 211–215.

ARCHARD, J. F., "Contact and Rubbing of Flat Surfaces," *Journal of Applied Physics,* Vol. 24, 1953, pp. 981–988.

BUCKLEY, D. H., "Definition and Effect of Chemical Properties of Surfaces in Friction, Wear, and Lubrication," in *Fundamentals of Tribology,* N. P. Suh and N. Saka, eds., MIT Press, Cambridge, Mass., 1980.

COOK, N. H., *Manufacturing Analysis,* Addison-Wesley, Reading, Mass., 1966.

GREEN, A. P., "Friction between Unlubricated Metals: A Theoretical Analysis of the Junction Model," *Proceedings of the Royal Society of London,* Series A, Vol. 228, 1955a, pp. 191–204.

GREEN, A. P., "The Plastic Yielding of Metal Junctions due to Combined Shear and Pressure," *Journal of the Mechanics and Physics of Solids,* Vol. 2, 1955b, pp. 197–211.

KRAGELSKII, I. V., "Friction Interaction of Solids," *Soviet Journal of Friction and Wear,* Vol. 1, 1980, pp. 7–20.

KRAMER, B. M., and SUH, N. P., "Tool Wear by Solution: Quantitative Understanding," *Journal of Engineering for Industry,* Transactions of the ASME, Vol. 102, 1980, pp. 303–309.

MCLAREN, K. G., and TABOR, D., "Friction of Polymers at Engineering Speeds: Influence of Speed, Temperature, and Lubrication," Paper No. 18, Lubrication and Wear Convention, Institution of Mechanical Engineers, Bournemouth, May 1963.

PINCHIBECK, P. H., "A Review of Plastic Bearings," *Wear,* Vol. 5, 1962, pp. 85–113.

RABINOWICZ, E., *Friction and Wear of Materials,* Wiley, New York, 1965.

RABINOWICZ, E., "Determination of Compatibility of Metals through Static Friction Test," *ASLE Transactions,* American Society of Lubrication Engineers, Vol. 14, 1971, pp. 198–205.

RABINOWICZ, E., "Wear Coefficients—Metals," *Wear Control Handbook,* American Society of Mechanical Engineers, New York, 1980.

SHAW, M. C., "Dimensional Analysis for Wear Systems," *Wear,* Vol. 43, 1977, pp. 263–266.

SHAW, M. C., and MACKS, E. F., *Analysis and Lubrication of Bearings,* McGraw-Hill, New York, 1949.

SIN, H.-C., SAKA, N., and SUH, N. P., "Abrasive Wear Mechanisms and the Grit Size Effect," *Wear,* Vol. 55, 1979, pp. 163–190.

STOWERS, I. F., "The Mechanisms of Fretting Wear," Ph.D. thesis, MIT, 1974.

SUH, N. P., "The Delamination Theory of Wear," *Wear,* Vol. 25, 1973a, pp. 111–124.

SUH, N. P., "Surface Treatment and Coating Techniques for Cemented Carbide Tools," *Proceedings of the North American Metalworking Conference,* Society of Manufacturing Engineers, Dearborn, Mich., 1973b, pp. 35–47.

SUH, N. P., "An Overview of the Delamination Theory of Wear," *Wear,* Vol. 44, 1977, pp. 1–16.

SUH, N. P., "New Theories of Wear Mechanisms and Their Implications for Tool Materials," *Wear,* Vol. 62, 1980, pp. 1–20.

WHITEHOUSE, D. J., "The Effects of Surface Topography on Wear," in *Fundamentals of Tribology,* N. P. Suh and N. Saka, eds., MIT Press, Cambridge, Mass., 1980.

APPENDIX 1.A

CLASSICAL MODEL FOR FRICTION AND WEAR BY ABRASIVE PARTICLES

The simplest idealized model for friction and wear due to abrasive particles as given by Rabinowicz (1965)* is presented in this appendix. A more detailed analysis of the abrasive wear process, together with a more exact model for abrasive wear, is given in Chapter 7.

An idealized model for abrasive wear is illustrated schematically in Fig. 1.11, which shows a conical abrasive particle removing a volume, V, of material by the cutting process. The normal load acting on the abrasive particle is L and the tangential force is F. As a very rough approximation, the normal load, L, is assumed to be equal to the indentation load in the absence of the tangential force (which obviously is an oversimplification since it violates the yield condition). Then the normal load, L, may be related to the hardness of the material, H, as

$$L = \frac{\pi w^2}{4} H \tag{1.A1}$$

where w is the projected diameter of the indentation.

Equation (1.A1) grossly overestimates L since there is no material behind the asperity and since the normal stress required for yielding is less

---

* All appendix references will be found in the chapter reference list immediately preceding the appendix.

than H due to the tangential load F. The tangential force, F, which is required for plowing, may similarly be expressed as

$$F = \frac{w^2 H}{4} \tan \theta \tag{1.A2}$$

Equation (1.A2) overestimates F since the tangential stress required for plowing is considerably less than H due to the lack of geometric constraints to the motion of the plastically displaced material. Using these approximate expressions for F and L, the friction coefficient, μ, due to the abrasive action may be expressed as

$$\mu_p = \frac{F}{L} = \frac{\tan \theta}{\pi} \tag{1.A3}$$

Equation (1.A3) states that sharp abrasives yield higher plowing friction, which is consistent with observed experimental results, and that μ_p is of the order of 1.

The wear coefficient, K_p, due to the abrasive process is defined as

$$K_p = \frac{3VH}{LS} \tag{1.A4}$$

Since V/S is equal to $(w^2 \tan \theta/4)$, K_p may be expressed as

$$K_p = \frac{3 \tan \theta}{\pi} = 3\mu_p \tag{1.A5}$$

The wear coefficient given by Eq. (1.A5) underestimates K_p since the estimated value L is larger than the actual value and since the actual wear volume is less than the volume displaced by the abrasive particle.

The specific energy expended, u, in removing a unit volume of material may be derived, using Eq. (1.A1), as

$$u = \frac{\text{external work done}}{\text{volume removed}} = \frac{\mu_p LS}{w^2 \tan \theta (S/4)} = H \tag{1.A6}$$

Equation (1.A6) states that if the simple cutting model for abrasive wear is correct, the specific energy expended in removing the material during the abrasive process is equal to the hardness of material. Comparison with experiments shows that u is normally much less than H. Even under metal-cutting conditions with a sharp tool u is about $\frac{1}{2}H$ (see Cook, 1966).

The abrasive (or plowing) wear coefficient, K_p, given by Eq. (1.A4), is a dimensionless quantity that relates two different physical processes. Therefore, there must be physical significance associated with K_p. One such physical meaning can be derived by rewriting Eq. (1.A4) as

$$K_p = 3\mu_p \frac{Vu}{\mu_p LS} = 3\mu_p \frac{\text{energy consumed in removing the material}}{\text{external work done}} \tag{1.A7}$$

In deriving Eq. (1.A7), Eq. (1.A6) was substituted. Since $3\mu_p = 0(1)$, K_p may be interpreted as being the ratio of the actual energy consumed in removing the material by the cutting mechanism to the total external work done. Conversely, since $K_p = 3\mu_p$ according to Eq. (1.A5), the classical model for abrasive wear states that the entire external work done is used to generate chips by the cutting process, which is self-consistent with the original assumption of the idealized model. In real situations, $K_p/3\mu_p <$ 0.3 and a great deal of external work is consumed to create ridges and to cause subsurface plastic deformation. The fraction of the external work done that is consumed in these noncutting processes increases with a decrease in the sharpness of the abrasive particles (i.e., as θ decreases). At small θ (i.e., $\tan \theta < 0.3$) there may not be any abrasive wear particles generated by the cutting process, but ridges will continue to form and subsurface deformation continues to take place, which may lead to loose wear particle formation through the delamination process discussed in Chapter 5.

2

CHEMICAL AND PHYSICAL STATE OF THE SOLID SURFACE

2.1 INTRODUCTION

The tribological behavior of a solid surface is directly affected by its chemical and physical states, which control the force transmission at the interface and which also control the wear processes. The importance of a comprehensive and definitive understanding of the surface physics and chemistry cannot be overstated.

Notwithstanding its importance, the science of surface phenomena is still in the early stages of development. The state of knowledge on surface properties is so inadequate that gross assumptions about surface phenomena have been necessary to solve tribological problems. The topics that require further development include dislocation mechanics near the surface, adhesion mechanisms, the formation and distribution of surface contaminants, and the flow strength of materials at the surface.

The properties of solid surfaces are usually different from those of the bulk. The differences stem from deviations in surface electronic and atomic structures from those of the bulk, the presence of contaminants, and the behavior of dislocations near the surface. These differences are manifest in unique electrical, chemical, and mechanical properties of the surface, which affect the tribological behavior of materials.

The purpose of this chapter is to review briefly some of the important surface properties. The bulk properties of metals, polymers, and ceramics are outlined first, followed by close examination of the atomic structure of metals and the mechanical behavior of solids. The atomic structure of the surface is also examined from the thermodynamic point of view. The information presented in this chapter provides a conceptual framework for later chapters.

2.2 BRIEF INTRODUCTION TO METALS, POLYMERS, AND CERAMICS

Metals are characterized by their metallic bonding, free electrons, and crystal structures. They also tend to react with nonmetallic elements to lower their free energy and form more stable compounds, such as oxides, chlorides, and carbides. The melting points of metals are, in general, higher than those of crystalline polymers but lower than those of ceramics. Metals are also, in general, tougher than ceramics and stronger than polymers. Because of these characteristics, metals are used in many tribological applications.

Most metals have body-centered-cubic (b.c.c.), face-centered-cubic (f.c.c.), or hexagonal-close-packed (h.c.p.) structures. Some have other structures, such as tetragonal. Since metals have crystalline structures, their mechanical and chemical behavior and properties are anisotropic and depend on their crystallographic orientation. Plastic deformation occurs along the planes of the closely packed structures. These crystallographic planes and directions of plastic deformation are called *slip planes* and *slip directions*, respectively. Similarly, the surface energy depends on the crystallographic plane; the surface energy of the close-packed planes is typically the lowest.

All metals have defects in the form of vacancies and dislocations. Dislocations that are line defects of the atomic arrangement lower the stress required to cause plastic deformation because of their high mobility under stress. Without dislocations, metals would be two to three orders of magnitude stronger than those typically measured. An isolated dislocation is extremely easy to move, but it takes a greater applied stress to move a number of dislocations and thus deform the solid plastically because as the number of dislocations increases, their mutual interactions impede their mobility. The increase in flow strength with deformation is called *work hardening*, which is a result of a greater number of dislocations interacting with each other as dislocations are generated due to plastic deformation. A well-annealed solid typically has a dislocation density of 10^6 cm^{-2}, whereas highly deformed metals have as many as 10^{12} dislocations per square centimeter. As the number of dislocations increases, they form dislocation cells that are a network of highly entangled dislocations surrounding a region of low dislocation density that is a few hundred angstroms in diameter. Dislocation behavior is of paramount importance in understanding the plastic deformation of crystalline solids. Although the dislocation behavior near the surface is extremely important in understanding all aspects of tribology, little is known about the structure and density of dislocations very near the surface (i.e., less than 100 Å).

Most metals used in engineering applications are polycrystalline, made up of many grains of all orientations. Therefore, polycrystalline metals have isotropic properties on a macroscopic scale, although individual grains are anisotropic. Grain boundaries are interfaces between two grains of different orientation. Most grain boundaries are regions of random atomic arrangement of finite thickness and therefore have higher free energy than the bulk. Thus impurities and solutes segregate on the grain boundaries. The yield strength and the toughness of polycrystalline metals increase with decreasing grain size, since the grain boundaries act as barriers to dislocation motion (thus requiring dislocation pile-up to break through the boundary) and to crack penetration.

Many metals used in engineering applications are alloys (both substitutional and interstitial) with multiphase structures rather than pure elements. The alloying elements are added for grain refinement, to control the rate of phase transformation during heat treatment, and to modify the mechanical and chemical properties. Typical metals have many phases, depending on the solubility of alloying elements and the presence of nonsoluble compounds such as oxides. Many metals used in tribological applications have hard ceramic phases (such as carbides) which are introduced to increase hardness and abrasion resistance. The effective spacing and size of the hard phase depends on specific tribological applications. For applications where toughness as well as hardness is important, the hard phase should be small in size and well dispersed. Both should be on the order of a few hundred angstroms for maximum effectiveness. On the other hand, for applications where abrasion resistance is important, it is desirable for the hard phase to be closely packed and to be larger in size than abrasive particles.

Polymers are covalently bonded solids. They are long-chain molecules which consist mostly of hydrocarbons and other nonmetallic elements. There are three kinds of engineering polymers: thermoplastic, thermoset, and elastomers. Thermoplastics are linear, long-chain polymers, most of which either melt or soften as the temperature is increased. One of the exceptions is Teflon (polytetrafluroethylene), which decomposes before melting. Certain thermoplastics, such as polyethylene and polypropylene, are partially crystalline (i.e., crystalline regions are separated by amorphous regions). The crystalline thermoplastics have melting points, as well as second-order glass transition temperatures, above which the plastic behaves in a "rubbery" fashion. Many crystalline plastics can undergo large plastic deformations at low temperatures. On the other hand, glassy thermoplastics such as polystyrene and polymethyl methacrylate (PMMA) are amorphous and brittle at room temperature. Thermosetting plastics are those with a three-dimensional network of covalent bonds. They are therefore amorphous and rigid at all temperatures, undergoing decomposition rather than melting at high temperatures. Elastomers differ from thermosetting plastics in that they have a smaller number of covalent bonds, thus allowing reversible deformation of the three-dimensional network without permanent displacement of molecules from their equilibrium positions. All these polymers are more corrosion resistant and chemically inert than most metals under normal operating conditions.

Typical thermoplastics have a broad molecular weight distribution. As a consequence, when these polymers are solidified, the molecular weight distribution near the surface or near an interface is substantially different from those of the bulk. Therefore, in certain crystalline thermoplastics the surface layer has different mechanical properties from those of the bulk. The existence of such a layer is attributed to the preferential nucleation of high-molecular-weight species during the crystallization process at certain nucleation sites (Schonhorn, 1969). Following this reasoning, low-molecular-weight species of the polymer are rejected at the molten plastic–air interface during the crystallization process, thus giving a weak surface layer. At a molten polymer–metal interface, a region of high cohesive strength is produced at the interface if the metal surface provides nucleation sites. This is caused by rejection of the low-molecular-weight species from the interface into the bulk. In the absence of nucleation sites, a weak plastic layer results at the plastic–metal interface.

Many thermoplastics with polar molecules [e.g., polyamides (commonly known as nylon), polyvinyl chloride, and polymethyl methacrylate] are hygroscopic. Water molecules diffuse into these plastics and form hydrogen bonds with the polar ends of these molecules. In the case of nylon, the volume expansion due to moisture absorption can be as much as 8% and thus significantly affect the dimensions of nylon bearings. When the condensation-polymerized polymers such as nylon and PET (polyethylene terephthalate) are wet, depolymerization occurs at high temperatures, thus lowering their mechanical properties.

The first tribological application of polymers was made about 50 years ago when phenol-formaldehyde resin (i.e., phenolic), reinforced with fibers and fabrics, was used to make bearings. Since then, particularly during the 1960s, a large number of new polymers have been introduced for engineering uses (Table 2.1). Polymeric materials are becoming increasingly important in tribology because of their low cost, light weight, quiet operation, ease of fabrication, and resistance to corrosion when used, for example, as plastic bearings. Plastic bearings can be particularly useful in applications where it is not possible to establish hydrodynamic lubrication due to the low sliding speeds (or oscillatory motions) or due to frequent stops and starts. Plastic bearings are also used in applications (e.g., food and textiles) where the presence of lubricant is not acceptable due to the risk of con-

TABLE 2.1 Introduction of Plastics Used in Tribology

1839	Vulcanization of natural rubber
1907	Phenol-formaldehyde resin
1926	Alkyd resins
1931	Neoprene synthetic rubber
1937	Styrene-butadiene, acrylonitrile-butadiene rubbers
1938	Nylon 616 (polyhexamethylene adipamide)
1941	Polyethylene
1942	Unsaturated polyesters for laminates
1943–1945	Silicones, flurocarbon resins, polyurethanes, styrene-butadiene
1947	Epoxy resins
1956	Linear polyethylene, acetals (polyoxymethylene)
1957	Polypropylene, polycarbonate
1959	*cis*-Polyisoprene and *cis*-polybutadiene
1960	Ethylene-propylene rubber
1962	Phenoxy resins, polyimide resins
1965	Polyphenylene oxide, polysulfones
1965	Kevlar fibers

tamination. Composite bearings with various fillers and fibers are better than normal polymeric bearings when high thermal stability and mechanical strength are required. In recent years acetals (polyoxymethylene), polyimide, reinforced teflon (PTFE), phenolics, and high-molecular-weight and ultrahigh-molecular-weight polyethylene have been commonly used in tribological applications. Many elastomers, such as silicone rubber, fluoroelastomers, and polyurethane, are often used as seals and in other tribological applications.

Ceramic is a material that is a combination of one or more metals with nonmetallic elements, usually oxygen, carbon, or nitrogen. The atoms of most ceramics are held together primarily by ionic bonding, while some ceramics have covalent and/or metallic bonds. Ceramics, especially oxides and nitrides, have a very low free energy of formation and therefore are very stable chemically. They do not readily react with other materials and have high melting points. Therefore, ceramics are used in high-temperature tribology applications. However, ceramics tend to be brittle and thus find limited use in applications that require toughness, although they have the best resistance to sliding wear.

2.3 GENERAL CHARACTERISTICS OF A SOLID SURFACE AND TRIBOLOGY

A solid surface is created when a solid fractures or when a liquid of a finite volume solidifies. A solid surface is an asymmetric boundary between two regions: one region where the interatomic forces are greater than the thermal energy of each atom so that the atoms are closely packed and the other region where there is gas with no "near-neighbor" interatomic interactions.

Because of the asymmetry, the atoms of the outer surface layer experience different forces from those in the bulk. Consequently, the electronic arrangement near the surface is considerably different from that of the bulk. In the case of many covalently bonded solids and a few metals, the atomic arrangement reconstructs itself to new equilibrium configurations (Estrup, 1975). An example of such reconstruction is shown in Fig. 2.1. One would not expect such a reconstruction of the surface of a perfect single crystal when the surface is a close-packed plane, because there is no atomic arrangement with a lower free energy. However, even a small amount of adsorbed gas atoms may affect the structure of the surface by displacing the surface atoms and forming a periodic array of substrate atoms and chemisorbed gas atoms (see Fig. 2.1) (Somorjai, 1972). There must be a certain degree of reconstruction on most polycrystalline solid surfaces because of the random orientation of grains, except when the surface deforms due to a sliding action which exposes close-packed slip planes.

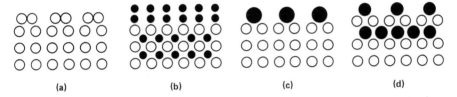

Figure 2.1 Atomic arrangement at the surface due to: (a) reconstruction of the surface atoms; (b) segregation of solute atoms; (c) chemisorption; (d) compound formation by mixing of absorbed atoms and the substrate atoms. (From Estrup, 1975.) Reprinted with permission.

To create a surface from the bulk, work must be done to break the atomic bonds. Consequently, the surface is at a state of higher free energy than the bulk; the surface energy is equal to the external work done in breaking the chemical bonds. The overall free energy of the surface and the surrounding is lowered when physisorption, chemisorption, and chemical reactions take place at the surface (see Fig. 2.1). In physisorption weak van der Waals–like forces act between an adsorbate and the substrate. Chemisorption bonds involve the mixing of the adsorbate and substrate wave functions, involving more reactive adsorbates such as H, O, CO, and NO (Schrieffer and Soven, 1975; Duke, 1968). Chemical reactions involve the formation of primary chemical bonds between the adsorbate and the substrate, resulting in a three-dimensional structure of a new compound. In tribology, chemisorption and chemical reactions at the surface are used to advantage in lowering frictional forces and minimizing the wear rates.

At the surface there is also an electrostatic potential difference associated with a charge double layer (Latanision, 1977; Herring, 1952). Because of the asymmetry of repelling forces acting on and between the electrons of the outermost shell of the surface atoms, one would expect to find excess electrons just outside the "original" surface and electron deficiency just inside the surface. Quantum mechanical calculation shows that there is a finite probability of finding an electron outside the metal surface in a vacuum. The difference in electronic density about the surface creates an electrostatic potential at the surface which is more negative outside than inside the surface. This is referred to as the *electrical double layer*. Such double layers are present at all interfaces, including the solid–electrolyte interface. Owing to the high mobility of conduction electrons, the electric field created by the double layer can penetrate only about 1 Å into the metal surface, but can penetrate in excess of 1 μm in the case of insulators.

It is not clear how the reconstruction of the atomic structure and the changes in the electronic state at the surface affect adhesion and tribological behavior. However, it may be reasonable to assume that the directionally bonded solids (i.e., covalently bonded and ionic solids) do not readily adhere to other solids during sliding. Even in the case of identical metals sliding against each other, the reconstructed surfaces cannot instantaneously re-establish the original chemical bond at asperity contacts. Even though the free energy is lowered by going back to the bulk state, there is a finite activation energy barrier that the atoms and electrons must overcome to return to the bulk state. Therefore, the welding of moving asperity junctions may not be a common occurrence for some materials when the temperature at the interface is low.

The change in chemical composition at the surface due to segregation of impurities and solutes can have a significant effect on the tribological characteristics of a surface. It has been shown experimentally that a freshly exposed surface of a solid solution establishes a new composition at the surface. Buckley (1980) showed that the surface of a 1% Al–Cu solid solution has a concentration of aluminum atoms 6.5 times higher at the surface than in the bulk (see Table 2.2), increasing the adhesion of these solid solutions to gold fivefold over pure copper. Surface enrichment by segregation was also observed in other systems: nickel in iron, silver in palladium, gold in copper, copper in nickel, silver in gold, tin in copper,

TABLE 2.2 Maximum Coverage of Minor Constituent on Alloy Surface

Alloy	Ratio of the Surface Concentration to Bulk Concentration
Cu–1 at % Al	6.5
Cu–5 at % Al	4.5
Cu–10 at % Al	3.1
Cu–1 at % Sn	15.0 ± 2
Fe–10 at % Al	8.0

Source: Buckley (1974).

aluminum in iron, and platinum in osmium. The segregation may occur as a result of the need to lower the strain energy created by the difference in atomic size (see Table 2.3).

Most surfaces are ordinarily covered with adsorbates and oxides that minimize the metal-to-metal contact at asperity contacts. In this sense adsorbates and oxides act as lubricants. Even heating metals in a vacuum at high temperatures cannot remove all adsorbates. One has to resort to argon ion bombardment and the like to remove all the extraneous elements (Buckley, 1980). One of the important consequences of a coherent oxide layer which is strongly bonded to substrate is that it alters the dislocation behavior near the surface, making the surface harder than the bulk upon plastic deformation. It should be noted that not all oxide layers formed on a metal surface are stable and adhere to the substrate, owing to the bulk density difference between the substrate and the oxide layer.

Even in the absence of adsorbates, the surface may be contaminated by migration of a few parts per million (ppm) of interstitial impurities to the surface. A few ppm of carbon in iron has been shown to change the surface chemistry (Buckley, 1980). Similar segregation has been seen in such systems as oxygen in platinum, phosphorus in iron, sulfur and carbon in nickel, sulfur in molybdenum, and sodium in lithium. In view of these contamination problems, one has to be extremely careful in using the surface energy data in tribological applications.

In boundary lubrication, it is essential that long-chain molecules be attracted to the solid surface and adhere for effective separation of metal surfaces. One possible clue as to how these molecules are attracted to the surface in the first place is given by an interesting correlation between the cohesive energy density (CED) of crystalline solids and the adsorption of albumin protein molecules (Suh and Nguyen, 1981). Experiments were performed to check the following hypotheses for protein adsorption: (1) the negatively charged albumin molecules accumulate on a solid surface

TABLE 2.3 Atomic Radius (Å)

Aluminum (f.c.c.)	1.431
Titanium (h.c.p.)	1.458
Vanadium (b.c.c.)	1.316
Chromium (b.c.c.)	1.249
Manganese (cubic comp.)	1.12
Iron (b.c.c.)	1.241
Cobalt (h.c.p.)	1.248
Nickel (f.c.c.)	1.245
Copper (f.c.c.)	1.278
Zinc (h.c.p.)	1.332
Silver (f.c.c.)	1.444
Cadmium (h.c.p.)	1.489
Tin (b.c. tetragonal)	1.509

due to the surface charges created as a consequence of proton diffusion into the solid; and (2) the solids with high CED, which have a higher interatomic potential, should make it more difficult for protons to diffuse into the solid and therefore, the higher the cohesive energy density, the smaller should be the protein accumulation. The results show that diamond, which has the highest CED, adsorbs the least amount of proteins, and pure metals such as gold, which have a very low CED, adsorb the most. The adsorption of proteins as a function of CED of the substrate is given in Table 2.4.

TABLE 2.4 Protein Adsorption of Various Materials as a Function of Their Cohesive Densities

Material	Cohesive Energy Density (cal/cm^3)	Surface Concentration $(\mu g/cm^2)$
Diamond	49,700	0.1
Aluminum oxide	34,200	0.1
Boron nitride	31,000	0.9
Calcium fluoride[a]	25,000	1.0, 3.4, 5.2
Magnesium oxide[a]	15,800	3.5, 4.2, 10.0
Glass	15,000	1.5
Platinum	14,800	18.5
Stainless steel	13,900	1.7[b]
Gold	8,600	66.7
Aluminum	7,800	3.8[b]
Silver	6,670	147.6

[a] Adsorption dependent on plane of cleavage.

[b] The surface concentration of protein is low because of the presence of chromium oxide and aluminum oxide on the surface of stainless steel and aluminum, respectively.

2.4 CHEMICAL INTERACTION OF METAL SURFACES WITH LUBRICANTS

Metal surfaces are often lubricated with gases, liquids, and solids. Liquid lubricants with additives are most commonly used to lower friction and wear. Lubricants minimize the metal-to-metal contacts either by forming a new chemical compound on the surface or by physically separating the surfaces, and thus lower the friction force. The mechanism of lubrication is in part dictated by the nature of interactions between the lubricant and the metal surface. There can be many different interactions: chemical reactions which form new compounds on the surface; physical interactions such as physisorption, which attracts long-chain molecules in the liquid to the surface due to van der Waals–type electrostatic forces; or chemisorption interactions involving strong chemical bonds between the liquid components and the metal surface.

Clean metal surfaces react with gases and long-chain molecules dissolved in the liquids (for a review of the subject, see Godfrey, 1980). The phenomenon has been used to advantage in developing lubricants that form a stable compound on the surface by inducing chemical reactions and that minimize adhesion at the asperity contacts. Oxygen is an important element in boundary lubricants. According to Klaus et al. (1975), dissolved oxygen reacts with super-refined mineral oil and the metal surface to form organometallics that reduce friction and wear.

Sulfur compounds are commonly used in boundary lubricants. These compounds decompose because of frictional heating and react with the metal surface to form metal sulfides. These compounds are typically iron sulfide and form on the plateaus of scuffed surfaces (Coy and Quinn, 1973). When oxygen is also present, a mixture of iron sulfide and iron oxide forms (Godfrey, 1962; Tomaru et al., 1977). In gaseous environment with oxygen, iron sulfide changes into iron oxide, indicating that iron oxide is more stable than iron sulfide (Buckley, 1974).

Phosphorus compounds such as tricresyl phosphate in mineral oil have been used in many lubricants. It is believed that they form iron phosphates. Similarly, chlorides form a stable surface layer.

The most commonly used additive is zinc organodithiophosphate (ZDTP). When this additive is added to a paraffin white mineral oil and a light hydraulic oil, brown and blue films are found to form on highly deformed steel surfaces (Jahanmir, 1981). Auger analysis and secondary-ion mass spectroscopy show that the blue film is 50 nm thick and is primarily a metastable protoxide of iron, $Fe_{1-x}O$ (Georges et al., 1977). The brown film contains P, S, and Zn, and is composed of minerals such as sulfides, sulfates, phosphates, and thiophosphates. Georges et al. claim that the brown film is the reaction product between zinc dithiophosphates and iron and contain some polymers of type $[Zn(PSO_2)]_n$.

These adsorptions of long-chain molecules on the surface also affect the mechanical properties of metals and other solids. The *Rebinder effect* (Rebinder, 1928) is the best known phenomenon that describes the adsorption-induced reduction of the surface hardness. The implication of such chemomechanical effects as the Rebinder effect on the tribological behavior of materials is not clearly known. A review article on surface effects in crystal plasticity by Latanision (1977) gives a good account of various surface phenomena under nontribological loading conditions.

It should be noted that the adsorption of long-chain molecules to the surface depends on the crystallographic orientation of the surface, especially in the case of ionic solids.

2.5 MECHANICAL PROPERTIES OF SOLID SURFACES

It was stated earlier that the chemical and physical properties of metallic and polymeric surfaces are different from the bulk. One of the physical properties tribologists are interested in is the flow strength of the materials near the surface. The question that has been investigated the most is whether the surface layer is softer or harder than the bulk when the specimen is subjected to uniform uniaxial deformation. This has been rather a controversial subject, since the experimental results support both views. According to Kramer and Demer (1961), the surface is harder, whereas Fourie (1968) holds the opposite view. Similar controversy also exists for highly worn surface layers (Savitskii, 1965; Kirk and Swanson, 1975). To understand the controversy, it is necessary to deal with dislocation dynamics near the surface, including generation and multiplication of dislocations and variation in dislocation density near the surface.

Kramer and Demer (1961) found that the original yield stress and the work-hardening behavior of an aluminum monocrystal could be recovered, as shown in Fig. 2.2, by chemically removing a 1-mm layer from the surface.

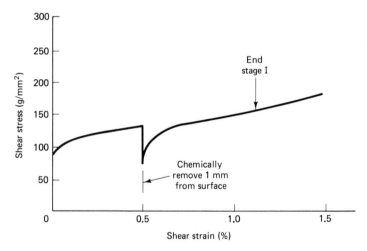

Figure 2.2 Kramer's experimental results showing that an aluminum monocrystal recovers the original yield stress upon reloading after removing a 1-mm surface layer electrochemically. (From Kramer and Demer, 1961.)

This experimental result was taken to mean that the stage I work hardening is confined to the near-surface layer (Kramer, 1963). Fourie (1968), on the other hand, found that the flow stress of plastically deformed copper monocrystal decreases near the surface, indicating that there was less work hardening. Fourie obtained his results by first deforming a large specimen, slicing it into thin sections ranging in thickness from 0.065 to 0.6 mm, and then reloading them. The results are shown in Fig. 2.3.

Many explanations have been advanced to deal with the soft–hard controversy based on dislocation theory. One of the explanations for Kramer's results is that dislocations are generated near the surface layer by a Frank–

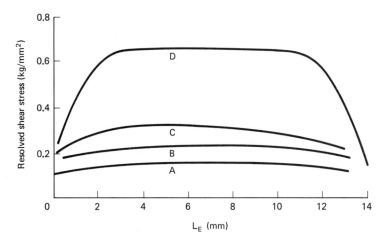

Figure 2.3 Fourie's experimental results, which show the flow stress distribution in copper monocrystals plotted as a function of the length of the glide path of edge dislocations, L_E, which has the value 0 and 14 mm at the original surfaces. Curve A is for an as-grown crystal; curves B, C, and D are for prestrains of 0.02, 0.029, and 0.053 respectively. (After Fourie, 1968.)

Read type of multiplication mechanism which can readily move into the interior of the solid but cannot egress out of the surface because the surface (or a contaminated surface layer) acts as a barrier to dislocation motion. Therefore, the dislocations accumulate at the surface faster than the interior (Latanision, 1977; Nabarro, 1977). Fourie's experimental results have been explained as follows. Near the surface, dislocations of opposite signs that are needed to form dislocation dipoles are not equally available. Since dislocation dipoles are responsible for some of the work hardening in stage II of f.c.c. metals, the surface layer does not work-harden as much as the interior.

Dislocations very near and parallel to the surface experience image forces due to their proximity to the surface. When there is no continuous, coherent, oxide layer adhering to the metal surface, such as when a fresh new surface is exposed due to deformation and/or wear, the image force attracts the dislocations to the surface. When the image force is greater than the resisting drag force (commonly referred to as the *dislocation friction stress*), these dislocations are attracted to the surface and disappear if the surface does not act as a barrier. Therefore, it is likely that the dislocations generated very near the surface during sliding may not be able to accumulate very near the surface, whereas the dislocations below this surface zone can cause work hardening. In this case the outermost surface layer where a fresh surface is just exposed may be softer than the interior. When a hard, continuous oxide layer is present on the metal surface, dislocations existing in the metal experience a force that repels them from the surface into the interior of the metal. However, in this case the surface layer will still have a higher dislocation density than the bulk, making it harder than the bulk.

The magnitude of the image shear stress τ_i acting on a dislocation parallel to the surface is given by

$$\tau_i = \frac{Gb}{4\pi(1 - \nu)h} \tag{2.1}$$

where G is the shear modulus, b the magnitude of the Burgers vector, ν Poisson's ratio, and h the distance from the surface. This force is opposed by the dislocation friction stress σ_f and by the change in the surface energy as the dislocations emerge from the surface. Assuming that the dislocation friction stress dominates, the maximum thickness of the surface layer in which the dislocation density is lower than the layer just below it may be expressed as

$$h = \frac{Gb}{4\pi(1 - \nu)\sigma_f} \tag{2.2}$$

For silicon iron, h is about 0.1 μm, whereas for pure copper it is about 10 to 20 μm.

The lower dislocation density zone given by Eq. (2.2) is valid only for single crystals whose slip planes intersect the surface. In polycrystalline metals, only the dislocations in the outermost grains would be able to egress to the surface since the dislocations in the subsurface grains will encounter grain boundaries that are barriers to dislocation motion. The outermost grain of the worn surface becomes very thin and aligns nearly parallel to the surface, due to the large plastic deformation. When the thickness of the grain is less than the effective depth of image force given by Eq. (2.2), the entire outer grain may be much softer than the subsurface grains.

To estimate the thickness of the outermost grain, consider the deformation of a spherical grain of diameter D due to the shear stress applied at the surface. The grain will then deform into an ellipsoid. The thickness of the grain c can be related to the equivalent strain $\bar{\varepsilon}$ (see Appendix 2.A for definitions, etc.) of the surface layer as (Dautzenberg and Zaat, 1973; see Appendix 2.B)

$$c = \frac{D}{\sqrt{3}\,\bar{\varepsilon}} \tag{2.3}$$

The experimentally measured maximum equivalent strain $\bar{\varepsilon}$ for f.c.c. metals is very large, being of the order of 100, while it is less for AISI 1020 steel, being of the order of 20 (Augustsson, 1974). Since the diameter of a typical f.c.c. metal is about 50 μm, the thickness of the deformed grain is about 0.3 μm, which is less than the effective depth of the image force

Sec. 2.5 Mechanical Properties of Solid Surfaces

for pure copper. Therefore, this grain may be softer than the subsurface grains if the surface is clean.

Definitive and reliable experimental results for hardness variation near the surface are not yet available. Perhaps the most reliable experimental results are those of Pethica and Oliver (1981), whose experimental technique is said to yield results reproducible within 10% at an indentation depth of 100 nm. The hardness measurements below an indentation depth of 50 nm varies a great deal, which is attributed to the variability of hardness over

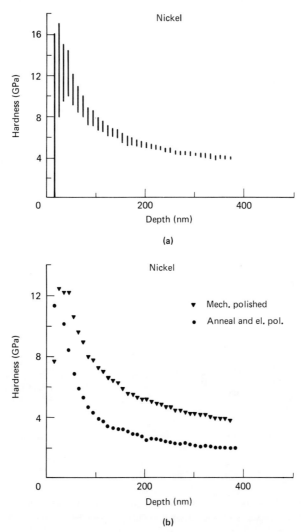

Figure 2.4 (a) Scatter of averaged data: bars two standard deviations high; (b) effect of vacuum anneal and electropolish on hardness. (From Pethica and Oliver, 1981.)

different regions of the same surface. Their results with mechanically polished nickel and with annealed and electropolished nickel are shown in Fig. 2.4. The experimental results show that the subsurface hardness of the mechanically polished nickel is greater than that of the bulk, but there are many regions of the outermost surface layer (<20 nm) where the hardness

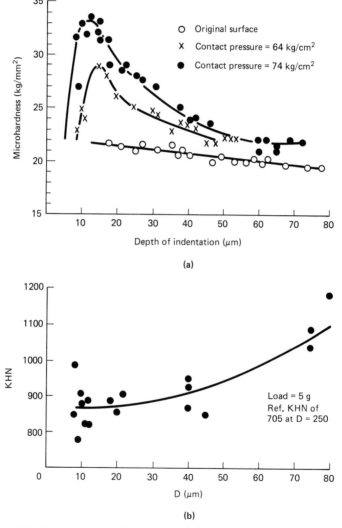

Figure 2.5 Measurements of the subsurface hardness variation: (a) indentation hardness of the worn surface of aluminum specimens under varying indentation load (b) indentation hardness of copper specimen wear tested under a normal load of 0.682 kg. [(a), From Sandor, 1972; (b), from Kirk, J. A., and Swanson, T. D., "Subsurface Effects During Sliding Wear," *Wear*, Vol. 35, 1975, pp. 63–67.]

Sec. 2.6   Thermodynamic Analysis of an Interface   41

is less than the maximum hardness value of the substrate. Savitskii (1965) and Kirk and Swanson (1975) also presented experimental results which show that the hardness is the maximum at subsurface and decreases toward the surface. These results are shown in Fig. 2.5. In general, however, there are more papers presented in the literature which claim that the hardness is the greatest at the surface (Gane, 1970; Pethica and Tabor, 1979; Ruff, 1976). None of the work published so far has treated the grain size effects and the maximum strain of the surface layer quantitatively. This controversy over the hardness of the outermost surface layer cannot be resolved until precise experiments are done using a reliable hardness-measuring technique that can measure a very thin surface layer. However, it may be reasonable to speculate that the metal surface can be either softer or harder than the substrate, depending on the cleanliness of the surface.

The actual hardness of the surface layer during sliding may also be different from the bulk because the surface layer undergoes cyclic loading. When a metal is subjected to cyclic loading, it either softens or hardens (Sandor, 1972; Morrow, 1965). Cold-worked metals usually undergo cyclic softening, while annealed metals may first undergo cyclic hardening followed by cyclic softening. Therefore, the actual flow stress of the metals at the surface during sliding may be lower than the bulk yield strength of the metal determined under a uniaxial tensile test, due to the extremely large number of cyclic loadings that occur at the contacting surfaces, especially the surface of the slider. There is also a residual permanent strain remaining after each cyclic loading, which is much smaller than the maximum strain amplitude of that cycle.

The significance of the hardness question in tribophysics is that if the outermost surface layer is softer than the substrate, the adhesion of the flat asperity contact at the surface cannot account for the observed friction coefficients in metals. Shearing will always occur through this soft layer, which yields at a lower stress than the bulk yield stress. So there must be other mechanisms that control the frictional force and eventually affect the wear behavior. Conversely, if the outermost surface layer is harder than the substrate, subsurface deformation is likely to occur.

2.6 THERMODYNAMIC ANALYSIS OF AN INTERFACE

In earlier sections it was stated that the atomic structure of the surface of most metals (and probably other solids as well) is highly heterogeneous. The causes for the heterogeneity are the following:

1. Reconstruction of the surface atoms of covalent solids and some metals
2. Adsorbates due to physisorption and chemisorption
3. Oxide layers due to chemical reaction
4. Segregation of solutes at the surface

5. Formation of compounds at the surface when adsorbates are present on the surface of metals, mixing the substrate atoms and adsorbate atoms

The fact that these layers on the surface do not simply sit on the substrate but are mixed with the substrate atoms has been determined experimentally and through quantum mechanical arguments (Duke, 1977). It can also be supported through thermodynamic arguments. In this section a thermodynamic analysis will be presented, following the treatment originally given by Cahn and Hilliard (1958).

For the purpose of analysis, consider a thick oxide layer formed on a metal substrate. Then the interface between the oxide layer and the substrate is the phase boundary between two different materials. The question we would like to answer is: How thick is the interface? If we assume that the interface has a finite thickness, the system consisting of the interface is locally a binary solution. If the mole fraction of oxygen, c, is a nonuniform property in the binary solution, the free energy per molecule, f, in the region of nonuniform composition will depend on both the local composition and the composition of the immediate environment. Assuming that the composition gradient is small compared with the reciprocal of the intermolecular distance, the concentration c and its derivatives may be taken as independent variables that can describe the local composition and the composition of the immediate environment. When f is a continuous function of these variables, it can be expanded in a Taylor series about f_0 (the free energy per molecule of a solution of uniform composition c) as

$$f(c, \nabla c, \nabla^2 c, \ldots) = f_0(c) + \sum_i L_i \left(\frac{\partial c}{\partial x_i}\right) + \sum_{i,j} K_{ij}^{(1)} \left(\frac{\partial^2 c}{\partial x_i \partial x_j}\right) \\ + \frac{1}{2} \sum_{i,j} K_{ij}^{(2)} \left(\frac{\partial c}{\partial x_i}\right)\left(\frac{\partial c}{\partial x_j}\right) + \ldots \quad (2.4)$$

where

$$L_i = \left[\frac{\partial f}{\partial (\partial c/\partial x_i)}\right]_0 \qquad K_{ij}^{(1)} = \left[\frac{\partial f}{\partial (\partial^2 c/\partial x_i \partial x_j)}\right]_0$$

$$K_{ij}^{(2)} = \left[\frac{\partial^2 f}{\partial (\partial c/\partial x_i) \partial (\partial c/\partial x_j)}\right]_0$$

i and j represent the successive substitution of spatial coordinates x, y and z, and the subscript 0 represents the value of the parameter of the uniform composition. In general, $K_{ij}^{(1)}$ and $K_{ij}^{(2)}$ are tensors reflecting the crystal symmetry and L_i's are components of a polarization vector in a polar crystal.

Sec. 2.6 Thermodynamic Analysis of an Interface

For a cubic crystal or an isotropic medium the free energy must be invariant due to the symmetry operators of reflection ($X_i \to -X_i$) and of rotation about a fourfold axis ($X_i \to X_j$). Therefore,

$$L_i = 0$$

$$K_{ij}^{(1)} = \begin{cases} K_1 = \left[\dfrac{\partial f}{\partial \nabla^2 c}\right]_0 & \text{if } i = j \\ 0 & \text{if } i \neq j \end{cases}$$

$$K_{ij}^{(2)} = \begin{cases} K_2 = \left[\dfrac{\partial^2 f}{\partial (|\nabla c|)^2}\right]_0 & \text{if } i = j \\ 0 & \text{if } i \neq j \end{cases} \quad (2.5)$$

Hence for a cubic lattice,

$$f(c, \nabla c, \nabla^2 c, \ldots) = f_0(c) + K_1 \nabla^2 c + K_2 (\nabla c)^2 + \cdots \quad (2.6)$$

Integrating over a volume V of the solution, we obtain for the total free energy F of this volume,

$$F = N_v \int f\, dv = N_v \int_v [f_0(c) + K_1 \nabla^2 c + K_2 (\nabla c)^2 + \cdots]\, dv \quad (2.7)$$

where N is the number of molecules per unit volume. Applying the divergence theorem yields

$$\int_v (K_1 \nabla^2 c)\, dv = -\int_v \left(\frac{dK_1}{dc}\right)(\nabla c)^2\, dv + \int_s (K_1 \nabla c\, n)\, ds \quad (2.8)$$

If we choose the boundary so that

$$\int_s (K_1 \nabla c\, n)\, ds = 0 \quad (2.9)$$

then

$$F = N_v \int \left[f_0(c) - \left(\frac{dK_1}{dc}\right)(\nabla c)^2 + K_2 (\nabla c)^2 + \cdots\right] dv$$

$$= N_v \int [f_0(c) + K(\nabla c)^2 + \cdots]\, dv \quad (2.10)$$

where

$$K = -\frac{dK_1}{dc} + K_2 = -\left[\frac{\partial}{\partial c}\left(\frac{\partial f}{\partial \nabla^2 c}\right)\right]_0 + \left[\frac{\partial^2 f}{\partial (|\nabla c|)^2}\right]_0$$

Equation (2.10) states that the free energy is a sum of two contributions: one being the free energy that this volume would have in a homogeneous solution, and the other the *gradient energy*, which is a function of the local composition.

At a flat oxide–metal interface where two isotropic phases c_α and c_β coexist, the energy of nonequilibrium material of composition intermediate between c_α and c_β may be represented as a continuous function $f_0(c)$ of the form shown in Fig. 2.6.

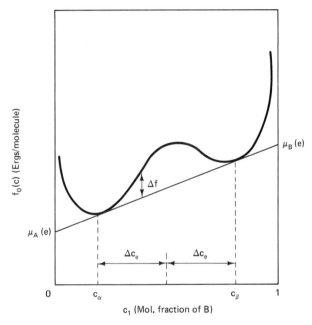

Figure 2.6 Free energy of a binary system as a function of composition. (From Cahn and Hilliard, 1958.) Reprinted with permission.

For a one-dimensional composition change across the interface, neglecting terms in derivatives higher than the second, the total free energy F of the system is

$$F = AN_v \int_{-\infty}^{\infty} \left[f_0(c) + K\left(\frac{dc}{dx}\right)^2 \right] dx \qquad (2.11)$$

The specific interfacial free energy, σ, is by definition the difference per unit area of interface between the actual free energy of the system and that which it would have if the properties of the phases are continuous throughout.

Hence

$$\sigma = N_v \int_{-\infty}^{\infty} \left[f_0(c) + K\left(\frac{dc}{dx}\right)^2 - c\mu_B(e) - (1-c)\mu_A(e) \right] dx \quad (2.12)$$

$$= N_v \int_{-\infty}^{\infty} \left[\Delta f(c) + K\left(\frac{dc}{dx}\right)^2 \right] dx$$

where

$$\Delta f(c) = f_0 - [c\mu_B(e) + (1-c)\mu_A(e)]$$
$$= c[\mu_B(c) - \mu_B(e)] + (1-c)[\mu_A(c) - \mu_A(e)]$$

To lower σ, the interfacial energy, a highly diffused layer [i.e., small $K(dc/dx)^2$] is desirable, but it increases the $\Delta f(c)$ term by increasing the volume of the nonequilibrium composition inhomogeneous layer. Therefore, an optimum exists where the $[\Delta f(c) + K(dc/dx)^2]$ term assumes a minimum value (i.e., the chemical potentials are constant through the system).

The foregoing thermodynamic treatment of the interface indicates that to minimize the free energy, the atoms of adsorbates and oxide layers form a mixed layer of finite thickness rather than simply sitting on the substrate. Therefore, most surfaces with contaminants are likely to be heterogeneous and thus may impede the egress of dislocations from the surface. Dislocations may still escape from the surface, however, if they can move to the surface before adsorbates and oxides can form on the clean surface following the deformation of the surface layer.

2.7 CONCLUDING REMARKS

In this chapter the chemical and physical states of the outermost layer of the surface were examined. It was shown that the geometric arrangement, the chemical state, and the physical properties are highly heterogeneous. The substrates are normally covered with adsorbates, oxides and other chemical reaction products, and segregated solutes. These contaminants are likely to prevent adhesion in many sliding situations.

It was also shown that the hardness of the surface layer is different from that of the substrate. The outermost surface layer can be either softer or harder than the substrate, depending on whether egress of dislocations is possible. Furthermore, it seems that the outermost grain of a polycrystalline metal can have a significantly different flow stress from that of the interior grains if dislocation egress is not impeded by surface contaminants and heterogeneous structures. Further research must be done to clarify the mechanical properties of the outermost surface layer.

Finally, the state of the contaminant–substrate interface is examined from a thermodynamic point of view. It is shown that the atoms of contaminants are mixed with the atoms of the substrate to lower the free energy of the system. Because they are mixed, it is difficult to remove these contaminants by such techniques as heating. Consequently, it is safe to assume that the surface of most engineering materials is always contaminated. This heterogeneous nature of solid surfaces can be used to minimize friction and wear. Additives can be added to lubricants to form stable compounds on the surface, which may lower the frictional force by preventing plowing in addition to adhesion. This is discussed further in Chapter 3.

REFERENCES

AUGUSTSSON, G., "Strain Field near the Surface due to Surface Traction," S.M. thesis, MIT, 1974.

BUCKLEY, D. H., "Oxygen and Sulfur Interactions with a Clean Iron Surface and the Effect of Rubbing Contact on These Interactions," *ASLE Transactions,* American Society of Lubrication Engineers, Vol. 17, 1974, pp. 206–212.

BUCKLEY, D. H., "Definition and Effect of Chemical Properties of Surfaces in Friction, Wear, and Lubrication," in *Fundamentals of Tribology,* N. P. Suh and N. Saka, eds., MIT Press, Cambridge, Mass., 1980.

CAHN, J. W., and HILLIARD, J. E., "Free Energy of a Non-uniform System: I. Interfacial Free Energy," *Journal of Chemical Physics,* Vol. 28, 1958, pp. 258–267.

COY, R. C., and QUINN, T. F. J., "An Application of Electron Probe Microanalysis and X-ray Diffraction to the Study of Surfaces Worn under Extreme Pressure Lubrication," *Tribology Convention 1972,* London, September 27–28, 1972, Institution of Mechancal Engineers, London, 1973, pp. 62–68.

DAUTZENBERG, J. H., and ZAAT, J. H., "Quantitative Determination of Deformation by Sliding Wear," *Wear,* Vol. 23, 1973, pp. 9–19.

DUKE, C. B., "Electronic Structure of Clean Metal Interfaces," General Electric Report 68-C-33, September 1968.

DUKE, C. B., "Atomic Geometry and Electronic Structure of Solid Surfaces," in *Surface Effects in Crystal Plasticity,* R. M. Latanision and J. T. Fourie, eds., Noordhoff, Leyden, 1977.

ESTRUP, P. J., "The Geometry of Surface Layers," *Physics Today,* April 1975, pp. 33–41.

FOURIE, J. T., "The Flow Stress Gradient between the Surface and the Center of Deformed Copper Single Crystals," *Philosophical Magazine,* Vol. 17, 1968, pp. 735–756.

GANE, N., "The Direct Measurement of the Strength of Metals on a Sub-micrometre Scale," *Proceedings of the Royal Society of London,* Series A, Vol. 317, 1970, pp. 367–391.

GEORGES, J. M., MEILLE, G., JACQUET, M., LAMY, B., and MATHIA, T., "A Study of the Durability of Boundary Films," *Wear*, Vol. 42, 1977, pp. 217–228.

GODFREY, D., "Chemical Changes in Steel Surfaces during Extreme Pressure Lubrication," *ASLE Transactions*, American Society of Lubrication Engineers, Vol. 5, 1962, pp. 57–66.

GODFREY, D., "Review of Usefulness of New Surface Analysis Instruments in Understanding Boundary Lubrication," in *Fundamentals of Tribology*, N. P. Suh and N. Saka, eds., MIT Press, Cambridge, Mass., 1980.

HERRING, C., "The Atomic Theory of Metallic Surfaces," in *Metal Interfaces*, American Society for Metals, Metals Park, Ohio, 1952, pp. 1–19.

HILL, R., *The Mathematical Theory of Plasticity*, Clarendon Press, Oxford, 1950.

JAHANMIR, S., "War of AISI 4340 Steel under Boundary Lubrication," *Wear of Materials 1981*, International Conference on Wear of Materials, San Francisco, March 30–April 1, 1981, ASME, New York, New York, pp. 648–655.

JOHNSON, W., and MELLOR, P. B., *Plasticity for Mechanical Engineers*, Van Nostrand, Reinhold, London, 1962.

KIRK, J. A., and SWANSON, T. D., "Subsurface Effects during Sliding Wear," *Wear*, Vol. 35, 1975, pp. 63–67.

KLAUS, E. E., TEWKSBURY, E. J., and BOSE, A. C., "Some Chemical Reactions in Boundary Lubrication," in *Proceedings of Japan Society of Lubrication Engineers—American Society of Lubrication Engineers*, International Lubrication Conference, Tokyo, June 9–11, 1975, T. Sakurai, ed., Elsevier, New York, 1976, p. 39.

KRAMER, I. R., "The Effect of Surface Removal on the Plastic Flow Characteristics of Metals," *Transactions of the American Institute of Mining, Metallurgical, and Petroleum Engineers*, Vol. 227, 1963, pp. 1003–1010.

KRAMER, I. R., and DEMER, L. J., "The Effect of Surface Removal on the Plastic Behavior of Aluminum Single Crystals," *Transactions of the American Institute of Mining, Metallurgical, and Petroleum Engineers*, Vol. 221, 1961, pp. 780–786.

LATANISION, R. M., "Surface Effects in Crystal Plasticity: General Review," in *Surface Effects in Crystal Plasticity*, R. M. Latanision and J. T. Fourie, eds., Noordhoff, Leyden, 1977.

McCLINTOCK, F. A., and ARGON, A. S., *Mechanical Behavior of Materials*, Addison-Wesley, Reading, Mass., 1966.

MORROW, J., "Low Cycle Fatigue Behavior of Quenched and Tempered SAE 1045 Steel," Report 277, Department of Theoretical and Applied Mechanics, University of Illinois, Urbana–Champaign, Ill., April 1965.

NABARRO, F. R. N., "Surface Effects in Crystal Plasticity—Overview from the Crystal Plasticity Standpoint," in *Surface Effects in Crystal Plasticity*, R. M. Latanision and J. T. Fourie, eds., Noordhoff, Leyden, 1977.

PETHICA, J. B., and OLIVER, W. C., "Ultra-microhardness Tests on Ion-Implanted Metal Surface," unpublished paper, 1981.

PETHICA, J. B., and TABOR, D., "Contact of Characterized Metal Surfaces at Very Low Loads: Deformation and Adhesion," *Surface Science*, Vol. 89, 1979, pp. 182–190.

REBINDER, P. A., Reports to the VIth Congress of Physicists, Moscow, No. 29, 1928.

RUFF, A. W., "Deformation Studies at Sliding Wear Tracks in Iron," *Wear*, Vol. 40, 1976, pp. 59–74.

SANDOR, B. I., *Fundamentals of Cyclic Stress and Strain*, University of Wisconsin Press, Madison, Wis., 1972.

SAVITSKII, K. B., cited in I. V. Kragelskii, *Friction and Wear*, Butterworth, Washington, D.C., 1965.

SCHONHORN, H., "Adhesion to Low Energy Polymers," in *Adhesion*, Gordon and Breach, New York, 1969.

SCHRIEFFER, J. R., and SOVEN, P., "Theory of Electronic Structure," *Physics Today*, April 1975, pp. 24–30.

SOMORJAI, G. A., *Principles of Surface Chemistry*, Prentice-Hall, Englewood Cliffs, N.J., 1972.

SUH, N. P., and NGUYEN, L. T., "Correlation of Protein Absorption with Cohesive Energy Density," to be published, 1985.

SUH, N. P., and TURNER, A. P. L., *Elements of the Mechanical Behavior of Solids*, McGraw-Hill, New York, Scripta Technica, Washington, D.C., 1975.

TOMARU, M., HIRONAKA, S., and SAKURAI, T., "Effects of Oxygen on the Load-Carrying Action of Some Additives," *Wear*, Vol. 41, 1977, pp. 117–140.

APPENDIX 2.A

ELEMENTS OF CONTINUUM MECHANICS

Since much of this book deals with continuum mechanics and the mechanical behavior of solids, it will be necessary for the reader to possess an essential background in the subject matter. It is suggested that the reader refer to reference books that deal with these topics extensively (e.g., Suh and Turner, 1975; Hill, 1950; Johnson and Mellor, 1962; McClintock and Argon, 1966). In this appendix, the basic governing relationships are given as a means of defining notations used throughout the book.

In a Cartesian coordinate system x_i the equation of motion for an isolated element in a body may be written as

$$\sum_{i=1}^{3} \frac{\partial \sigma_{ij}}{\partial x_i} + X_j = \rho \frac{\partial^2 u_j}{\partial t^2} \qquad (2.\text{A}1)$$

where σ_{ij} is the stress acting on the x_i plane along the x_j direction (i.e., the ratio of the force component F_j acting on an infinitesimal plane A_i), X_j the body force along the x_j axis, ρ the mass density of the element, and u_j the displacement along the x_j axis. In the absence of couple stresses, $\sigma_{ij} = \sigma_{ji}$. In many applications of tribology the body force and the acceleration are negligible. Therefore, the equilibrium equation may be written as

Chap. 2   Appendix 2.A

$$\sum_{j=1}^{3} \frac{\partial \sigma_{ij}}{\partial x_j} = 0 \tag{2.A2}$$

The stress σ_{ij} is a second-order tensor and therefore can be transformed into σ_{ij}'' of the x_i' coordinate system. Mohr's circle is often used for this purpose, especially for the plane stress case.

Second-order tensors have invariants. The two important quantities derived from the invariants of σ_{ij} are

$$\sigma = \frac{1}{3} \sum_{k=1}^{3} \sigma_{kk}$$

$$\bar{\sigma} = \left[\frac{1}{2}[(\sigma_{11} - \sigma_{22})^2 + (\sigma_{22} - \sigma_{33})^2 + (\sigma_{33} - \sigma_{11})^2] \right.$$

$$\left. + 3(\sigma_{12}^2 + \sigma_{13}^2 + \sigma_{23}^2) \right]^{1/2} = \left(\frac{3}{2} \sigma_{ij}' \sigma_{ij}' \right)^{1/2} \tag{2.A3}$$

The first invariant σ is the hydrostatic stress (i.e., negative hydrostatic pressure p) and the second invariant $\bar{\sigma}$ is approximately equal to twice the maximum shear stress. σ does not affect the plastic deformation of metals, whereas the magnitude of $\bar{\sigma}$ controls the plastic deformation. $\bar{\sigma}$ is called the equivalent (or effective) stress. The other important invariants are the maximum principal stress σ_I and the minimum principal stress σ_III. The difference between the maximum and minimum principal stresses is equal to twice the maximum shear stress, which is also an invariant.

The deformation of a body is often described in terms of strain ε_{ij}, which is defined as

$$\varepsilon_{ij} = \frac{1}{2}\left(\frac{\partial u_i}{\partial x_j} + \frac{\partial u_j}{\partial x_i} \right) \tag{2.A4}$$

When the deformation is small (such as in typical elastic deformation), the strain ε_{ij} is also a second-order tensor and follows the same transformation rule as σ_{ij}. However, when the deformation is large, such as in plastic deformation, only the incremental strain $d\varepsilon_{ij}$ can be treated as a second-order tensor. For geometric compatibility to be satisfied, three displacement components u_i and the nine strain components ε_{ij} must be compatible so as not to permit material overlap or creation of holes, violating the continuity relationship. This compatibility condition can be stated mathematically as

$$\frac{\partial^2 \varepsilon_{ii}}{\partial x_j^2} + \frac{\partial^2 \varepsilon_{jj}}{\partial x_i^2} = 2 \frac{\partial^2 \varepsilon_{ij}}{\partial x_i \, \partial x_j} \tag{2.A5}$$

$$\frac{\partial^2 \varepsilon_{ii}}{\partial x_j \, \partial x_k} = \frac{\partial}{\partial x_i}\left(-\frac{\partial \varepsilon_{jk}}{\partial x_i} + \frac{\partial \varepsilon_{ki}}{\partial x_j} + \frac{\partial \varepsilon_{ij}}{\partial x_k} \right)$$

Two important quantities derived from the invariants of the strain tensor $d\varepsilon_{ij}$ are

$$d\varepsilon = \sum_{k=1}^{3} d\varepsilon_{kk} = \Delta(\text{volume})$$

$$(d\bar{\varepsilon})^2 = \frac{4}{9}\left[\frac{1}{2}[(d\varepsilon_{11} - d\varepsilon_{22})^2 + (d\varepsilon_{22} - d\varepsilon_{33})^2 + (d\varepsilon_{11} - d\varepsilon_{33})^2] \right. \quad (2.A6)$$
$$\left. + 3(d\varepsilon_{12}^2 + d\varepsilon_{13}^2 + d\varepsilon_{23}^2)\right]$$

The first invariant, $d\varepsilon$, is equal to the incremental volume change, whereas $d\bar{\varepsilon}$ represents distortion, approximately being equal to twice the maximum shear strain and is proportional to distortion energy. $\bar{\varepsilon}$ is commonly referred to as equivalent or effective strain.

The relationship given by Eqs. (2.A1) through (2.A6) must be satisfied by all solids and liquids. However, the constitutive relationship between σ and ε depends on the specific material. For linear elastic solids, Hooke's law is used, which may be written as

$$\varepsilon'_{ij} = \frac{1}{2G}\sigma'_{ij} = \frac{1+\nu}{E}\sigma'_{ij} \quad (2.A7)$$

where G is the shear modulus, E Young's modulus, and ν Poisson's ratio of the solid. The deviator stress σ'_{ij} and the deviator strain ε'_{ij} are defined as

$$\sigma'_{ij} = \sigma_{ij} - \delta_{ij}\sigma = \sigma_{ij} - \frac{1}{3}\delta_{ij}\sum_{k=1}^{3}\sigma_{kk} \quad (2.A8a)$$

$$\varepsilon'_{ij} = \varepsilon_{ij} - \frac{1}{3}\delta_{ij}\varepsilon = \varepsilon_{ij} - \frac{1}{3}\delta_{ij}\sum_{k=1}^{3}\varepsilon_{kk} \quad (2.A8b)$$

where the Kronecker delta δ_{ij} is equal to 1 when $i = j$ and zero when $i \neq j$. Equation (2.A7) is often written as

$$\varepsilon_{xx} = \frac{1}{E}[\sigma_{xx} - \nu(\sigma_{yy} + \sigma_{zz})]$$
$$\varepsilon_{xy} = \frac{1}{2G}\sigma_{xy} \quad (2.A9)$$

Most metals yield plastically (which is defined as time-independent permanent deformation) when the critical shear stress reaches a critical value. Two different types of yield criteria are used to denote the onset of plastic deformation. The *Tresca yield criterion* states that when the maximum shear stress reaches the shear yield strength, the metal deforms plastically, which may be expressed as

$$\sigma_I - \sigma_{III} = 2k \tag{2.A10}$$

where σ_I and σ_{III} are the maximum and the minimum principal stresses, respectively, and k is the yield strength of the material in shear. The other criterion is the *von Mises criterion,* which states that the yielding occurs when the distortion energy reaches a critical value. This condition may be written as

$$\bar{\sigma} = \sqrt{3}\,k \tag{2.A11}$$

The strain of an elastoplastic solid is made of elastic components, ε_{ij}^e, and plastic components, ε_{ij}^p:

$$\varepsilon_{ij} = \varepsilon_{ij}^e + \varepsilon_{ij}^p \tag{2.A12}$$

When the plastic strain is much larger than the elastic strain component, it is often assumed that

$$\varepsilon_{ij} = \varepsilon_{ij}^p \tag{2.A13}$$

to simplify the mathematical analysis.

The stress–strain relationship of a plastic solid is given by the Prandtl–Reuss relationship, which may be written as

$$d\varepsilon_{ij}^p = \frac{3}{2}\frac{d\bar{\varepsilon}^p}{\bar{\sigma}}\sigma_{ij}' \tag{2.A14}$$

where $\bar{\varepsilon}^p$ and $\bar{\sigma}$ are effective plastic strain and effective stress, respectively. Equation (2.A14) may be written as

$$d\varepsilon_{11}^p = \frac{d\bar{\varepsilon}^p}{\bar{\sigma}}\left[\sigma_{11} - \frac{1}{2}(\sigma_{22} + \sigma_{33})\right]$$

$$d\varepsilon_{12}^p = \frac{3}{2}\frac{d\bar{\varepsilon}^p}{\bar{\sigma}}\sigma_{12} \tag{2.A15}$$

Note that the equations above are similar to Hooke's law, except that ν is replaced by $\frac{1}{2}$ (because plastic deformation does not involve any volume change) and $1/E$ is replaced by $d\bar{\varepsilon}/\bar{\sigma}$. To solve the Prandtl–Reuss equation, the relationship between $\bar{\sigma}$ and $\bar{\varepsilon}$ must be known. One commonly used relationship is the *power law relationship,* which may be written as

$$\bar{\sigma} = C(\bar{\varepsilon}^p)^n \tag{2.A16}$$

where C and n are material constants. n is sometimes called the work-hardening index. Another commonly used relationship is the *rigid–perfectly plastic relationship,* which states that the material remains rigid until a critical stress is reached and then deforms plastically without any work hardening.

The incremental plastic work done per unit volume is

$$dW^p = \sum_{i=1}^{3}\sum_{j=1}^{3} \sigma'_{ij}\, d\varepsilon'^p_{ij} = \bar{\sigma}\, d\bar{\varepsilon}^p \tag{2.A17}$$

APPENDIX 2.B

DETERMINATION OF THE EQUIVALENT STRAIN OF DEFORMED GRAINS

Under a typical sliding condition the material near the surface undergoes large plastic deformation. The strain distribution near the surface can be determined experimentally by sectioning the specimen perpendicular to the surface and by measuring the distortion of the subsurface grains (Dautzenberg and Zaat, 1973). From these approximate measurements of the distorted grain shape the *equivalent* (or sometimes called *effective*) strain distribution can be determined as a function of the depth from the surface.

Consider a spherical grain subjected to shear deformation as shown in Fig. 2.B1. The incremental strains $d\varepsilon_{ij}$ associated with the deformation

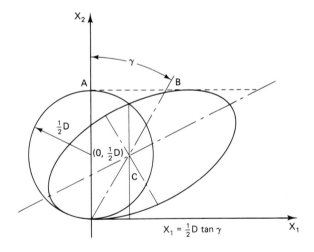

Figure 2.B1 Deformation of an ideal spherical grain under pure shear.

are all equal to zero except $d\varepsilon_{12}$. Then, from the Levy–von Mises equation,

$$d\varepsilon_{ij} = \frac{3}{2}\frac{d\bar{\varepsilon}}{\bar{\sigma}}\sigma'_{ij} \tag{2.B1}$$

where the deviator stress σ_{ij}, the equivalent stress $\bar{\sigma}$, and the effective strain $\bar{\varepsilon}$ are defined as

$$\sigma'_{ij} = \sigma_{ij} - \frac{1}{3}\delta_{ij}\sum_{k=1}^{3}\sigma_{kk}$$

$$\bar{\sigma} = \left[\sum_{i=1}^{3}\sum_{j=1}^{3}\frac{3}{2}\sigma'_{ij}\sigma'_{ij}\right]^{1/2}$$

$$= \left[\frac{1}{2}[(\sigma_{11}-\sigma_{22})^2 + (\sigma_{22}-\sigma_{33})^2 + (\sigma_{33}-\sigma_{11})^2]\right.$$

$$\left. + 3(\sigma_{12}^2 + \sigma_{23}^2 + \sigma_{31}^2)\right]^{1/2} \quad (2.B2)$$

$$d\bar{\varepsilon} = \frac{2}{3}\left[\frac{1}{2}[(d\varepsilon_{11}-d\varepsilon_{22})^2 + (d\varepsilon_{33}-d\varepsilon_{22})^2 + (d\varepsilon_{11}-d\varepsilon_{33})^2]\right.$$

$$\left. + 3(d\varepsilon_{12}^2 + d\varepsilon_{23}^2 + d\varepsilon_{31}^2)\right]^{1/2}$$

The corresponding stress components are obtained as

$$\sigma_{11} = \sigma_{22} = \sigma_{33} \quad (2.B3)$$

$$\sigma_{23} = \sigma_{31} = 0$$

Then the equivalent strain and the equivalent stress are

$$d\bar{\varepsilon} = \frac{2}{\sqrt{3}}d\varepsilon_{12} = \frac{1}{\sqrt{3}}\frac{dl_1}{l_2}$$

$$\bar{\sigma} = \sqrt{3}\,k \quad (2.B4)$$

where l_1 and l_2 are the lengths of two elements parallel to the x_1 and x_2 axes, respectively, and k is the critical shear strength of the material. The total equivalent strain $\bar{\varepsilon}$ may be obtained by integrating $d\bar{\varepsilon}$ as

$$\bar{\varepsilon} = \frac{1}{\sqrt{3}}\tan\gamma \quad (2.B5)$$

where γ is the angle of shear deformation shown in Fig. 2.B1.

In the (x_1, x_2) coordinate system the circle is represented by

$$x_1^2 + \left(x_2 - \frac{D}{2}\right)^2 = \frac{D^2}{4} \quad (2.B6)$$

After pure shear deformation by an angle γ the point (x_1, x_2) of the circle is displaced to (x'_1, x'_2) of the ellipse. The relationship between (x_1, x_2) and (x'_1, x'_2) is

$$x'_1 = x_1 + x'_2\tan\gamma \quad (2.B7)$$

$$x'_2 = x_2$$

Substitution of Eq. (2.B7) into Eq. (2.B6) yields

$$x_1'^2 + x_2'^2(1 + \tan^2\gamma) - 2x_1'x_2'\tan\gamma - x_2'D = 0 \qquad (2.B8)$$

The length of the cord c can be determined by finding the x_2' coordinates of the intersections which the line $x_1' = \tfrac{1}{2}D\tan\gamma$ makes with the ellipse, and then subtracting the smaller x_2' from the larger x_2'. Then, from Eq. (2.B8) we obtain

$$c = D\cos\gamma \qquad (2.B9)$$

The equivalent strain $\bar{\varepsilon}$ can be obtained in terms of c from Eqs. (2.B5) and (2.B9) as

$$\bar{\varepsilon} = \frac{1}{\sqrt{3}}\tan\gamma = \frac{1}{\sqrt{3}}\left(\frac{1}{\cos^2\gamma} - 1\right)^{1/2} = \frac{1}{\sqrt{3}}\left(\frac{D^2}{c^2} - 1\right)^{1/2} \qquad (2.B10)$$

When $c \ll D$, the equation above becomes

$$\bar{\varepsilon} = \frac{1}{\sqrt{3}}\frac{D}{c} \qquad (2.B11)$$

Since metal grains rarely have a spherical shape, an approximate method of determining the strain from nonspherical grains is needed. We will use an approximate method called the *linear intercept method*, where a linear intercept is a line that lies in the (x_1, x_2) plane parallel to the x_2 axis. The length of the linear intercept is the distance between the two intersections which the line makes with the grain boundary (see Fig. 2.B2).

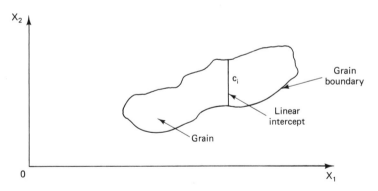

Figure 2.B2 Linear intercept.

By measuring a certain number of linear intersections and using the fact that the average linear intercept possesses a constant ratio to the smallest axis of the ellipsoid generated due to the pure shear deformation of a sphere, the effective strain of the grain can be determined. The mathematical derivation of the linear intercept method is given below.

The volume V of an ellipsoid is

$$V = \frac{4}{3} \frac{\pi abd}{8} \qquad (2.B12)$$

where a, b, and d are the main axes of the ellipsoid (Fig. 2.B3). Since a small areal element in the $x_1 x_3$ plane is $(dx_1 dx_3)$, the volume V may be also expressed as

$$V = \Delta x_1 \Delta x_3 \sum_{i=1}^{n} c_i \qquad (2.B13)$$

where n is equal to the number of small elements.

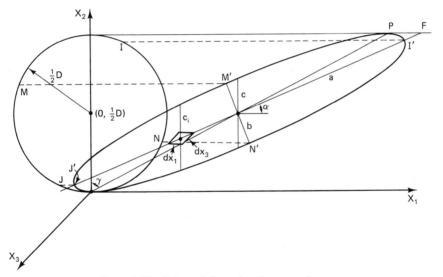

Figure 2.B3 Sphere deformed under pure shear.

The average linear intercept $\bar{c}$ is defined as

$$\sum_{i=1}^{n} c_i = n\bar{c} \qquad (2.B14)$$

The area of the ellipse with a as the major axis and d as the minor axis is

$$\frac{\pi da}{4} \qquad (2.B15)$$

The plane containing the major axis and the minor axis makes the angle α with the $x_1 x_3$ plane, so the projection of the small area element $(\Delta x_1 \Delta x_3)$

on the ad's plane is

$$\frac{\Delta x_1 \, \Delta x_3}{\cos \alpha} \tag{2.B16}$$

Equation (2.B13) can be written as

$$\frac{\Delta x_1 \, \Delta x_3}{\cos \alpha} \cos \alpha \sum_{i=1}^{n} c_i = V \tag{2.B17}$$

Substituting Eqs. (2.B12), (2.B14), and (2.B16) into (2.B17), we obtain

$$\bar{c} \cos \alpha = \frac{2}{3} b \tag{2.B18}$$

Now a new coordinate system is introduced into Fig. 2.B3 with its center at ($\frac{1}{2} D \tan \gamma$, $\frac{1}{2} D$) and axis parallel to the major and minor axes of the ellipse in the $x_1 x_2$ plane. The major axis extends from J' to I' and the minor axis from N' to M'. If this ellipse is transferred back to the original circular shape, point M' will go to M, N' to N, I' to I, and J' to J. Representing these new points in the (x_1, x_2) coordinate system yields $b = D \tan \alpha$ and $a = D/\tan \alpha$.

Then Eq. (2.B18) may be written as

$$\bar{c} \cos \alpha = \frac{2}{3} D \tan \alpha \tag{2.B19}$$

Since $c = D \cos \alpha$, Eq. (2.B19) may be expressed as

$$\bar{c} \cos \alpha = \frac{2}{3} c \frac{\tan \alpha}{\cos \alpha} \tag{2.B20}$$

From Fig. 2.B3, γ may be related to α as

$$\tan \alpha = \frac{\frac{1}{2}D}{\frac{1}{2}D \tan \gamma + PF} \tag{2.B21}$$

which may be expressed as

$$\tan \gamma = \frac{1}{\tan \alpha} - \tan \alpha = \frac{1 - \tan^2 \alpha}{\tan \alpha} = \frac{2}{\tan 2\alpha} \tag{2.B22}$$

Substitution of Eq. (2.B22) into Eq. (2.B20) yields

$$\bar{c} = \frac{2}{3} c (1 + \tan^6 \alpha)^{1/2} \tag{2.B23}$$

For $\alpha < \pi/8$ (i.e., for $\gamma > \pi/3$) c may be approximated within an accuracy of 1% as

$$\bar{c} = \frac{2}{3} c \tag{2.B24}$$

For a sphere the following holds:

$$V = \frac{1}{6}\pi D^3 = \frac{\pi D^2}{4\overline{D}} \tag{2.B25}$$

where $\overline{D}$ is the average diameter of the sphere and $\pi D^2/4$ is the area of the plane which goes through the center of the sphere. Therefore, Eq. (2.B10) may be expressed as

$$\overline{\varepsilon} = \left[\frac{(\overline{D}/\overline{c})^2 - 1}{3}\right]^{1/2} \tag{2.B26}$$

APPENDIX 2.C

ANALYSIS OF PLASTIC DEFORMATION BY THE SLIP-LINE FIELD METHOD

Introduction

Many physical processes are described by partial differential equations. These differential equations can be classified into three groups: hyperbolic, parabolic, and elliptic equations. Hyperbolic equations describe physical processes where discontinuities exist, such as in wave propagation and the deformation of a rigid–perfectly plastic solid along the maximum shear direction. These hyperbolic equations can be solved by the *method of characteristics*. By the use of this method, partial differential equations can be converted to total differential equations which are valid along the characteristic lines. Then the total differential equation can be solved either analytically or numerically along these characteristic lines.

The governing equations for deformation of a rigid–perfectly plastic solid are hyperbolic. When these equations are solved by the method of characteristics, it is called the slip-line field method. The characteristics are lines of the maximum shear stress along which the deformation takes place. These lines are called slip lines.

The method of characteristics will be reviewed first, followed by specific discussions on the slip-line field method.

Consider a partial differential equation

$$P(x, y, z)\frac{\partial z}{\partial x} + Q(x, y, z)\frac{\partial z}{\partial y} = R(x, y, z) \tag{2.C1}$$

where z is a function of x and y and z is specified along the boundary. Let the integral that satisfies Eq. (2.C1) be

$$u(x, y, z) = c \tag{2.C2}$$

Then the variation of z on the integral surface may be written as

$$dz = \frac{\partial z}{\partial x} dx + \frac{\partial z}{\partial y} dy \tag{2.C3}$$

Solving Eqs. (2.C1) and (2.C3) for one of the unknowns, $\partial z/\partial x$, we obtain

$$\frac{\partial z}{\partial x} = \frac{|N|}{|D|} \tag{2.C4}$$

where

$$|N| = \begin{vmatrix} R & Q \\ dz & dy \end{vmatrix}$$

$$|D| = \begin{vmatrix} P & Q \\ dx & dy \end{vmatrix}$$

A unique solution exists if

$$|D| \neq 0 \tag{2.C5}$$

If the denominator is equal to zero (i.e., $|D| = 0$), an infinite number of solutions exist when $|N| = 0$. That is, along the characteristic line which is given by

$$|D| = 0 \qquad \frac{dy}{dx} = \frac{Q}{P} \tag{2.C6}$$

a functional relation given by the following equation is valid:

$$|N| = 0 \qquad Q\, dz = R\, dy \tag{2.C7}$$

Equation (2.C7), which describes the variation of z along the family of curves given by Eq. (2.C6), is completely equivalent to the original differential equation, Eq. (2.C1). Equation (2.C6) is sometimes called the *compatibility equation* and Eq. (2.C7) is known as the *characteristic equation*. The curves given by Eq. (2.C6) are known as *characteristics*.

Now consider a nonhomogeneous second-order partial differential equation with constant coefficients of the following kind:

$$a \frac{\partial^2 z}{\partial x^2} + b \frac{\partial^2 z}{\partial x\, \partial y} + c \frac{\partial^2 z}{\partial y^2} = e \tag{2.C8}$$

For this problem the values of z, $\partial z/\partial x$, and $\partial z/\partial y$ are specified along a certain known curve Σ. The variations of $d(\partial z/\partial x)$ and $d(\partial z/\partial y)$ along the curve Σ on the integral surface may be written as

$$d\left(\frac{\partial z}{\partial x}\right) = \frac{\partial^2 z}{\partial x^2} dx + \frac{\partial^2 z}{\partial x\, \partial y} dy$$

$$d\left(\frac{\partial z}{\partial y}\right) = \frac{\partial^2 z}{\partial x\, \partial y} dx + \frac{\partial^2 z}{\partial y^2} dy \tag{2.C9}$$

Solving Eqs. (2.C8) and (2.C9) for $\partial^2 z/\partial x^2$, one obtains

$$\frac{\partial^2 z}{\partial x^2} = \frac{|N|}{|D|} \qquad (2.C10)$$

where N and D are given in terms of the coefficients. Equation (2.C10) has a unique solution if $|D| \neq 0$ and an infinite number of solutions if $|D| = 0$. Setting $|D| = 0$, we find that:

1. If $b^2 - 4ac < 0$, there are no real roots for dy/dx, indicating that the curve Σ cannot be specified (i.e., elliptic equations).
2. If $b^2 - 4ac > 0$, two real roots of dy/dx exist and therefore there are an infinite number of characteristic curves and solutions (i.e., hyperbolic equations).
3. If $b^2 = 4a$, there is only one real root of dy/dx (i.e., parabolic equations).

Slip-Line Field Method

Now consider the deformation of a rigid–perfectly plastic solid in plane strain (Hill, 1950). The governing equilibrium equations are

$$\frac{\partial \sigma_{xx}}{\partial x} + \frac{\partial \sigma_{xy}}{\partial y} = 0$$

$$\frac{\partial \sigma_{xy}}{\partial x} + \frac{\partial \sigma_{yy}}{\partial y} = 0 \qquad (2.C11)$$

The material yields when maximum shear stress reaches a critical value k, that is,

$$\sigma_I - \sigma_{III} = 2k \qquad (2.C12)$$

where σ_I is the maximum principal stress and σ_{III} is the minimum principal stress. The failure (i.e., yield) envelope and Mohr's circle are shown in Fig. 2.C1. The stresses acting on an infinitesimal element corresponding to a specific shear angle, ϕ, is shown in Fig. 2.C2. In terms of the hydrostatic pressure, p, and the shear angle, ϕ, the stresses may be expressed as

$$\sigma_{xx} = -p - k \sin 2\phi$$
$$\sigma_{yy} = -p + k \sin 2\phi \qquad (2.C13)$$
$$\sigma_{xy} = k \cos 2\phi$$

where $p = (\sigma_{xx} + \sigma_{yy})/2$.

Substituting Eq. (2.C13) into Eq. (2.C11) and noting that the variations of p and ϕ are

$$dp = \frac{\partial p}{\partial x} dx + \frac{\partial p}{\partial y} dp$$

$$d\phi = \frac{\partial \phi}{\partial x} dx + \frac{\partial \phi}{\partial y} dy \qquad (2.C14)$$

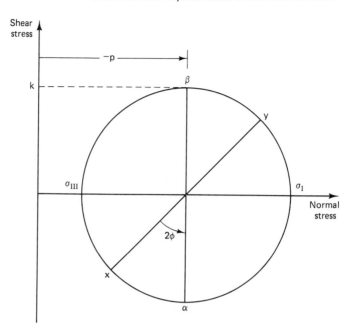

Figure 2.C1 Mohr's circle. Plastic deformation occurs when the shear stress is equal to k.

the equations can be solved for $\partial p / \partial x$ and so on. Solving the compatibility equation $|D| = 0$, we obtain

$$\frac{dy}{dx} = \tan \phi \quad (\alpha \text{ line})$$
$$\frac{dy}{dx} = -\cot \phi \quad (\beta \text{ line})$$
(2.C15)

Equation (2.C15) states that the maximum shear stress directions are the characteristic directions. From the characteristic equation (i.e., $|N| = 0$) we obtain

$$p + 2k\phi = \text{constant} \ldots \text{along the } \alpha \text{ line}$$
$$p - 2k\phi = \text{constant} \ldots \text{along the } \beta \text{ line}$$
(2.C16)

Equation (2.C16) states that along the maximum shear stress directions α and β, the hydrostatic pressure, p, and the shear angle, ϕ, must satisfy Eq. (2.C16). The use of these equations is illustrated in Appendix 3.A.

Equation (2.C16) was obtained considering only the equilibrium condition and the yield criterion. However, we must also satisfy the continuity relationship.

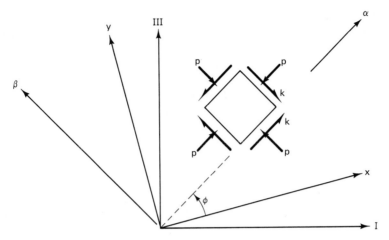

Figure 2.C2 Stresses acting on an element oriented along the α and β directions. I and III are the maximum and minimum principal directions, respectively. ϕ is the shear angle which defines the α direction with respect to the x axis.

From the Prandtl–Reuss equations

$$\dot{\varepsilon}_{ij} = \frac{3}{2} \frac{\dot{\bar{\varepsilon}}}{\bar{\sigma}} \sigma'_{ij} \tag{2.C17}$$

it can be shown that

$$\frac{\dot{\varepsilon}_{xx} - \dot{\varepsilon}_{xy}}{\dot{\varepsilon}_{xy}} = \frac{\sigma_{xx} - \sigma_{yy}}{2\sigma_{xy}} = -\tan 2\phi \tag{2.C18}$$

σ'_{ij} is the deviator stress defined in Eq. (2.A8), and $\bar{\varepsilon}$ and $\bar{\sigma}$ are the effective strain and the effective stress, respectively, which is defined in Eq. (2.B2). In terms of velocity gradients, Eq. (2.C18) may be written as

$$\cos 2\phi \left(\frac{\partial v_y}{\partial y} - \frac{\partial v_x}{\partial x} \right) - \sin 2\phi \left(\frac{\partial v_x}{\partial y} + \frac{\partial v_y}{\partial x} \right) = 0 \tag{2.C19}$$

where v_x and v_y are the x and the y components of velocity u.

The incompressibility condition may be written as

$$\frac{\partial v_x}{\partial x} + \frac{\partial v_y}{\partial y} = 0 \tag{2.C20}$$

The variations of v_x and v_y may be expressed as

$$dv_x = \frac{\partial v_x}{\partial x} dx + \frac{\partial v_x}{\partial y} dy$$

$$dv_y = \frac{\partial v_y}{\partial x} dx + \frac{\partial v_y}{\partial y} dy \tag{2.C21}$$

If U and V are the α and β components of the resultant velocity u, respectively, the particle velocity along the x and y axes may be written as (see Fig. 2.C3)

$$v_x = U \cos \phi - V \sin \phi \qquad (2.C22)$$
$$v_y = U \sin \phi + V \cos \phi$$

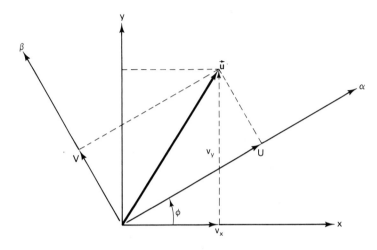

Figure 2.C3 Velocity vector u in the (x, y) coordinate system. U and V are the velocity components along the α and the β directions, respectively. v_x and v_y are the velocity components along the x and y axes, respectively.

Substituting Eq. (2.C22) into Eqs. (2.C19) through (2.C21), and solving the characteristic equation, we obtain

$$dU - V \, d\phi = 0 \ldots \text{along the } \alpha \text{ line} \qquad (2.C23)$$
$$dV + U \, d\phi = 0 \ldots \text{along the } \beta \text{ line}$$

Equation (2.C23) states that the velocity components and the shear angle must vary according to these characteristic equations.

3

GENERATION AND TRANSMISSION OF FORCE AT THE INTERFACE: THE GENESIS OF FRICTION

3.1 INTRODUCTION

One of the three fundamental aspects of tribology is the genesis of friction between sliding surfaces, as discussed in Chapter 1. Frictional force must be present whenever energy-consuming mechanisms exist between and near the sliding surfaces. According to the first law of thermodynamics, the external work done by the frictional force must be equal to the sum of the internal energy increase and the energy dissipated at and near the interface. Internal energy can change through the creation of new surfaces and the residual elastic stress field. Among the energy-consuming mechanisms are plastic and viscous deformation, hysteresis loss (including mechanical, electrical, and magnetic), creation of dislocations, and so on. Fortunately for tribologists, only a few of these energy-consuming mechanisms play important roles at sliding or rolling contacts. In this chapter the genesis of friction is examined by considering the most important energy-consuming mechanisms, since the internal energy change is smaller than the energy dissipated.

Study of frictional behavior dates back to prehistoric times, when humankind used the frictional behavior of materials to their advantage. We still see monuments such as the Egyptian pyramids which could not have been built without a good understanding of frictional behavior. The first comprehensive scientific investigation of frictional behavior is often attributed to Amontons (1699), who established modern phenomenological concepts on friction. Since then, many towering figures of science have worked on tribological problems. Coulomb (1785) advanced the roughness theory and

Euler (1748) tried to explain the difference between the static and dynamic coefficients of friction.

Beginning in the 1930s, Bowden and Tabor (1950) initiated modern pioneering research in tribology, which has affected tribologists for many decades. Although various theories have been advanced to explain the friction mechanism, the adhesion theory they advanced has dominated other theories for many decades (Bowden and Tabor, 1950; Ernst and Merchant, 1940; Rabinowicz, 1965; Shaw and Macks, 1949). Unfortunately, the experimental results often deviate from the adhesion theory. The experimentally observed coefficients of friction are generally much larger than those predicted; the prediction is worst when the experiment is done in vacuum or in an inert environment which best simulates the conditions assumed in deriving the friction coefficient based on the adhesion theory. Furthermore, the chemical compatibility argument that the sliding surfaces of metals with greater mutual solubility have a higher coefficient of friction than a pair of less soluble metals cannot explain the large variations in the frictional behavior of metals with little difference in chemical solubility. There are other basic questions on adhesion, such as those discussed in Chapter 2, which cast doubt on the thesis that friction is entirely due to adhesion. It is clear that adhesion *does* exist in some sliding situations, but it cannot explain all the observed frictional phenomena adequately and satisfactorily in terms of basic principles of nature.

As discussed in Chapter 1, the essence of the adhesion theory is that asperities of sliding surfaces come in contact with opposing asperities and form welded junctions which must be sheared to satisfy the kinematic requirements for sliding. Therefore, the frictional force depends directly on the actual area of contact, which is a function of the applied normal and tangential load. To explain the discrepancy between the experimentally measured and the theoretically predicted friction coefficients, the adhesion theory has relied on the argument that the real area of contact is larger when some of these junctions are under tensile loading, requiring a correspondingly larger force to shear the interface.

Because the adhesion theory of friction emphasizes the importance of adhesion between asperities, a great deal of attention in the past has been devoted to the role of surface energy and the mutual solubility of the contacting materials (Rabinowicz, 1965). It is reasonable to assume that metals with greater solubility will more readily form strong junctions and thus have higher friction and wear coefficients. Many experimental results were presented in support of the argument. However, questions have been raised on the validity of the argument as to whether or not under typical sliding conditions friction can be entirely attributed to adhesion due to mutual dissolution. Furthermore, Buckley (1980) pointed out that many of the surface energy data may be incorrect, since surfaces are easily contaminated by chemisorption and physisorption and the chemical composition of the surface is different from that of the bulk. A more realistic model appears to be that the frictional behavior is controlled by plowing, adhesion, and asperity deformation (Suh and Sin, 1981).

The frictional force depends on the history of sliding. The frictional force undergoes significant changes during the early stages of sliding before reaching steady-state frictional behavior. The time-dependent nature of frictional behavior is a rich source of information in understanding frictional behavior. The difference between the static and kinetic coefficients of friction is well known, but the time-dependent nature of the kinetic coefficient of friction is less well known (Abrahamson et al., 1975). This phenomenon is very important. For example, when the interface is lubricated, the time scale is so expanded that the important frictional phenomenon of interest between well-lubricated surfaces may in fact be a time-expanded version of the early stages of the frictional behavior observed under the dry sliding conditions.

The frictional behavior of materials is important in tribology not only because the frictional force between sliding surfaces is of interest, but also because it generally affects the wear behavior (Rabinowicz, 1977; Suh and Sin, 1981). The delamination wear, which as shown in Chapters 4 and 5 is a predominant wear mechanism for sliding of metal surfaces, is clearly affected by surface traction due to its effect on plastic deformation, crack nucleation, and crack propagation at the subsurface. Also, plowing by wear particles and asperity deformation affect both the wear process and frictional behavior by creating new wear particles and increasing the surface traction.

This chapter describes a somewhat different theory to explain the frictional behavior of materials than those given in other text and reference books. It will be demonstrated that adhesion does not play a significant role at the onset of sliding in typical sliding situations and that the frictional force is generated as a consequence of three different mechanisms: asperity deformation, plowing by wear particles, and adhesion. That is, the coefficient of friction is not a given material property, because it also depends on the mechanical properties of the opposing surface and the environment. Experimental results will first be summarized before presenting theoretical support for the postulated genesis of friction. Finally, the friction space concept is presented as a means of representing the frictional behavior of materials. Various experimental techniques used to measure friction are described in Appendix 3.

3.2 TYPICAL FRICTION TESTS AND EXPERIMENTAL OBSERVATIONS OF FRICTIONAL BEHAVIOR OF METALS

In Chapter 1 the phenomenological aspects of friction were discussed. In studying the fundamental mechanisms of friction, however, it is necessary to investigate beyond the general phenomenological aspects of friction in great detail. The results of such a study will be described in this section before presenting a friction theory in a later section.

A series of experiments were conducted at MIT to study the friction and wear behavior of various combinations of the following materials: Armco iron and AISI 1020, 1045, and 1095 steel.[1] These iron-based metals with differing carbon contents have large differences in hard-phase concentrations and hardness. These materials were chosen to minimize chemical differences, although one would expect the surface of Armco iron to be substantially different chemically from the steel specimens because of the absence of interstitial carbon atoms in Armco iron.

Armco iron was recrystallized at 973 K for 1 hour. AISI 1020, 1045, and 1095 steels were austenitized at 1173 K for 15 minutes, oil quenched, and then tempered at 673 K for 1 hour to obtain a spherodized microstructure. The hardness and the volume fraction of cementite are listed in Table 3.1.

TABLE 3.1 Experimental Materials

Material	Heat Treatment	Vickers Hardness (MPa)	Volume Fraction of Cementite
Armco iron	973 K, 1 hr; air-cooled	980 ± 50	0.0004
AISI 1020 steel	Austenitized at 1173 K, 15 min; oil-quenched; 673 K, 1 hr; air-cooled	1710 ± 100	0.020
AISI 1045 steel	Spheroidized: 1173 K, 15 min; oil-quenched; 673 K, 1 hr; air-cooled	4120 ± 130	0.067
AISI 1095 steel	Spheroidized: 1173 K, 15 min; oil-quenched; 673 K, 1 hr; air-cooled	6080 ± 350	0.142

Some tests were also conducted with OFHC copper, which was polished with 4/0 abrasive paper. These specimens were tested in air sliding against AISI 1020 steel. The initial coefficient of friction, μ_i, was about the same as those obtained with steel and iron specimens.

Samples of 6.35 mm in diameter were tested for friction and wear using crossed-cylinder geometry (see Appendix 3.D). The specimen (rotating cylinder) was rotated by the spindle of a lathe, and the slider (stationary

---

[1] These material specific experiments are described here as a means of isolating the mechanisms that cause friction. Different experimental results may be obtained with different sets of materials, since these fundamental mechanisms affect materials in different ways depending on the testing conditions and material properties.

cylinder) was held stationary in a holder attached to a lathe tool dynamometer which was mounted on the carriage of the lathe. Both normal and tangential forces were measured by a dynamometer–recorder assembly.

Tests were conducted in a purified argon atmosphere except for AISI 1020 steel, where some samples were also tested in air under both lubricated and unlubricated conditions. Water and light machine oil were used as lubricants. The experimental results were obtained under the following conditions: normal load of 1 kg (9.8 N), sliding speed of 0.02 m/s, total sliding distance of 36 m, and at room temperature.

Some of the specimens were sectioned along the sliding direction to measure the slope of the asperity by taking micrographs of the asperities, since the asperities of the machined surfaces were orientation dependent.

Extremely well polished surfaces were slid against each other to investigate the nature of the surface damage after predetermined amounts of sliding. These surfaces were observed using scanning electron microscopy.

The friction and wear coefficients of the iron–carbon system are tabulated in Tables 3.2 and 3.3, respectively. There are several important results worth considering in detail. First, the coefficient of friction changes as a function of the distance slid, especially at the early stage of sliding. It usually has a low initial value and gradually increases until reaching a steady-state value. After it reaches a maximum value the friction coefficient sometimes drops down if the stationary slider is much harder than the moving specimen. The same pair of materials do not show the drop in the

TABLE 3.2 Friction Coefficients

Slider (stationary cylinder)	Coefficient[a]	Specimen (rotating cylinder)			
		Armco Iron	1020 Steel	1045 Steel	1095 Steel
Armco iron	μ_i	0.13	0.20	0.24	0.20
	μ_s	0.71	0.75	0.69	0.76
	μ^*	—	—	—	—
1020 steel	μ_i	0.18	0.20	0.13	0.12
	μ_s	0.55	0.68	0.57	0.65
	μ^*	0.80	—	—	—
1045 steel	μ_i	0.16	0.17	0.17	0.12
	μ_s	0.52	0.53	0.71	0.69
	μ^*	0.77	0.71	—	—
1095 steel	μ_i	0.17	0.17	0.14	0.17
	μ_s	0.51	0.54	0.58	0.67
	μ^*	0.76	0.73	—	—

[a] μ_i initial coefficient of friction; μ_s, steady-state coefficient of friction; μ^*, peak value of the friction coefficient.

TABLE 3.3 Wear Coefficients

Slider (stationary cylinder)	Coefficient[a]	Specimen (rotating cylinder)			
		Armco Iron	1020 Steel	1045 Steel	1095 Steel
Armco iron	K_{sp}	46.0 ± 37.2	127.0 ± 45.0	331.0 ± 268.0	1210.0 ± 120.0
	K_{sl}	25.1 ± 17.5	6.94 ± 2.33	16.4 ± 14.0	17.1 ± 0.5
1020 steel	K_{sp}	6.10 ± 2.75	143.0 ± 88.0	25.3 ± 5.0	42.3 ± 17.6
	K_{sl}	5.23 ± 2.60	85.4 ± 48.4	2.57 ± 2.27	1.80 ± 1.33
1045 steel	K_{sp}	3.59 ± 1.69	15.6 ± 5.9	94.2 ± 17.8	565.0 ± 269.0
	K_{sl}	2.89 ± 2.37	13.9 ± 7.2	49.4 ± 7.5	60.0 ± 17.4
1095 steel	K_{sp}	2.11 ± 0.89	8.83 ± 4.04	7.32 ± 5.10	15.1 ± 7.6
	K_{sl}	2.35 ± 0.95	4.97 ± 2.72	5.24 ± 3.78	14.7 ± 10.3

[a] K_{sp}, wear coefficient of specimen; K_{sl}, wear coefficient of slider. All K values are multiplied by 10^{-4}.

Sec. 3.2 Typical Friction Tests and Experimental Observations

coefficient of friction when their roles are reversed. The initial coefficient of friction is always in the range of about 0.1 to 0.2, regardless of the materials tested and whether or not lubricants are used. Second, the steady-state coefficient of friction and the wear rates are higher when identical metals are slid against each other than when a harder stationary slider is slid against a softer moving specimen. However, when a softer stationary slider is slid against a harder moving specimen, the steady-state coefficients of friction are nearly the same as those of the identical materials sliding against each other. In this case the wear rates of dissimilar pairs of metals are much greater than those of identical metals.

These changes in the friction and wear behavior are related to the changes in the surface topography as shown in Figs. 3.1 and 3.2, which are the micrographs of the slider surface and the specimen surface, respectively. These figures show that when the stationary slider is harder than the specimens, the hard surface is polished to a mirror finish and the high spots of the softer surface acquire the same mirror finish. When the material underneath the polished surface fails, new high spots are created, which become polished to a mirror finish again. This does not happen when the stationary slider is softer than the specimen or when identical metals are slid against each other, because the soft metal wears continuously, creating many wear particles which get entrapped between the sliding interface. In these cases many plowing grooves are observed and the surface always stays rough.

From these experimental results the following observations are made:

1. The coefficients of friction versus sliding distance (or time) may be summarized by using the two typical plots shown in Fig. 3.3. The behavior shown in Fig. 3.3(a) always holds when identical metals are slid against each other. The drop in the coefficient of friction in Fig. 3.3(b) is associated with mutual polishing of the mating surfaces (Suh et al., 1980). The behavior shown in Fig. 3.3(b) results primarily when the hardness of the stationary slider is much greater than the moving specimen. In this case the slider does not wear very rapidly and the asperities of the hard slider are gradually removed, creating a polished hard surface.
2. When wear particles are brushed from the sliding interface, the coefficient of friction decreases to a low value and gradually reaches a steady-state value again, as schematically illustrated in Fig. 3.4 [The effect of wear particles on the friction coefficient was also reported by Suh et al. (1980) and Kuwahara and Masumoto (1980).]
3. The coefficient of friction can differ by as much as 0.2, even for the same pair of chemically identical materials, depending on which is a stationary slider and which is a moving specimen (see Table 3.2).

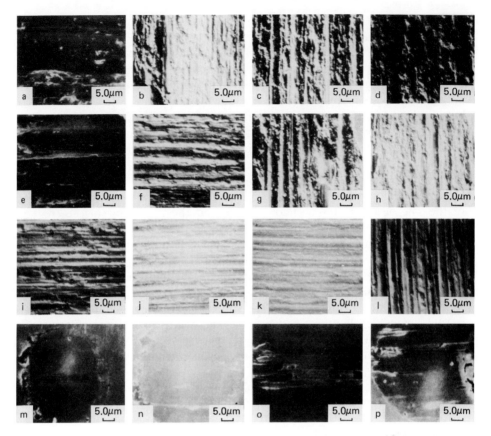

Figure 3.1 Scanning electron micrographs of the surfaces of worn sliders: (a)–(d), iron on iron, 1020, 1045, and 1095 steel, respectively; (e)–(h), 1020 on iron, 1020, 1045, and 1095 steel, respectively; (i)–(l) 1045 on iron, 1020, 1045, and 1095 steel, respectively; (m)–(p) 1095 on iron, 1020, 1045, and 1095 steel, respectively. Corresponding friction and wear coefficients are given in Tables 3.2 and 3.3, respectively. The micrographs of the worn surfaces of the opposing specimens are shown in Figure 3.2. (From Suh, N. P., and Sin, H.-C., "The Genesis of Friction," *Wear*, Vol. 69, 1981, pp. 91–114.)

4. The initial value of the kinetic coefficient of friction is in the neighborhood of 0.1 to 0.2 (but mainly in the range 0.12 to 0.17) for many materials tested—gold on gold, steel on steel, brass on steel, and so on (Suh and Sin, 1981; Suh et al., 1980)—and also regardless of whether or not lubricants are used.
5. When the friction test is performed with extremely well polished surfaces, plowing grooves are formed from the onset of testing.
6. Frictional behavior depends very much on experimental conditions.

Sec. 3.2 Typical Friction Tests and Experimental Observations 71

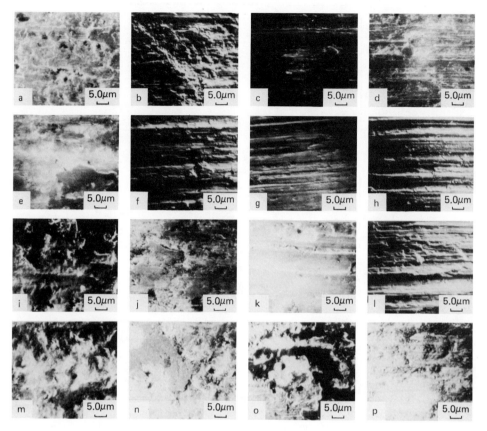

Figure 3.2 Scanning electron micrographs of the surfaces of worn specimens: (a), (e), (i), (m) iron on iron, 1020, 1045, and 1095 steel, respectively; (b), (f), (j), (n) 1020 on iron, 1020, 1045, and 1095 steel, respectively; (c), (g), (k), (o) 1045 on iron, 1020, 1045, and 1095 steel, respectively; (d), (h), (l), (p) 1095 on iron, 1020, 1045, and 1095 steel, respectively. The micrographs of the worn surfaces of the sliders (i.e., counterface) are given in Figure 3.1 and the corresponding friction and wear coefficients are given in Tables 3.2 and 3.3, respectively. (From Suh, N. P., and Sin, H.-C., "The Genesis of Friction," *Wear*, Vol. 69, 1981, pp. 91–114.)

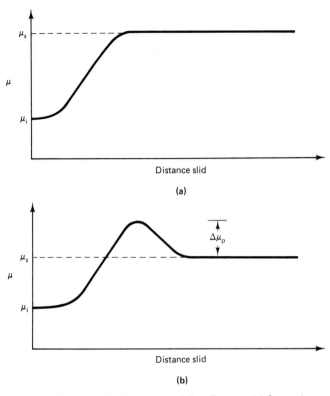

Figure 3.3 Coefficient of friction versus sliding distance: (a) for an Armco iron slider sliding against an Armco iron specimen ($\mu_i = 0.13$, $\mu_s = 0.71$); (b) for an AISI 1095 steel slider sliding against an Armco iron specimen ($\mu_i = 0.17$, $\mu_s = 0.51$, $\Delta\mu_p = 0.25$). (From Suh, N. P., and Sin, H.-C., "The Genesis of Friction," *Wear*, Vol. 69, 1981, pp. 91–114.)

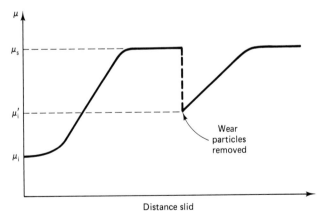

Figure 3.4 Effect of removing wear particles for an Armco iron slider sliding against an Armco iron specimen ($\mu_i = 0.13$, $\mu_s = 0.71$, $\mu_i' = 0.4$). (From Suh, N. P., and Sin, H.-C., "The Genesis of Friction," *Wear*, Vol. 69, 1981, pp. 91–114.)

3.3 GENESIS OF FRICTION

The experimental results clearly indicate that the observed friction coefficients cannot be explained solely in terms of the adhesion theory. For example, the effect of entrapped wear particles and the existence of μ_i, which is independent of environmental conditions and materials tested, cannot be explained by the adhesion theory. The theory is further defied by the dramatic changes in the coefficient of friction when the role of the slider and the specimen is reversed.

The experimental results discussed in the preceding section indicate that the coefficient of friction between the sliding surfaces is due to the various combined effects of asperity deformation, μ_d; plowing by wear particles and hard surface asperities, μ_p; and adhesion between the flat surfaces, μ_a. The relative contribution of these components depends on the condition of the sliding interface, which is affected by the history of sliding, the specific materials used, the surface topography, and the environment.

To clarify the time-dependent friction behavior of the materials tested, the plot of friction versus distance slid will be subdivided into the following stages (see Fig. 3.5) and each stage will be discussed qualitatively.

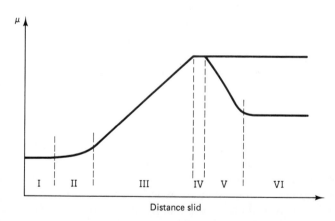

Figure 3.5 Six stages in the frictional force versus distance slid relationship. (From Suh, N. P., and Sin, H.-C., "The Genesis of Friction," *Wear*, Vol. 69, 1981, pp. 91–114.)

Stage I. In this early stage the frictional force is largely a result of plowing of the surface by asperities. Adhesion does not play any significant role in this stage, due to the contaminated nature of the surface. The deformation of asperities does take place at the onset of sliding, which affects the static coefficient of friction. However, in stage I, asperity deformation is not the major factor that determines the coefficient of friction, since the asperities in contact deform as soon as sliding commences and the surface is easily polished. Consequently, the coefficient of friction in this initial stage, μ_i, is largely independent of material combinations, the surface conditions, and the environmental conditions.

Stage II. In this second stage, the frictional force begins to rise slowly due to increase in adhesion. When the interface is lubricated, stage I persists for a long time and stage II may not be present. The slope in stage II can be steeper if the wear particles generated by the asperity deformation and fracture are entrapped between the sliding surfaces and plow the surfaces.

Stage III. This stage is characterized by a steep increase in slope due to the rapid increase in the number of wear particles entrapped between the sliding surfaces as a consequence of higher wear rates. The slope is also affected by the increase in adhesion due to the increase in clean interfacial areas. The force required to deform the asperities will continue to contribute to the frictional force in this stage as long as surface asperities are present. The wear particles are generated when the process of wear particle formation by subsurface deformation, crack nucleation, and crack propagation postulated by the delamination theory of wear is completed (Suh, 1973). Some of the wear particles get entrapped between the surfaces, causing plowing. The plowing will be the greatest when the wear particles are entrapped between metals of nearly equal hardness, because they will penetrate into both surfaces, preventing any slippage between the particle and the surface.

Stage IV. This stage is reached when the number of wear particles entrapped between the interface remains constant. This occurs when the number of the newly entrapped particles equal the number of entrapped particles leaving the interface. The adhesion contribution to friction also remains constant in stage IV. The asperity deformation continues to be important, since the wear by delamination creates new rough surfaces with asperities. However, in most cases asperity deformation is not as important as plowing since asperities deform readily and the frequency of new asperity generation is slow. When two like metals are slid against each other or when the mechanism responsible for stage V does not play a significant role, the coefficient of friction in stage IV is the steady-state frictional coefficient between the two metals.

Stage V. In some cases, such as when a very hard stationary slider is slid against a soft specimen, the asperities of the hard surface are gradually removed, creating a mirror finish, as shown in Fig. 3.6. In this case the frictional force decreases, due to the decrease in plowing and asperity deformation. Plowing decreases since wear particles cannot anchor to a polished hard surface.

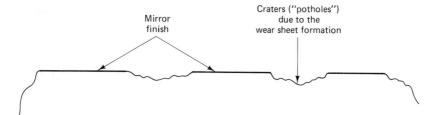

Figure 3.6 Hard stationary surface polished by a soft surface. (From Suh, N. P., and Sin, H.-C., "The Genesis of Friction," *Wear*, Vol. 69, 1981, pp. 91–114.)

Stage VI. Eventually, when the hard surface becomes mirror smooth to a maximum extent, the softer surface also acquires the same mirror finish and the frictional force levels off. The surfaces are never completely smooth since there are always "potholes" due to the creation of delamination wear particles. These craters provide anchoring points for wear particles. When the hard surface is not stationary but moving against the softer surface, the hard surface remains rough, because the number of wear particles entrapped at the interface is large due to the more rapid rate of the stationary soft slider. In this case, stages V and VI are not present.

It should be noted that the sequence by which the six stages of friction shown in Fig. 3.5 occurs is specific to the materials tested, the experimental arrangement, and the environmental conditions. In other cases, with differing materials and conditions, the sequence by which the frictional force changes may be different from that shown in Fig. 3.5. For example, if the surface of a sliding material with low ductility is initially very rough (unlike those used to generate the results discussed in the preceding section) and if the normal load is high, the frictional force may reach a high value as soon as sliding commences due to the rapid generation of wear particles as a result of the immediate fracture of asperities. Regardless of the specific shape of the friction versus sliding distance curve, the interfacial traction is caused primarily by a combined effect of the three basic mechanisms discussed in this chapter.

3.4 ANALYSIS OF THE FRICTION-GENERATING MECHANISMS

The three basic mechanisms (i.e., asperity deformation, plowing, and adhesion) that are responsible for the generation of the friction force in metals will be analyzed in this section. The asperity deformation determines the static coefficient of friction and also affects the dynamic coefficient of friction, since asperities are continuously generated due to delamination of wear sheets. However, the contribution of the asperity deformation to the dynamic coefficient of friction is not large relative to those by plowing and adhesion, since new asperities are generated only with the formation of delaminated wear particles, which often requires a large number of cyclic loading by the asperities of the opposing surface. On the other hand, plowing takes place continuously whenever wear particles are entrapped between the sliding surfaces or when the asperities of the counterface plow clean flat surfaces that come into contact during steady-state sliding. The relative magnitude of these components will be determined approximately by using the slip-line field.

3.4.1 Analysis of the Asperity Deformation

Consider two representative asperities approaching each other as shown in Fig. 3.7. When these asperities come into contact with each other, they

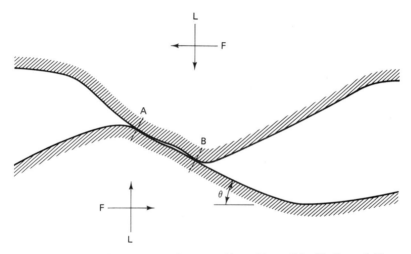

Figure 3.7 Two interacting surface asperities. (From Suh, N. P., and Sin, H.-C., "The Genesis of Friction," *Wear*, Vol. 69, 1981, pp. 91–114.)

have to deform in such a manner that the resulting displacement field is compatible with the sliding direction and that the sum of the vertical components of the surface traction at the contacting asperities must be equal to the applied normal load. A possible slip-line field that satisfies the kinematic condition is given in Fig. 3.8. The solution demands that the shear stress along OA be whatever is necessary to satisfy the condition that $\theta = \alpha$, which is necessary to constrain the resulting deformation in the sliding direction. This would be possible even under the lubricated conditions if the interface OA is not perfectly smooth but rough enough to allow mechanical interlocking. The derivation of the normal and tangential force corresponding to the slip-line field shown in Fig. 3.8 is given in Appendix 3.A.

The general solution is sketched in Fig. 3.9. If it is assumed that the

Sec. 3.4 Analysis of the Friction-Generating Mechanisms

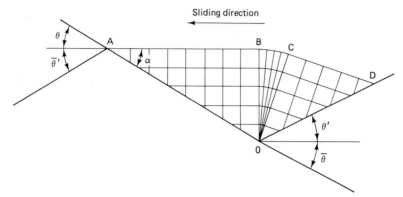

Figure 3.8 Geometrically compatible slip-line field. It can be seen that $\theta > \bar{\theta}$ and $\theta = \alpha$. (From Suh, N. P., and Sin, H.-C., "The Genesis of Friction," *Wear*, Vol. 69, 1981, pp. 91–114.)

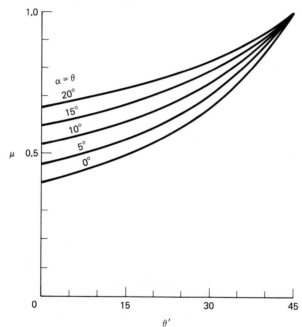

Figure 3.9 Slip-line field solution for friction as a function of the slope of asperities. (From Suh, N. P., and Sin, H.-C., "The Genesis of Friction," *Wear*, Vol. 69, 1981, pp. 91–114.)

asperity deformation is the only phenomenon that takes place at the interface and is entirely responsible for the frictional force under a given load, the coefficient of friction due to asperity deformation varies from 0.39 to 1 as the slope of asperities increases from 0 to 45°. These values are closer to the static coefficient of friction than the dynamic friction coefficient measured during the early stage of sliding, stage I (Kato et al., 1980).

The contribution of the asperity deformation to friction is expected to be the largest when two identical metals are slid against each other since the surfaces always remain rough. In a dynamic situation where the surfaces become smooth, most of the normal load is carried by the entrapped wear particles and the flat contacts. Therefore, the actual contribution of the asperity deformation to the frictional force is expected to be a small fraction of the estimated value in a dynamic situation.

It should be noted that the slip-line analysis done to determine μ_d is similar to Green's results, which were discussed in Chapter 1 (Green, 1955). However, in this case, μ_d does not depend on adhesion since in the analysis the asperity deformation was constrained kinematically. μ_d represents the frictional force due to the deformation of all interacting asperities. Before the onset of sliding between two surfaces, μ_d largely controls the static coefficient of friction.

3.4.2 Analysis of Adhesion Component of the Friction Coefficient

A frictional force can arise due to the adhesion of two nearly flat surfaces. Unlike the deformation of asperities this frictional force is a function of the adhesion between the two opposing surfaces. The adhesion force arises either due to the welding of two nearly flat portions of the surface or when the atoms are brought together in close proximity for interatomic interactions but without welding. The adhesion can also arise at the slopes of two interacting asperities, but its contribution to friction has already been included in deriving μ_d as a subset of kinematically constrained asperity deformation problems. The specific experimental results presented in this chapter show that μ_a is not present (or is at least negligible) at the onset of sliding, probably due to the presence of contaminants on the surface. With the deformation of asperities and exposure of fresh new surfaces, the adhesion between nearly flat surfaces is expected to increase. The exact flat adhesion area cannot be determined a priori, since the applied normal load may also be carried by interacting asperities and entrapped wear particles, although the limiting cases can be analyzed.

Consider two nearly flat surfaces coming into contact as shown in Fig. 3.10. (Sometimes this type of contact is called a "rubbing" contact.) Depending on the nature of adhesion along the interface *ED*, the force required to move the rubbing surfaces with respect to each other varies. When there is no adhesion the force will be zero and when there is complete adhesion it will reach a maximum. The solution to this problem can be obtained again using the slip-line fields, similar to the solution shown in Fig. 3.10. The exact geometric shape of the slip-line field will depend on the boundary condition at ED. The solution sought can be adapted from the work of Challen and Oxley (1979), who derived an expression for the

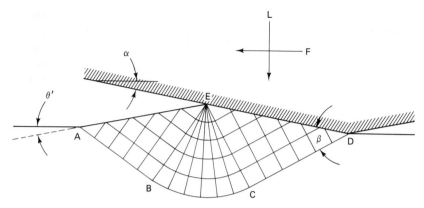

Figure 3.10 Slip-line field for a rubbing contact. (From Suh, N. P., and Sin, H.-C., "The Genesis of Friction," *Wear*, Vol. 69, 1981, pp. 91–114.)

friction coefficient as (see Appendix 3.B)

$$\mu_a = \frac{A \sin \alpha + \cos (\cos^{-1} f - \alpha)}{A \cos \alpha + \sin (\cos^{-1} f - \alpha)} \quad (3.1)$$

where

$$A = 1 + \frac{\pi}{2} + \cos^{-1} f - 2\alpha - 2 \sin^{-1} \frac{\sin \alpha}{\sqrt{1 - f}}$$

f = strength of the adhesion at *ED* as expressed as a fraction of the shear flow strength of the softer material (3.2)

α = slope of the hard asperity

For nearly flat surfaces $\alpha \to 0$. Therefore, Eqs. (3.1) and (3.2) reduce to

$$\mu_a = \frac{f}{A + \sin (\cos^{-1} f)} \quad (3.3)$$

where

$$A = 1 + \frac{\pi}{2} + \cos^{-1} f \quad (3.4)$$

μ_a varies from 0 to 0.39 as f changes from 0 to 1.

The friction coefficient determined by Eqs. (3.1) and (3.2) is based on the assumption that all the applied normal load is carried by the flat interfaces. However, since part of the normal load is also carried by purely elastic contacts, the interacting asperity junctions discussed in the preceding section, and the entrapped wear particles, μ_a under typical sliding conditions should be less than 0.4. The experimental results obtained with the hard AISI

1095 steel slider and the soft Armco iron specimen showed that the steady-state coefficient of friction reached a value of 0.51 when both surfaces were polished smooth, and thus the friction was caused primarily by adhesion. The agreement between the theory and the experiment is reasonable, since asperity interactions and plowing by wear particles must have also contributed to the frictional force.

3.4.3 Analysis of Plowing Component of the Friction Coefficient

The plowing component of the frictional force can be due to the penetration of hard asperities or due to the penetration of wear particles. The plowing due to wear particles is illustrated schematically in Fig. 3.11. When two surfaces are of equal hardness, the particle can penetrate both surfaces. As the surfaces move with respect to each other, grooves will be formed in one or both of the surfaces. When one of the surfaces is very hard and smooth, the wear particle will simply slide along the hard surface and no plowing can occur. However, when the hard surface is very rough, wear particles can anchor in the hard surface and plow the soft surface.

The friction due to plowing was investigated by Sin et al. (1979), who showed that the contribution of plowing to the friction coefficient is very

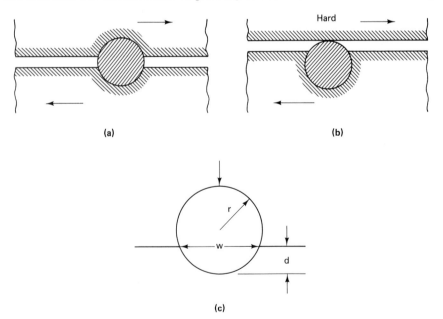

Figure 3.11 Idealized model of wear particle interaction between two sliding surfaces: (a) surfaces of equal hardness; (b) one smooth very hard surface; (c) geometry of the wear particle. (From Suh, N. P., and Sin, H.-C., "The Genesis of Friction," *Wear*, Vol. 69, 1981, pp. 91–114.)

Sec. 3.4 Analysis of the Friction-Generating Mechanisms

sensitive to the ratio of the radius of curvature of the particle to the depth of penetration. The friction coefficient by plowing μ_p is given by (Appendix 3.C):

$$\mu_p = \frac{2}{\pi}\left\{\left(\frac{2r}{w}\right)^2 \sin^{-1}\frac{w}{2r} - \left[\left(\frac{2r}{w}\right)^2 - 1\right]^{1/2}\right\} \quad (3.5)$$

where w is the width of the penetration and r is the radius of curvature of the particle. For the experiments discussed in Section 3.2, the ratio w/r measured by sectioning the worn specimen was in the neighborhood of 0.8. Substituting this value into Eq. (3.5), the plowing coefficient of friction is found to be 0.2. This value is in the same range as the decrease in the friction coefficient observed by removing the wear particles from the Armco iron–Armco iron and Armco iron–AISI 1095 steel interfaces, which were 0.31 and 0.16, respectively. The range of possible values of μ_p as a function of the ratio $w/2r$ is shown in Fig. 3.12. The minimum and the maximum values of μ_p are 0 and 1, respectively.

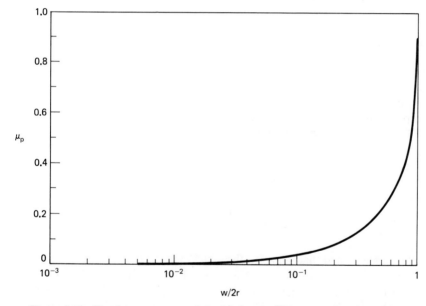

Figure 3.12 Plowing component of the friction coefficient as a function of the ratio of the width to the diameter of the entrapped wear particle. (From Suh, N. P., and Sin, H.-C., "The Genesis of Friction," *Wear*, Vol. 69, 1981, pp. 91–114.)

Plowing not only increases the total frictional force and delamination wear, but also creates small wear particles, which in turn affect the subsequent wear of sliding surfaces. Plowing action forms ridges along the sides of plowed grooves. When these ridges are deformed flat and subjected to repeated loading, some of them become loose wear particles with continued sliding. This is illustrated schematically in Fig. 3.13.

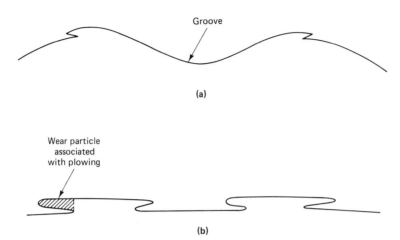

Figure 3.13 Schematic illustration of wear particle formation due to plowing: (a) ridges formed along the sides of the plowed grooves; (b) flattened ridges. (From Suh, N. P., and Sin, H.-C., "The Genesis of Friction," *Wear*, Vol. 69, 1981, pp. 91–114.)

3.5 OTHER FRICTION COMPONENTS

In addition to these three friction components (i.e., adhesion, plowing, and asperity deformation), there can be other contributions to friction. For example, when viscoelastic surfaces with high hysteresis (i.e., high internal damping) loss are slid against each other, external work must be done by the tangential component of the surface traction to overcome the cyclic energy loss due to the hysteresis loss. The fact that such a frictional force component exists can be demonstrated by rolling a hard sphere on an elastomer and by comparing the coefficient of friction with the internal damping of the elastomer. The frictional force change due to the change in the rolling speed correlates with the internal damping fluctuation of the elastomer due to the frequency change, as shown in Fig. 3.14.

Frictional force can also arise when the wear debris is a viscoelastic or plastic substance that sticks to the sliding interface and undergoes repeated deformation consuming energy. This was indeed one of the major contributors to the frictional force when graphite-reinforced polyurethane was slid against a steel slider. The wear debris, consisting of sticky matrix material, spread on the interface and continued to deform, increasing the frictional force whenever the slider passed over this region. (See Chapter 6 for further details.)

Frictional force is a result of the work done at the interface, as discussed earlier in this chapter. Therefore, when any energy-consuming mechanism exists between and near the interface, frictional force must be present to supply the energy, according to the first law of thermodynamics. Therefore, there can be many mechanisms that contribute to the friction, but the

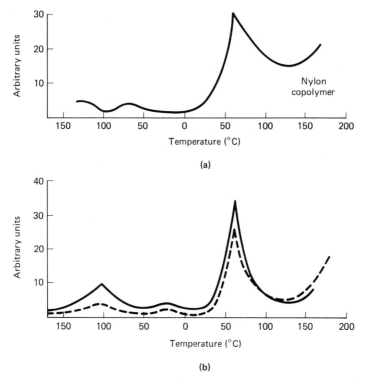

Figure 3.14 (a) Rolling friction μ_r of $\frac{3}{16}$-in. steel ball over the surface of a nylon copolymer, as a function of temperature (load 1050 g). (b) Low-frequency viscoelastic loss data for the same polymer as function of temperature; solid line, damping loss or internal friction; dashed line, damping loss corrected for change in modulus ($-150°C < T < +200°C$; damping loss, 200–1200 cps). (From Ludema, K. C., and Tabor, D., "The Friction and Viscoelastic Properties of Polymeric Solids," *Wear*, Vol. 9, 1966, pp. 329–348.)

primary mechanisms between common metals in normal environment seems to be plowing, adhesion, and asperity deformation.

3.6 RELATIVE CONTRIBUTIONS OF μ_d, μ_a, AND μ_p TO THE OVERALL FRICTION FORCE

Based on the experimental results obtained using the iron–carbon system and the analysis of the adhesion, plowing, and asperity deformation, the relative values of various friction components may be summarized as follows:

μ_d: The friction coefficient due to asperity deformation can be as large as 0.43 to 0.75 when the entire applied normal load is carried by typical surface asperities with a slope of 4 to 20°. It appears that μ_d is responsible for the static coefficient of friction. The reason μ_d is not a major factor in stage I is that once the original asperities deform, asperity interactions cannot take place. This friction component can contribute partially to the steady-state coefficient of friction if new asperities are continuously generated as a consequence of the wear process.

μ_a: The adhesion component of the friction coefficient between metals varies from about 0 to 0.4, depending on the nature of adhesion between the flat part of the interacting surfaces. The low value is for a well-lubricated surface with light lubricant, while the high value is for identical metals sliding against each other without any surface contaminants and oxide layers.

μ_p: The plowing component of the friction coefficient between metals varies from nearly 0 to 1.0 from a theoretical point of view, depending on the depth of penetration, but normally by less than 0.4 in a typical situation. The high values are associated with two identical metals sliding against each other with deep penetration by wear particles, while the low value is obtained when either wear particles are totally absent from the interface or a soft surface is slid against a hard surface with a mirror finish.

The determination of the total friction coefficient in a given condition is complex. It is difficult to determine the relative contributions of μ_d, μ_a, and μ_p to the total friction coefficient, because analyses for μ_d and μ_a were done assuming that the total normal load is carried by either asperities or flat contact areas. In real situations the normal load is apportioned among the asperity contacts, flat adhesion junctions, and the entrapped particles. Therefore, the relative contribution of these friction components to total friction must also be dependent on the normal load distribution. In this case the maximum value of the friction coefficient cannot greatly exceed 1. The friction coefficient can be much larger than 1, however, if each of the mechanisms that contribute to friction can take place sequentially as well as concurrently. Consider, for example, a flat junction and an asperity in contact, as shown in Fig. 3.15. When the flat areas come into contact first and form an adhesion junction, the analysis performed for μ_a is strictly valid. When the asperities also come into contact with further sliding, the normal load must be shared between the asperity junction (as in the results given in Fig. 3.9) and the flat welded junction. Since the flat junctions have higher normal "stiffness" (i.e., the force required to cause unit displacement along the vertical direction), the normal load carried by the flat welded junction may decrease without the corresponding decrease in the tangential force. In this case the friction coefficient can exceed 1 and the asperities will shear along the dashed line if the materials are identical or along the crossed line if the top slider is much harder than the bottom slider.

An important conclusion of this chapter is that the friction coefficient

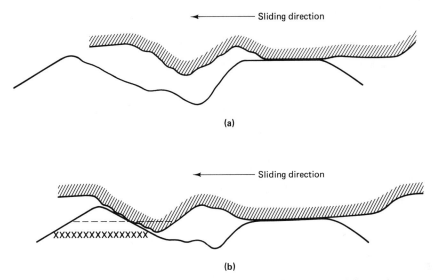

Figure 3.15 Two sliding surfaces in contact: (a) first contact of flat surfaces; (b) flat surface contact and asperity contact. (From Suh, N. P., and Sin, H.-C., "The Genesis of Friction," *Wear*, Vol. 69, 1981, pp. 91–114.)

can vary from test to test even for the same pair of materials, depending on the test conditions. For example, when abrasive particles are entrapped between the sliding surfaces from the very beginning of the sliding action, the initial coefficient of friction can be very large, due to the plowing contribution. Therefore, it no longer suffices to give friction coefficients simply as material properties. Therefore, Suh et al. (1981) proposed a new concept that can properly represent the frictional behavior.

This concept is the *friction space,* which represents the coefficient of friction as a function of adhesion, plowing, and roughness, as shown in Fig. 3.16. The adhesion is expressed in terms of the nondimensional interfacial shear strength, f, of the flat contacts, which is defined in Eq. (3.2). The roughness is plotted in terms of the slope of the surface asperities, while the plowing is given in terms of the ratio of the width of wear particle (or hard asperity) penetration to its radius. The lowest surface in Fig. 3.16 represents the possible values of friction coefficients when there are no asperities (i.e., $\theta = 0$, which forms the lower bound). The θ_i friction surface in the figure represents friction when it is affected by the initial roughness of the machined surface, while the θ^* surface corresponds to the steady-state case, where the roughness is that of delaminated surfaces. As the surface gets rougher and/or the number of the steady-state asperities increase, μ will increase and the friction value will move in the friction space along the μ axis. The important thing to emphasize here is that the friction value can be anywhere within the friction space bound by the $\theta = 0$ surface and the $\theta = \theta^*$ surface, depending on the initial test conditions and material

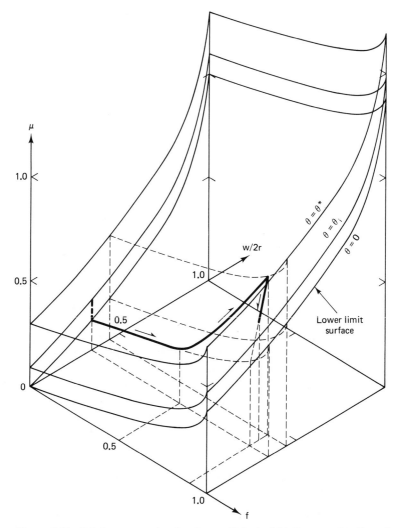

Figure 3.16 Friction space showing the coefficient of friction as a function of adhesion between flat contacts, wear particle penetration, and surface roughness; $f = \tau_s/k$, where τ_s is the shear stress at the interface and k is the shear flow strength of the soft metal; θ is the slope of asperities; $w/2r$ is the ratio of the width of particle penetration to the diameter of the particle. (From Suh, N. P., and Sin, H.-C., "The Genesis of Friction," *Wear*, Vol. 69, 1981, pp. 91–114.)

properties. The friction surface θ^* is plotted from the actual experimental results obtained with Armco iron sliding against AISI 1095 steel. In this case the asperity contribution to friction was 0.3.

Although this figure is not precise, it gives a reasonable picture of what happens in a given situation. The paths of the friction coefficient change shown in Fig. 3.3 are also plotted in this friction space (indicated by a heavy line in Fig. 3.16). The friction coefficient starts from the initial roughness plane θ_i toward a nearly flat surface (i.e., $\theta = 0$) and traces along the paths indicated in Fig. 3.3. In some cases the θ_i surface may be above the θ^* friction surface, depending on the initial surface finish relative to the steady-state surface roughness.

The foregoing argument may be applied to the specific case of gold sliding against gold. When unlubricated gold specimens (which do not have any oxide layer) are slid against each other, the frictional force is the sum of μ_a and μ_p or the sum of a fraction of μ_d, μ_a, and μ_p, depending on the situation. For example, if the normal load is first borne by flat contacts only, friction will be entirely due to adhesion, μ_a, which will reach a maximum value. Then if asperities of the opposing surfaces come in contact, an additional frictional force will be required to deform the asperities. In addition to these frictional forces, the third frictional component may also affect the frictional behavior if the wear particles become wedged in between the sliding surfaces. Therefore, the coefficient of friction between gold on gold may be as high as 1.4 to 1.6 and fluctuate between a maximum and a minimum value.

Lubricated surfaces can have a coefficient of friction whose magnitude will be determined by the degree of plowing and asperity interaction. Lubricated well-polished surfaces are found to have a coefficient of friction of approximately 0.04 for a hard surface and 0.12 for a soft-iron surface. In this case the frictional force is due primarily to plowing, as evidenced by small ploughed grooves left on the surface. However, when wear particles are entrapped between the sliding surfaces of similar hardness, the plowing component of the frictional force can increase, raising the friction coefficient.

In the past the high friction coefficient between like metals has been explained in terms of greater adhesion due to their greater solubility (Rabinowicz, 1965). However, the evidence presented in this chapter shows that the *compatibility* of metals is dictated more by their mechanical behavior than by their chemical behavior. This is quite reasonable since the diffusion rate at typical sliding junctions at room temperature is so low that the solubility between the metals cannot account for the observed wear rates (Suh, 1980).

The fact that plowing of the surfaces by wear particles is a major cause of friction can be illustrated using specially prepared surfaces. Figure 3.17 shows the modulated surface of a copper specimen. A checkerboard surface pattern was created on the copper surface by etching away every other block of the checkerboard. The dimples created on the surface by the etching process provide recessed space of 50 microns deep into which wear particles can be dropped and removed from the sliding contacts. This modulated surface of the copper specimen is slid against a pin of 0.25 inches in diameter without using any lubricants. The pin was made of three different materials: copper, iron, and chromium. These materials have a wide range

Figure 3.17 Scanning electron micrographs of the surfaces of (a) flat Cu disk and (b) modulated Cu disk.

of solubility in copper, from complete to nil. The frictional behavior of these materials is shown in Figure 3.18. The coefficient of friction of the modulated copper surface remains at about 0.2 and does not change with the sliding distance regardless of the specific pin material used. This indicates that chemical compatibility has little effect on the friction coefficient at low sliding speeds. However, when the surface of the copper is flat without dimples and thus cannot remove wear particles from the interface, the results are very different. The coefficient of friction of the flat surface increases with the sliding distance in all cases and the degree of increase

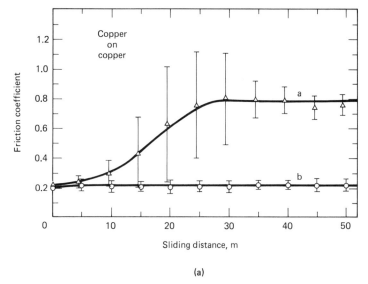

(a)

Figure 3.18a Coefficient of friction *vs.* sliding distance: for Cu pin sliding on (a) flat Cu and (b) modulated Cu disks.

Sec. 3.6 Relative Contributions of μd, μa, and μp to the Overall Friction Force 89

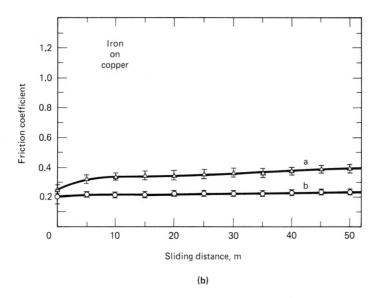

Figure 3.18b Coefficient of friction *vs.* sliding distance: for Fe pin sliding on (a) flat Cu and (b) modulated Cu disks.

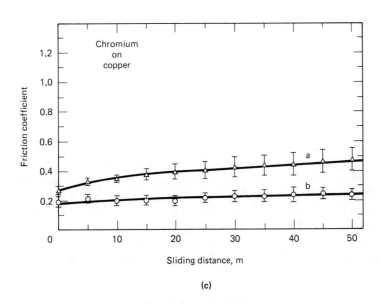

Figure 3.18c Coefficient of friction *vs.* sliding distance: for Cr pin sliding on (a) flat Cu and (b) modulated Cu disks.

depends on the relative hardness of the pin and the specimen, because the degree of plowing is a function of the relative hardness. These results clearly illustrate that plowing by wear particles plays a significant role in determining the frictional force, that the plowing is a strong function of the relative hardness, and that chemical compatibility has little effect at low speed sliding.

3.7 CONCLUDING REMARKS

In this chapter it is shown that the classical adhesion theory of friction cannot explain experimental results by itself and that the frictional coefficient is not an inherent material property. The coefficient of friction depends very much on the sliding conditions, material combinations, and geometry. The coefficient of friction between typical engineering materials is composed of three components: μ_a, due to the deforming asperities; μ_p, due to plowing by wear particles entrapped between sliding surfaces and hard surface asperities; and μ_p, due to adhesion. Typical values of μ_a can range from 0 to 0.4 under typical conditions. However, μ_p can be as large as 1.0 when the depth of penetration by wear particles is large. The friction coefficient due to asperity deformation dictates the static friction coefficient and can range from 0.43 to 0.75, depending on the slope of asperities. Asperity deformation also contributes to the dynamic friction, but its contribution is smaller than indicated since asperities are deformed as soon as they are formed.

The mechanisms responsible for the genesis of friction presented in this chapter should be operative in the case of ceramics, most thermosetting plastics and thermoplastics, and most metals. The frictional behavior of highly crystalline thermoplastics such as Teflon and high-density polyethylene are affected also by the molecular orientation of molecules and their transfer to the counterface. Certain plastics and elastomers such as polyurethane can have high friction due to the deformation of wear debris which is stuck at the sliding interface. The frictional behavior of crystalline thermoplastics is treated in Chapter 6.

Now that we know what causes friction, we should be able to apply this knowledge and control the frictional behavior of materials. Indeed, it is shown in Chapter 9 that friction and wear behavior, even including electrical contact problems, can be manipulated by geometric and other means!

REFERENCES

ABRAHAMSON, E. P., II, JAHANMIR, S., and SUH, N. P., "The Effect of Surface Finish on Wear of Sliding Surfaces," *CIRP Annals,* International Institution for Production Engineering Research, Vol. 24, 1975, pp. 513–514.

AMONTON, G., "De la résistance causée dans les machines," *Mémoires de l'Académie Royale A*, 1699, pp. 275–282.

BOWDEN, F. P., and TABOR, D., *Friction and Lubrication of Solids*, Clarendon Press, Oxford, Part I, 1950, pp. 90–121, Part II, 1964, pp. 52–86.

BOWDEN, F. P., MOORE, A. J. W., and TABOR, D., "The Ploughing and Adhesion of Sliding Metals," *Journal of Applied Physics*, Vol. 14, 1943, pp. 80–91.

BUCKLEY, D. H., "Definition and Effect of Chemical Properties of Surfaces in Friction, Wear, and Lubrication," in *Fundamentals of Tribology*, N. P. Suh and N. Saka, eds., MIT Press, Cambridge, Mass., 1980.

CHALLEN, J. M., and OXLEY, P. L. B., "An Explanation of the Different Regimes of Friction and Wear Using Asperity Deformation Models," *Wear*, Vol. 53, 1979, pp. 229–243.

COULOMB, C. A., "Théorie des machines simples," *Mémoires de Mathématique et de Physics de l'Académie Royale*, 1785, pp. 161–342.

ERNEST, H., and MERCHANT, M. E., "Surface Friction between Metals—A Basic Factor in the Metal Cutting Process," *Proceedings of the Special Seminar on Friction and Surface Finish*, 1940, pp. 76–101 (Publ. by MIT Press, 1969).

EULER, L., "Sur le frottement des corps solides," *Histoire de l'Académie Royale à Berlin*, Vol. iv, 1748, pp. 122–132.

GODDARD, J., and WILMAN, H., "A Theory of Friction and Wear during the Abrasion of Metals," *Wear*, Vol. 5, 1962, pp. 114–135.

GREEN, A. P., Friction between Unlubricated Metals: A Theoretical Analysis of the Junction Model," *Proceedings of the Royal Society of London*, Series A, Vol. 228, 1955, pp. 191–204.

HISAKADO, T., "On the Mechanism of Contact between Solid Surfaces," *Bulletin of the JSME*, Vol. 13, No. 55, 1970, pp. 129–139.

KATO, S., MARUI, E., KOBAYASHI, A., and MATSUBAYUSHI, T., "Characteristics of Surface Topography and Static Friction on Scraped Surface Slideway, Part I and Part II," *Transactions of the American Society of Mechanical Engineers*, Vol. 103, 1980, pp. 97–108.

KOMVOPOULOS, K., SAKA, N., and SUH, N. P., "The Mechanism of Friction in Boundary Lubrication," *Journal of Tribology, ASME Trans.*, to be published in 1985.

KUWAHARA, K., and MASUMOTO, H., "Influence of Wear Particles on the Friction and Wear between Copper Disk and Pin of Various Kinds of Metal," *Journal of Japan Society of Lubrication Engineers*, Vol. 25, 1980, pp. 126–131.

LUDEMA, K. C., and TABOR, D., "The Friction and Viscoelastic Properties of Polymeric Solids," *Wear*, Vol. 9, 1966, pp. 329–348.

RABINOWICZ, E., *Friction and Wear of Materials*, Wiley, New York, 1965, pp. 51–108.

RABINOWICZ, E., *Proceedings of the International Conference on Wear of Materials*, St. Louis, Mo., 1977, ASME, New York, pp. 36–40.

SHAW, M. C., and MACKS, E. F., *Analysis and Lubrication of Bearings*, McGraw-Hill, New York, 1949, pp. 457–461.

SIN, H.-C., SAKA, N., and SUH, N. P., "Abrasive Wear Mechanisms and the Grit Size Effect," *Wear,* Vol. 55, 1979, pp. 163–190.

SUH, N. P., "The Delamination Theory of Wear," *Wear,* Vol. 25, 1973, pp. 111–124.

SUH, N. P., "New Theories of Wear and Their Implications on Tool Materials," *Wear,* Vol. 60, 1980, pp. 1–20.

SUH, N. P., and SIN, H.-C., "The Genesis of Friction," *Wear,* Vol. 69, 1981, pp. 91–114.

SUH, N. P., SIN, H.-C., TOHKAI, M., and SAKA, N., "Surface Topography and Functional Requirements for Dry Sliding Surfaces," *CIRP Annals,* International Institution for Production Engineering Research, Vol. 29, 1980, pp. 413–418.

TZUKIZOE, T., and SAKAMOTO, T., "Friction in Scratching without Metal Transfer," *Bulletin of the JSME,* Vol. 18, No. 115, 1975, pp. 65–72.

APPENDIX 3.A

SLIP-LINE FIELD SOLUTION FOR DEFORMATION OF ASPERITY CONTACTS AT THE SLIDING CONTACT

The deformation of asperities can be analyzed using the method of the slip-line field discussed in Appendix 2.C. For the purpose of slip-line analysis, it is assumed that the material is rigid–perfectly plastic and the asperities undergo plane strain deformation.

Figure 3.A1 gives a possible slip-line field for the asperity contact between sliding surfaces. The interface OA between asperities and the stress-free surfaces OD are both assumed to be straight, with their directions defined by the angles θ and θ' measured from the sliding direction. From the figure it can be noticed that the slip line $ABCD$ is a β line. Using the Hencky relations, the stresses along the slip line can be obtained.

At D,

$$\phi = \theta' + \frac{\pi}{4}$$
$$p = k \tag{3.A1}$$

Along AB,

$$\phi = \alpha - \theta + \frac{\pi}{2}$$
$$p_{AB} = k\left(1 + \frac{\pi}{2} + 2\alpha - 2\theta - 2\theta'\right) \tag{3.A2}$$

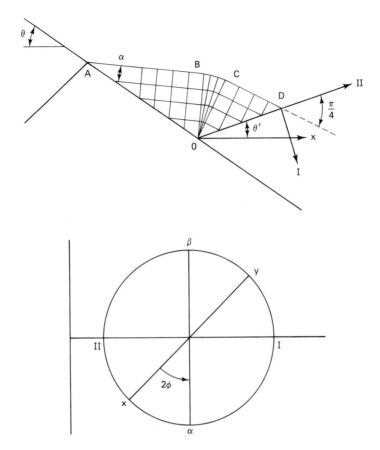

Figure 3.A1 Slip-line field for an asperity contact. (From Suh, N. P., and Sin, H.-C., "The Genesis of Friction," *Wear*, Vol. 69, 1981, pp. 91–114.)

Isolating the junction along *ABO* as shown in Fig. 3.A2, we can find the resultant forces as

$$-F_y = L = (AB)p \cos(\theta - \alpha) - (AB)k \sin(\theta - \alpha)$$
$$+ (OB)k \cos(\theta - \alpha) - (OB)p \sin(\theta - \alpha)$$
$$-F_x = F = (AB)p \sin(\theta - \alpha) + (AB)k \cos(\theta - \alpha)$$
$$+ (OB)k \sin(\theta - \alpha) + (OB)p \cos(\theta - \alpha)$$

(3.A3)

Using the geometric relations, *L* and *F* are expressed as

$$L = (OA)[p \cos \theta + k \sin(2\alpha - \theta)]$$
$$F = (OA)[p \sin \theta + k \cos(2\alpha - \theta)]$$

(3.4A)

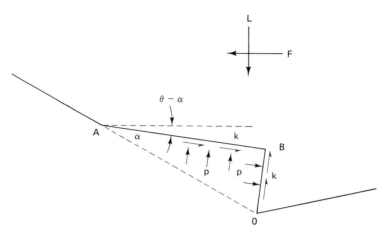

Figure 3.A2 Isolation of the junction of the asperity contact along *ABO* in Fig. 3.A1. (From Suh, N. P., and Sin, H.-C., "The Genesis of Friction," *Wear*, Vol. 69, 1981, pp. 91–114.)

Therefore, the coefficient of friction is

$$\mu_d = \frac{F}{L} = \frac{(1 + \pi/2 + 2\alpha - 2\theta - 2\theta') \sin \theta + \cos (2\alpha - \theta)}{(1 + \pi/2 + 2\alpha - 2\theta - 2\theta') \cos \theta + \sin (2\alpha - \theta)} \quad (3.A5)$$

When, θ, θ', and α are small (θ, θ', $\alpha \to 0$),

$$\mu_d = \frac{1}{1 + \pi/2} = 0.39$$

For several values of α, θ, and θ', the coefficient of friction is plotted in Fig. 3.9.

If the junction does not weld along *AO*, the shear stress along *OA* will be much less than k. The interfacial shear stress is related to the angle α as $\tau = k \cos 2\alpha$. When there is no shear stress along *OA* (i.e., $\alpha = \pi/4$), the junction will slide along *OA* until the junction can deform under the influence of the normal load alone. During this sliding the coefficient of friction can be obtained by substituting $\alpha = \pi/4$ into Eq. (3.A5) as

$$\mu_d = \frac{(1 + \pi - 2\theta - 2\theta') \sin \theta + \cos (\pi/2 - \theta)}{(1 + \pi - 2\theta - 2\theta') \cos \theta + \sin (\pi/2 - \theta)} \quad (3.A6)$$

or

$$\mu = \tan \theta$$

which is the same expression as that derived from the roughness theory of friction.

The slip-line field solution derived above is a general solution. When the sliding occurs, the slip-line field should satisfy the kinematic condition, which corresponds to the case of the slip line AB being parallel to the sliding direction. Therefore, α is equal to θ.

APPENDIX 3.B

ADHESION COMPONENT OF FRICTION

According to the rubbing model of Challen and Oxley (1979), the straight line joining A and D is parallel to the free surface (see Fig. 3.10). Also, the shear force acting on the soft asperity at the interface must act in the direction DE to oppose the motion. These two conditions together with the condition that the slip line must be inclined at an angle of $\pi/4$ to the free surface AE define the slip-line field. It follows from a geometry that

$$AE \sin \theta' = ED \sin \alpha \qquad (3.B1)$$

and

$$AE \sin \frac{\pi}{4} = ED \sin \beta \qquad (3.B2)$$

where θ' is the angle of slope of AE and β is the angle EDC.

The shear force on DE is obtained as

$$\tau_{DE} = fk = k \cos 2\beta \qquad (3.B3)$$

or

$$f = \cos 2\beta \qquad (3.B4)$$

Using the relation $\cos 2\beta = 1 - 2 \sin^2 \beta$, and Eqs. (3.B1) through (3.B3), θ' may be expressed as

$$\theta' = \sin^{-1} \frac{\sin \alpha}{(1 - f)^{1/2}} \qquad (3.B5)$$

The friction coefficient can be obtained by substituting the proper angles ($\theta \to \alpha$, $\alpha \to \beta$) into Eq. (3.A5) in Appendix 3.A as

$$\mu = \frac{(1 + \pi/2 + 2\beta - 2\alpha - 2\theta') \sin \alpha + \cos (2\beta - \alpha)}{(1 + \pi/2 + 2\beta - 2\alpha - 2\theta') \cos \alpha + \sin (2\beta - \alpha)} \qquad (3.B6)$$

or

$$\mu = \frac{A \sin \alpha + \cos (\cos^{-1} f - \alpha)}{A \cos \alpha + \sin (\cos^{-1} f - \alpha)} \qquad (3.B7)$$

where

$$A = 1 + \frac{\pi}{2} + 2\beta - 2\alpha - 2\theta'$$

$$= 1 + \frac{\pi}{2} + \cos^{-1} f - 2\alpha - 2\sin^{-1}\frac{\sin\alpha}{(1-f)^{1/2}} \quad (3.B8)$$

APPENDIX 3.C

PLOWING COMPONENT OF FRICTION

Bowden et al. (1943) were among the first to attempt to model the plowing mechanism in dry sliding. Assuming a spherical or a cylindrical shape for the hard asperity, they obtained simple expressions for the plowing force. Bowden et al. found that when a small hemispherical slider of steel slides on unlubricated indium, plowing accounts for as much as one-third of the total friction. Theoretical calculations for such simple shapes as cones, spheres, and pyramids have also been made by many investigators (Goddard and Wilman, 1962; Hisakado, 1970; Tzukizoe and Sakamoto, 1975), and the experimental results agree fairly closely with the predictions of these models (Hisakado, 1970; Sin et al., 1979; Komvopoulos et al., 1984).

For a conical asperity [Fig. 3.C1(a)], the normal and tangential forces acting on an elemental area, dA, respectively, are

$$dL = p \cos\theta \, dA \quad (3.C1)$$

$$dF = p \sin\theta \cos\gamma \, dA + s \sin\gamma \, dA \quad (3.C2)$$

where $dA = \sec\theta r \, dr \, d\gamma$, and p and s are the normal pressure and the shear stress, respectively, acting on the elemental area as the cone plows. Integration of Eqs. (3.C1) and (3.C2) over the front half of the conical surface where contact occurs yields the total normal and tangential forces, respectively, as

$$L = \frac{\pi w^2}{8} p \quad (3.C3)$$

$$F = \frac{w^2}{4} p \left(\tan\theta + \frac{s}{p} \sec\theta \right) \quad (3.C4)$$

where w is the diameter of indentation. The coefficient of friction can be obtained, by dividing Eq. (3.C4) by Eq. (3.C3), as

$$\mu = \frac{2}{\pi} \left(\tan\theta + \frac{s}{p} \sec\theta \right) \quad (3.C5)$$

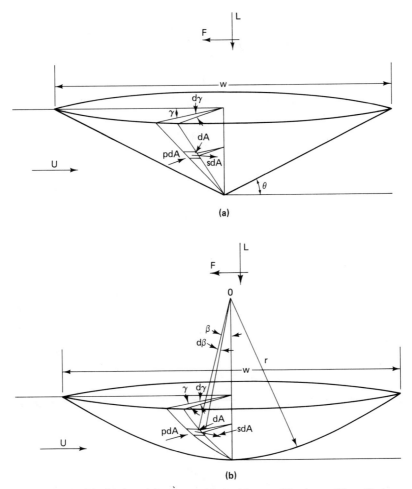

Figure 3.C1 Idealized models of wear debris: (a) cone; (b) sphere. (From Komvopoulos et al., 1984.)

For a spherical asperity [Fig. 3.C1(b)] the normal and tangential forces acting on an infinitesimal area, dA, can be expressed as

$$dL = p \cos \beta \, dA \tag{3.C6}$$

$$dF = p \sin \beta \cos \gamma \, dA + s \sin \gamma \, dA \tag{3.C7}$$

where $dA = r^2 \sin \beta \, d\beta \, d\gamma$, and r is the radius of the sphere. Integrating the two equations above over the front half of the sphere where contact occurs, the total normal and tangential forces can be expressed as

$$L = \frac{\pi w^2}{8} p \tag{3.C8}$$

$$F = pr^2\left\{\sin^{-1}\frac{w}{2r} - \frac{w}{2r}\left[1 - \left(\frac{w}{2r}\right)^2\right]^{1/2}\right\} + 2sr^2 \quad (3.C9)$$
$$\left\{1 - \left[1 - \left(\frac{w}{2r}\right)^2\right]^{1/2}\right\}$$

The coefficient of friction is obtained, by dividing Eq. (3.C9) by Eq. (3.C8), as

$$\mu = \frac{2}{\pi}\left(\frac{2r}{w}\right)^2 \left(\sin^{-1}\frac{w}{2r} - \frac{w}{2r}\left[1 - \left(\frac{w}{2r}\right)^2\right]^{1/2}\right.$$
$$\left. + 2\frac{s}{p}\left\{1 - \left[1 - \left(\frac{w}{2r}\right)^2\right]^{1/2}\right\}\right) \quad (3.C10)$$

where w is the diameter of indentation.

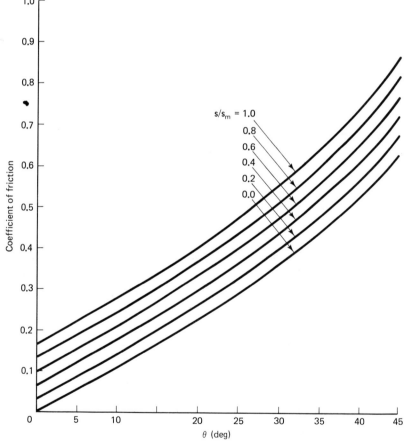

Figure 3.C2 Coefficient of friction for a conical asperity as a function of the angle θ for different boundary "frictional" conditions. (From Komvopoulos et al., 1984.)

It is reasonable to assume that the normal pressure p is approximately equal to the hardness, p_m, of the material being plowed, which is equal to $6s_m$, where s_m is the shear strength. Using this assumption, Eqs. (3.C5) and (3.C10) may be rewritten as

$$\mu = \frac{2}{\pi}\left(\tan\theta + \frac{1}{6}\frac{s}{s_m}\sec\theta\right) \quad (3.\text{C}11)$$

$$\mu = \frac{2}{\pi}\left(\frac{2r}{w}\right)^2\left(\sin^{-1}\frac{w}{2r} - \frac{w}{2r}\left[1 - \left(\frac{w}{2r}\right)^2\right]^{1/2} \right. \quad (3.\text{C}12)$$
$$\left. + \frac{1}{3}\frac{s}{s_m}\left\{1 - \left[1 - \left(\frac{w}{2r}\right)^2\right]^{1/2}\right\}\right)$$

Depending on the interfacial conditions, the ratio s/s_m can assume values between zero and unity. When the ratio s/s_m approaches a value

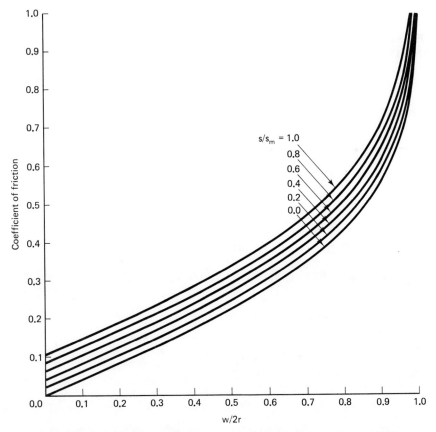

Figure 3.C3 Coefficient of friction for a spherical asperity as a function of the ratio of groove width to diameter of sphere for different boundary "frictional" conditions. (From Komvopoulos et al., 1984.)

close to zero, Eqs. (3.C11) and (3.C12) can be rewritten as

$$\mu = \frac{2}{\pi} \tan \theta \tag{3.C13}$$

$$\mu = \frac{2}{\pi} \left(\frac{2r}{w}\right)^2 \left\{ \sin^{-1} \frac{w}{2r} - \frac{w}{2r} \left[1 - \left(\frac{w}{2r}\right)^2 \right]^{1/2} \right\} \tag{3.C14}$$

Equation (3.C14) is identical to Eq. (3.5).

Figure 3.C2 shows several friction curves obtained from Eq. (3.C12) (conical asperity) as a function of the angle θ and for different interfacial "frictional" conditions. Figure 3.C3 shows the friction curves obtained from Eq. (3.C13) (spherical asperity) as a function of the ratio $w/2r$ for different interfacial frictional conditions.

APPENDIX 3.D

MEASUREMENT OF THE COEFFICIENT OF FRICTION AND WEAR RATES

In many experiments involving friction and wear, a stationary member (i.e., the slider) is rubbed against a moving member (i.e., the specimen). The experimental arrangement for measuring the frictional force and wear rate can be:

1. Cylinder on cylinder (Fig. 3.D1)
2. Annular disk on annular disk (Fig. 3.D2)
3. Pin on disk (Fig. 3.D3)
4. Pin on cylinder (Fig. 3.D4)
5. Two rotating disks (Fig. 3.D5)
6. Two oscillating flat pads, often for fretting wear (Fig. 3.D6)

The frictional force can be measured using any suitable transducers,

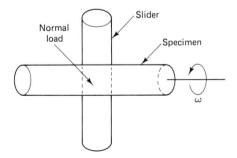

Figure 3.D1 Cylinder on cylinder experiment.

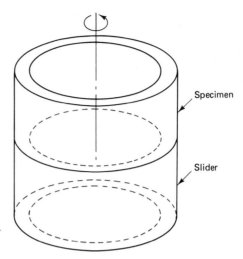

Figure 3.D2 Annular disk on annular disk experiment.

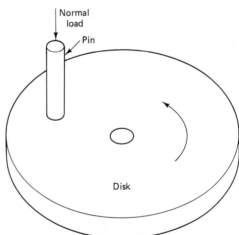

Figure 3.D3 Pin-in-disk experiment.

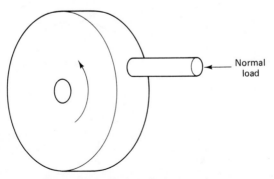

Figure 3.D4 Pin-in-cylinder experiment.

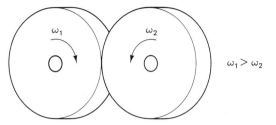

Figure 3.D5 Two rotating disks.

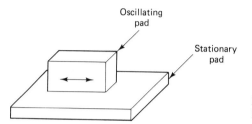

Figure 3.D6 Two flat pads, one of which is under oscillatory motion.

while the normal load is usually applied using dead weights. A strain gage arrangement for the pin-on-disk experiment is shown in Fig. 3.D7.

The wear rate can be measured by determining the weight changes of the specimen and the slider, or by measuring the topological profile of the worn surface, or by measuring the change in the length of a soft pin rubbing against a much more wear-resistant material. One should use the simplest technique that yields the results within the desired accuracy.

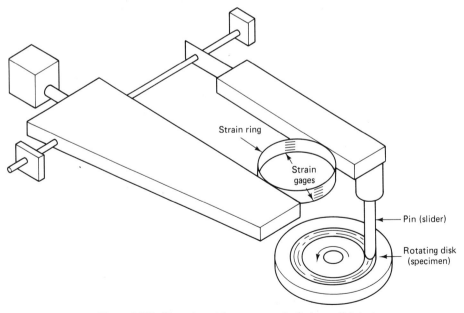

Figure 3.D7 Experimental arrangement of pin-on-disk tests.

4

RESPONSE OF MATERIALS TO SURFACE TRACTION

4.1 INTRODUCTION

In Chapter 3 the genesis of friction between sliding metal surfaces was presented by examining the nature and the magnitude of surface traction at asperity contacts. It was shown that the friction force at an interfacial contact of an elastoplastic solid is due primarily to one of the following mechanisms: plowing of the surface by wear particles and hard asperities, adhesion of the flat interfacial contacts, and asperity deformation. Because these forces are transmitted from the surface to the bulk, the material at the surface and subsurface undergoes physical changes. This chapter examines these changes and how they occur in various materials. From our understanding of materials behavior, we can definitively state that the material response to the externally applied cyclic load must differ from material to material depending on their microstructures and whether the material is elastic, elastoplastic, or viscoelastic. In this chapter these material responses will be characterized by relating the applied stress field to the resulting strain–strain rate–temperature field, including their fracture behavior. To achieve this goal, the stress distribution in elastic and elastoplastic solids will be analyzed first and their effect on the damage process will be assessed.

Materials can exhibit two basic responses when an external force is applied:

1. *Deformation of the surface and subsurface:* Solids can deform elastically, plastically, and in a viscous manner. Many solids exhibit all these deformation characteristics (i.e. elastoplastic, viscoelastic, and vis-

coplastic), depending on the stress and temperature. Elastic and plastic deformation is time independent, whereas viscous deformation is time dependent.
2. *Fracture of Solids:* New surfaces are created when solids fracture, which occurs in two steps. Cracks and voids can be nucleated at or below the surface. From these nucleated or preexisting flaws, cracks can propagate, creating long cracks.

Each of these phenomena constitutes a separate class of mechanics problems and, therefore, will be treated individually in the following sections.

In the analysis the material will be assumed to be homogeneous and isotropic. These are not necessarily realistic assumptions for metals and crystalline polymers, because the surface becomes highly inhomogeneous and anisotropic during sliding, due to large plastic deformation. However, the need to simplify the mathematical analysis justifies these assumptions. On the other hand, these assumptions may not be too unreasonable in studying the subsurface crack nucleation and propagation phenomenon, since the amount of deformation at the subsurface where cracks nucleate and propagate is much less than that at the outermost layer of the surface.

The load applied to the sliding surface by the moving asperity contact is cyclic. Typically, the sliding material experiences a large number of cycles since the asperity contact length is of the order of 10 to 100 μm and the distance between the asperities is of the order of 1000 μm. The material response to this cyclic loading differs depending on whether the material behaves elastically, elastoplastically, or viscoelastically. Under cyclic loading, metals yield at lower yield stress and cracks extend due to the opening or the sliding of the crack tips.

This chapter treats only those situations where the interfacial temperature rise is negligible. This is the case when the sliding speed is low or when the applied load is small. In Chapter 8 the case of high interfacial temperature rise is considered.

4.2 RESPONSE OF ELASTIC SOLIDS TO THE CYCLIC LOAD APPLIED AT SLIDING CONTACTS

When a perfectly elastic solid[1] is slid against a slider, the material at a given position experiences a cyclic loading and elastic deformation as the asperities of the counterface move over the surface. In elastic solids with

---

[1] A *perfectly elastic solid* is defined as that material which undergoes only reversible deformation. It should be noted, however, that many elastic solids which are nominally elastic may involve small local plastic deformation when cracks propagate through these materials or when the stress in a region exceeds the yield stress. The inelastic effects contribute to hysteresis during cyclic loading.

Sec. 4.2 Response of Elastic Solids to the Cyclic Load

negligible anelastic effects, the stresses and strains are completely reversible and therefore the solid returns to its original state after each cyclic loading without leaving any residual stresses and strains. When the tensile stress in the elastic solid exceeds the fracture strength, however, cracks start propagating from preexisting surface or internal cracks, or from cracks nucleated underneath a sharp indenter.[2] When these cracks generated from various parts of the solid link together through crack propagation, loose wear particles are formed. This type of wear process occurs when brittle solids (such as glass, carbides, etc., at room temperature) are slid against hard abrasive surfaces under a load, or when solid particles impinge against the surface at high speeds (Oh and Finnie, 1967, 1970). The underlying basic mechanics that control these crack nucleation and propagation phenomena in a perfectly elastic solid are treated in this section. This is done by determining the stress distribution and the mechanics of crack formation under the load applied at the surface of an elastic solid.

4.2.1 Stress Distribution in Perfectly Elastic Solids Due to Various Interfacial Loading

Beginning in the late nineteenth century, a number of leading mathematicians analyzed stress distributions in a homogeneous, isotropic, linear elastic semi-infinite solid due to various contact loads. The shape of the indenter was idealized as being infinitely sharp or surfaces with various curvatures. In this section the loading of a half-space by four different indenters is considered: a point indenter, a spherical indenter, a line indenter, and a cylindrical indenter. The reason for considering these four different types of asperity geometry is that in sliding situations or in abrasive and erosive environments, some of the contacts are extremely sharp, while in many cases they may be approximated as being spherical or cylindrical. The shape of the indenter is very important because it controls the magnitude of the stress and the volume of the material under stress. The behavior of nominally glassy solids is very sensitive to the volume of the highly stressed material since the failure of these solids depends on the probability of finding a preexisting flaw in the stressed region.

Stress distribution due to a point force. The indentation of a solid by a sharp indenter may be idealized as shown in Fig. 4.1. A concentrated point force P is applied at the origin of the cylindrical coordinate system (r, θ, z) which is located at the surface of a semi-infinite elastic solid. Due to the applied load, the elastic solid undergoes elastic deformation and a stress field is established, which will be determined in this section (Boussinesq, 1885).

---

[2] The fracture criterion for brittle solids is discussed in Appendix 4.A.

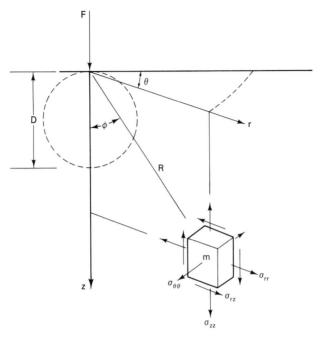

Figure 4.1 Point force P is applied at the origin of the cylindrical coordinate system (r, θ, z). (Adapted from Timoshenko and Goodier, 1970.)

If the displacement components of an infinitesimal element along the r, θ, and z axes are denoted by u, v, and w, respectively, the v component is equal to zero and the u and w components are independent of θ due to symmetry. Furthermore, all the derivatives with respect to θ are also equal to zero. Consequently, the shear stresses $\sigma_{z\theta}$ and $\sigma_{r\theta}$ are also equal to zero. Other stress components are given by Boussinesq (1885) as

$$\sigma_{rr} = \frac{P}{2\pi}\left[(1-2\nu)\frac{1}{r^2} - \frac{z}{r^2(r^2+z^2)^{1/2}} - \frac{3r^2z}{(r^2+z^2)^{5/2}}\right]$$

$$\sigma_{zz} = -\frac{3P}{2\pi}\frac{z^3}{(r^2+z^2)^{5/2}}$$

$$\sigma_{\theta\theta} = \frac{(1-2\nu)P}{2\pi}\left[-\frac{1}{r^2} + \frac{z}{r^2(r^2+z^2)^{1/2}} + \frac{z}{(r^2+z^2)^{3/2}}\right]$$

$$\sigma_{rz} = -\frac{3P}{2\pi}\frac{rz^2}{(r^2+z^2)^{5/2}}$$

(4.1)

The details of these derivations can be found in Appendix 4.B, which is adapted from Timoshenko and Goodier (1970).

If we isolate an infinitesimal element m, the resultant stress acting on the plane perpendicular to the z axis (i.e., the bottom surface of the element

Sec. 4.2 Response of Elastic Solids to the Cyclic Load

m of Fig. 4.1) has a magnitude σ^R which is a function of the applied force F and its position as

$$\sigma^R = (\sigma_{zz}^2 + \sigma_{rz}^2)^{1/2} = \frac{3P}{2\pi} \frac{z^2}{(r^2 + z^2)^2} = \frac{3P}{2\pi} \frac{\cos^2\phi}{r^2 + z^2} \qquad (4.2)$$

Since the ratio $(r^2 + z^2)/\cos^2\phi$ is equal to the square of the diameter D of a circle passing through the origin as shown in Fig. 4.1, Eq. (4.2) states that the resultant stress is constant along the circle, being equal to $3P/2\pi D^2$. The direction of the resultant stress passes through the origin, as the ratio σ_{zz}/σ_{rz} is equal to z/r.

The displacements can be found by substituting Eq. (4.1) into Hooke's law, which may be expressed as

$$u = \frac{(1 - 2\nu)(1 + \nu)P}{2\pi Er} \left[\frac{z}{(r^2 + z^2)^{1/2}} - 1 + \frac{r^2 z}{(1 - 2\nu)(r^2 + z^2)^{3/2}} \right]$$

$$v = 0 \qquad (4.3)$$

$$w = \frac{P}{2\pi E} \left[\frac{(1 + \nu) z^2}{(r^2 + z^2)^{3/2}} + \frac{2(1 - \nu^2)}{(r^2 + z^2)^{1/2}} \right]$$

At $z = 0$, the radial and the vertical components of displacement become

$$u = -\frac{(1 - 2\nu)(1 + \nu)P}{2\pi Er} \qquad (4.4)$$

$$w = \frac{(1 - \nu^2)P}{\pi Er}$$

According to Eq. (4.4), the radial and vertical displacements at the surface decrease inversely with distance from the origin.

Equations (4.1) through (4.3) indicate that the stresses (including the resultant stress σ^R) and the displacement approach infinity at the origin. Obviously, this singularity is a result of the assumption that the cross-sectional area of the point force indenter is equal to zero. In real situations, no physical indenter has a nonzero area of contact. Furthermore, the stress cannot indefinitely increase since the material will either yield or fracture when the stresses exceed the inherent strength of the material, relieving the stress concentration at the point of loading.

Important features of the Boussinesq solution are given in Figs. 4.2 through 4.4 for the case $\nu = 0.25$. Figure 4.2 shows trajectories of the three principal normal stress σ_{11}, σ_{22}, and σ_{33}, which are curves of tangent directions of the principal stresses at each point (Lawn and Swain, 1975). The contours of these stresses are plotted in Fig. 4.3. In these figures the contours of σ_{11} and σ_{33} lie in the rz plane passing through the origin, whereas σ_{22} is perpendicular to the plane, being the hoop stress. σ_{11} is tensile everywhere, whereas σ_{33} is compressive everywhere. The hoop stress σ_{22}

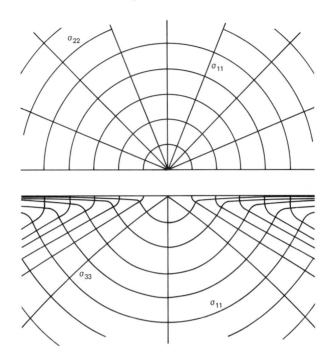

Figure 4.2 Stress trajectories in Boussinesq field. The surface view (top) shows one-half of the surface and the side view (bottom) shows the rz plane passing through the origin. Plotted for $\nu = 0.25$. (After Lawn and Swain, 1975.)

is tensile below the region $\phi < 51.8$ and compressive near the surface. The trajectories of the σ_{22} component are circumferential (i.e., along the θ axis) even in the tensile region below the surface. The trajectories of the σ_{11} component are radial at the surface and circular about the origin below the surface near the z axis. The magnitudes of the tensile stresses, σ_{11} and σ_{22}, are nearly the same at the point of loading. Therefore, under a sharp indenter cracks may run along the z axis below the surface and along the circumferential direction at the surface due to the σ_{11} component, whereas subsurface cracks may propagate radially due to the σ_{22} component from the crack generated along the z axis due to the σ_{11} component. Figure 4.4 shows the magnitudes of the principal stresses, the maximum shear stress, and the hydrostatic pressure as a function of the angular position from the z axis. The stress magnitudes are normalized with respect to $P/\pi R^2$. The tensile stress terms contributed by σ_{rr} and $\sigma_{\theta\theta}$ components of Eq. (4.1) are a sensitive function of Poisson's ratio, completely disappearing when ν reaches the maximum value of 0.5 (i.e., incompressible solid). Therefore, solids with smaller ν develop larger tensile stress and thus are more likely to be brittle (Lawn and Swain, 1975).

The implications of the Boussinesq solution will be discussed later in

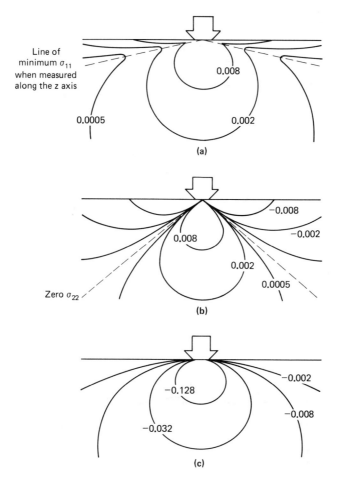

Figure 4.3 Contours of principal normal stresses, (a) σ_{11}, (b) σ_{22}, (c) σ_{33}, in Boussinesq field, shown in plane containing contact axis. Plotted for $\nu = 0.25$. Unit of stress is $P/\alpha \pi a^2$, where contact "diameter" (arrowed) is $2a\sqrt{\alpha}$. Note that the solution is valid only at $R \gg a$. (After Lawn and Swain, 1975.)

this section when we consider the experimental results after other stress fields due to other indenters are reviewed.

Hertzian stress distribution due to a spherical indenter. Consider the case of load being applied to a semi-infinite elastic solid by a smooth spherical indenter. To solve this problem, the semi-infinite solid will be treated as a sphere with an infinite radius. Figure 4.5 shows the cylindrical coordinate system (r, θ, z) and the two spheres in contact. The stress distribution and the displacement field in these spheres can be determined by the superposition of the Boussinesq solution for a point load once the

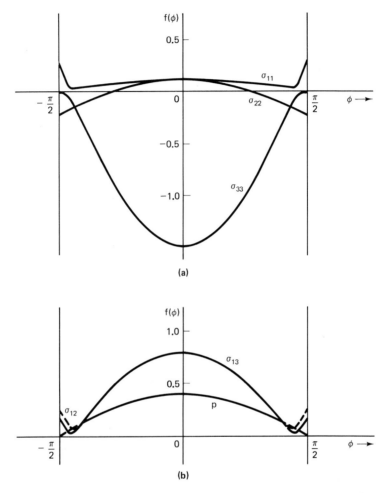

Figure 4.4 Angular variation of principal stress components in Boussinesq field, plotted in terms of dimensionless function $(\sigma_{ij}\pi R^2/\nu P)$: (a) principal normal stresses, σ_{11}, σ_{22}, σ_{33}; (b) maximum shear stress, σ_{13} or σ_{12} (dashed line), and hydrostatic compression, p. Plotted for $\nu = 0.25$. (After Lawn and Swain, 1975.)

interfacial stresses acting at the contact area are known. Hertz (1895) determined the interfacial stresses which were later used by Huber (1904) to determine the stresses and the displacement field in the spheres.

When the contact area is free of friction, the interfacial stress consists of only the normal pressure acting between the two spheres. The pressure distribution is hemispheric, which may be expressed as (see Appendix 4.B for details)

$$(\sigma_{zz})_{z=0} = -p = -p_0 \frac{(a^2 - r^2)^{1/2}}{a} \tag{4.5}$$

Sec. 4.2 Response of Elastic Solids to the Cyclic Load

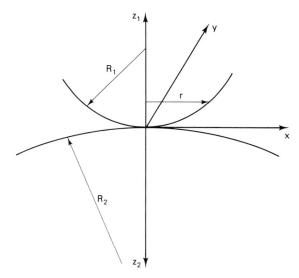

Figure 4.5 Two spherical bodies in contact. At zero load the contact occurs at a point $x = y = z_1 = z_2 = 0$.

where a is the radius of the contact area and p_0 is the maximum pressure acting at $r = 0$ and $z = 0$ (i.e., the center of the contact area). The radius a and the peak pressure p_0 are

$$a = \left[\frac{3P}{16}\frac{k_1 + k_2}{\rho_1 + \rho_2}\right]^{1/3} = \left[\frac{3P}{4}\frac{(k_1 + k_2)R_1R_2}{R_1 + R_2}\right]^{1/3} \quad (4.6)$$

$$p_0 = \frac{3P}{2\pi a^2}$$

where

$$k_i = \frac{1 - \nu_i^2}{E_i} \quad i = 1 \text{ or } 2$$

$$\rho_i = \frac{1}{R_i} = \text{curvature of the sphere } i \quad (4.7)$$

The magnitude of p_0 indicates that the maximum pressure at the center of the contact zone is 1.5 times the average pressure between the spherical contacts. The magnitudes of the other principal stresses, σ_{rr} and $\sigma_{\theta\theta}$, are equal to $(1/2 + \nu)\sigma_{zz}$.

Using the stress distribution at the contact area given by Eq. (4.5) and adopting the Boussinesq solution for a point load, the stress distribution can be found. From such an analysis it can be shown that the maximum tensile stress is caused by the radial stress component at the circular boundary of the contact zone. Its magnitude is given by

$$(\sigma_{rr})_{max} = \frac{1 - 2\nu}{3} p_0 \tag{4.8}$$

Therefore, in the case of perfectly elastic solids, which eventually fracture when the tensile stress exceeds the fracture strength, the failure is expected to occur right around the periphery of the contact area, whereas under a sharp indenter the crack is expected to occur at the point of loading. The results presented so far assumed that there is no friction between the contacting spheres. When there is friction at the contact area, the maximum tensile stress occurs outside the contact area, as shown in Fig. 4.6 (Johnson et al., 1973).

It is also interesting to note that the maximum shear stress due to the Hertzian load is not maximum at the surface but a finite distance below the surface, indicating that yielding occurs first at the subsurface. For example, when $\nu = 0.3$, the maximum shear stress is located on the z axis at a depth of about half of the contact radius, and its magnitude is about $0.31 p_0$. When there is a frictional (i.e., tangential) force in addition to the normal load, the location of the maximum shear stress moves toward the

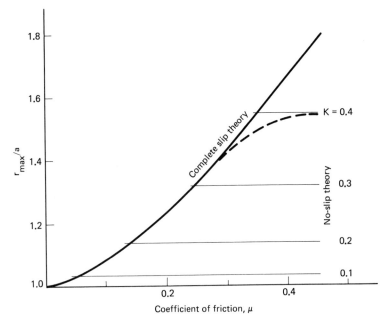

Figure 4.6 Position of the maximum radial tensile stress: (a) as a function of K (no-slip theory); (b) as a function of μ (complete slip theory). An estimated curve (chain line) is interpolated for $K = 0.4$. (From Johnson et al., 1973.)

Sec. 4.2 Response of Elastic Solids to the Cyclic Load

surface. This point will be discussed in greater detail when we consider the plastic deformation of elastoplastic solids under sliding conditions.

Stress distribution due to a line load with normal and tangential components. In many tribological applications the interfacial contact load consists of both normal and tangential components. For these and other problems, the analysis becomes much simpler if a plane strain condition is assumed. This assumption is equivalent to assuming that a uniform load is applied by an infinitely long indenter. Although a point contact and spherical indenters are better approximations of the contact geometry, the plane strain assumption is a reasonable approximation which will not cause much loss in generalization.

Figure 4.7 shows a semi-infinite elastic solid loaded by a concentrated line indenter which is infinitely long along the y direction. The force per unit depth F has a vertical component P and a tangential component Q. The stress distribution in the solid due to the concentrated load is given by (Michell, 1902; Timoshenko and Goodier, 1970; Smith and Liu, 1953)

$$\sigma_{xx} = -\frac{2P}{\pi z}\cos^2\theta \sin^2\theta + \frac{2Q}{\pi z}\sin^3\theta \cos\theta = \Lambda \sin^2\theta$$

$$\sigma_{zz} = -\frac{2P}{\pi z}\cos^4\theta + \frac{2Q}{\pi z}\cos^3\theta \sin\theta = \Lambda \cos^2\theta \quad (4.9)$$

$$\sigma_{xz} = -\frac{2P}{\pi z}\sin\theta \cos^3\theta + \frac{2Q}{\pi z}\cos^2\theta \sin^2\theta = \Lambda \sin\theta \cos\theta$$

where the angle θ and the function Λ are given by

$$\theta = \cos^{-1}\frac{z}{(x^2 + z^2)^{1/2}} \qquad \Lambda = \frac{2}{\pi z}(-P\cos^2\theta + Q\sin\theta \cos\theta)$$

The ratio Q/P is the friction coefficient μ. It should be noted that the contribution of P to σ_{xx} is exactly the same as the contribution of Q to σ_{xz}

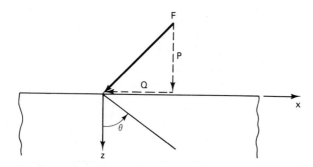

Figure 4.7 Semi-infinite elastic solid loaded by a concentrated load.

except for the sign. Similarly, the contribution of P to σ_{xz} is exactly the same as that of Q on σ_{zz}. In the case of plane strain, the normal strain σ_{yy} is zero. Therefore, from Hooke's law, σ_{yy} can be found as

$$\sigma_{yy} = \nu(\sigma_{xx} + \sigma_{zz}) \tag{4.10}$$

Not surprisingly, the stress components have singularities along the line of loading, just as the concentrated point force had at the origin. The stress fields are similar to those of the point-force case, except that the contours of constant stress are tilted along the direction of the resultant force F due to the frictional force.

Stress distribution due to two-dimensional cylindrical indenter with friction. The force transmission at most asperity contacts does not occur along a line. The load is typically distributed over a contact, as shown in Fig. 4.8. If we make the assumption that the normal and the tangential loads at the interfacial contact are distributed elliptically, as predicted by the Hertzian solution for elastic solids, the surface traction at the asperity contact may be expressed as

$$\sigma_{zz} = \begin{cases} 0 & \text{for } |x| > a \\ -p_0\left(1 - \dfrac{x^2}{a^2}\right)^{1/2} & \text{for } |x| \leq a \end{cases} \tag{4.11}$$

$$\sigma_{xz} = \begin{cases} 0 & \text{for } |x| > a \\ q_0\left(1 - \dfrac{x^2}{a^2}\right)^{1/2} & \text{for } |x| \leq a \end{cases} \tag{4.12}$$

where p_0 and q_0 are, respectively, the maximum normal and tangential stresses acting at the origin (i.e., $x = 0$, $z = 0$) of the coordinate system, and $2a$ is the width of the asperity contact. The stress distribution in the solid due to the distributed load given by Eqs. (4.11) and (4.12) can be obtained by rewriting Eq. (4.9) for the case where the concentrated load is acting at $x = \xi$ and by integrating the resulting equation with respect to

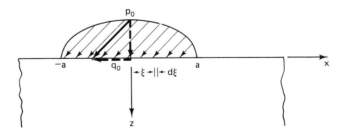

Figure 4.8 Semi-infinite elastic solid loaded by elliptically distributed load. The maximum normal and tangential stresses are p_o and q_o, respectively.

Sec. 4.2 Response of Elastic Solids to the Cyclic Load

ξ from $-a$ to a as

$$\sigma_{xx} = \frac{2q_0}{\pi z a} \int_{-a}^{a} (a^2 - \xi^2)^{1/2} \frac{(x-\xi)^3 z}{[z^2 + (x-\xi)^2]^2} d\xi$$

$$- \frac{2P_0}{\pi a z} \int_{-a}^{a} (a^2 - \xi^2)^{1/2} \frac{z^2(x-\xi)^2}{[z^2 + (x-\xi)^2]^2} d\xi$$

$$\sigma_{zz} = \frac{2q_0}{\pi a z} \int_{-a}^{a} (a^2 - \xi^2)^{1/2} \frac{(x-\xi)z^3}{[z^2 + (x-\xi)^2]^2} d\xi$$

$$- \frac{2P_0}{\pi a z} \int_{-a}^{a} (a^2 - \xi^2)^{1/2} \frac{z^4}{[z^2 + (x-\xi)^2]^2} d\xi \quad (4.13)$$

$$\sigma_{xz} = \frac{2q_0}{\pi a z} \int_{-a}^{a} (a^2 - \xi^2)^{1/2} \frac{(x-\xi)^2 z^2}{[z^2 + (x-\xi)^2]^2} d\xi$$

$$- \frac{2P_0}{\pi a z} \int_{-a}^{a} (a^2 - \xi^2)^{1/2} \frac{z^3(x-\xi)}{[z^2 + (x-\xi)^2]^2} d\xi$$

Integrating Eq. (4.13), the stresses in the semi-infinite solid due to the elliptically distributed normal and tangential contact load are obtained as (Liu, 1950)

$$\sigma_{xx} = \frac{q_0}{\pi}\left[(2x^2 - 2a^2 - 3z^2)\psi + 2\pi\frac{x}{a} + 2(a^2 - x^2 - z^2)\frac{x}{a}\bar{\psi}\right]$$

$$- \frac{P_0}{\pi} z\left(\frac{a^2 + 2x^2 + 2z^2}{a}\bar{\psi} - \frac{2}{a} - 3x\psi\right) \quad (4.14)$$

$$\sigma_{zz} = \frac{q_0}{\pi} z^2\psi - \frac{P_0}{\pi} z(a\bar{\psi} - x\psi)$$

$$\sigma_{xz} = \frac{q_0}{\pi}\left[(a^2 + 2x^2 + 2z^2)\frac{z}{a}\bar{\psi} - 2\pi\frac{z}{a} - 3xz\psi\right] - \frac{P_0}{\pi} z^2\psi$$

in which

$$\psi \equiv \frac{\pi}{k_1} \frac{1 - (k_2/k_1)^{1/2}}{(k_2/k_1)^{1/2}\{2(k_2/k_1)^{1/2} + [(k_1 + k_2 - 4a^2)/k_1]\}^{1/2}}$$

$$\bar{\psi} \equiv \frac{\pi}{k_1} \frac{1 + \left(\frac{k_2}{k_1}\right)^{1/2}}{(k_2/k_1)^{1/2}\{2(k_2/k_1)^{1/2} + [(k_1 + k_2 - 4a^2)/k_1]\}^{1/2}} \quad (4.15)$$

$$k_1 \equiv (a + x)^2 + z^2$$

$$k_2 \equiv (a - x)^2 + z^2$$

The maximum normal and tangential stresses at the boundary p_0 and q_0 are related to the resultant force per unit length P and Q of Eq. (4.9) as

$$p_0 = \frac{2P}{\pi a} \qquad (4.16)$$

$$q_0 = \frac{2Q}{\pi a}$$

The coefficient of friction μ is related to p_0 and q_0 as

$$\mu = \frac{q_0}{p_0} \qquad (4.17)$$

When two elastic cylinders whose longitudinal axes are parallel are in contact, half of the contact area a is related to the resultant normal load per unit length P, the material properties of the cylinders in contact (modulus E and Poisson's ratio ν), and the radius of each cylinder R as

$$a = \left(\frac{2P\Delta}{\pi}\right)^{1/2} \qquad (4.18)$$

where

$$\Delta = \frac{1}{\frac{1}{2}(1/R_1 + 1/R_2)}\left(\frac{1 - \nu_1^2}{E_1} + \frac{1 - \nu_2^2}{E_2}\right) \qquad (4.19)$$

The subscripts 1 and 2 refer to the first and second cylinders. It should be noted that the stress distribution in a semi-infinite solid given by Eq. (4.14) corresponds to the case $R_1 \to \infty$.

The σ_{xx} component of the stress at the boundary $z = 0$ of the semi-infinite solid can be derived from Eq. (4.13) by setting $z = 0$ and evaluating the integral (Liu, 1950). The stress is given as

$$\sigma_{xx} = \begin{cases} 2q_0\left[\frac{x}{a} - \left(\frac{x^2}{a^2} - 1\right)^{1/2}\right] & \text{for } x \geq a \\ 2q_0\left[\frac{x}{a} + \left(\frac{x^2}{a^2} - 1\right)^{1/2}\right] & \text{for } x \leq -a \\ 2q_0\frac{x}{a} - p_0\left(1 - \frac{x^2}{a^2}\right)^{1/2} & \text{for } |x| \leq a \end{cases} \qquad (4.20)$$

The stress distributions at an asperity contact given by Eqs. (4.11), (4.12), and (4.20) are valid only for an elastic semi-infinite solid in contact with an elastic cylinder. However, it is often assumed that such an elliptical stress distribution exists even between elastoplastic solids, although the stress distribution tends to be more uniform across the interface in elastoplastic solids when yielding occurs at the contact area (Johnson, 1968; Hardy, 1971).

The stress distribution due to the elliptically distributed surface traction is plotted for various cases in Figs. 4.9 through 4.15. Figures 4.9 through 4.11 show the distribution of the principal stresses σ_I and σ_{III} and the maximum shear stress τ_{max} at various depths due to the normal and tangential load at the surface. The contours of these stress distributions in the xz plane are also plotted in Figs. 4.12 through 4.14. All of these stresses lie on the xz plane. The contours of these stresses are significantly different from when the surface is loaded by either normal or tangential load alone, as shown in Fig. 4.15, which is a plot of the maximum shear stress distribution when only the normal load is applied. It should be noted that the depth of the maximum shear stress becomes more shallow with the increase in the tangential load.

The stress distribution in a semi-infinite elastic solid due to the cylindrical indenter is similar in its general characteristics to that caused by a spherical indenter. When the coefficient of friction is equal to zero, the maximum tensile stress at the surface is caused by the σ_{xx} component (which is also equal to the maximum principal stress at the surface) and is located at both outer peripheries of the contact area. As the indenter begins to exert a tangential force along the x direction, in addition to the normal load along the z axis, the magnitude of the maximum tensile stress at the trailing edge of the contact area increases, while the tensile stress σ_{xx} at the leading edge decreases and eventually becomes compressive. As shown in Fig. 4.12 there is a large tensile stress σ_{xx} behind the slider (i.e., behind the moving asperity contact). The tensile stress behind the slider is maximum at the surface and decays with depth. When this maximum tensile stress is sufficiently large to initiate a crack from an existing surface flaw, cracks will propagate perpendicular to the surface. However, away from the surface, the direction of these cracks will change due to the change in the principal tensile stress direction as shown in Fig. 4.9. Once these cracks reach a finite depth, they will cease to propagate since the magnitude of the maximum tensile stress decays rapidly away from the surface. This results in the formation of a large number of semi-circular ring cracks behind the slider. When these cracks link up, loose particles are generated.

As was the case with a spherical indenter, the maximum shear stress under a cylindrical indenter also occurs at a finite depth below the surface, when only the normal pressure is applied at the contact area. This is shown in Fig. 4.15. The location of the maximum shear stress is at $z = 0.78a$, for $\nu = 0.3$. As the tangential force is increased at the asperity contact, the location of the maximum shear stress moves toward the surface, reaching the surface when the coefficient of friction is $\frac{1}{9}$. In the case of elastoplastic solids, plastic deformation occurs whenever the maximum shear stress reaches the shear strength of the material.

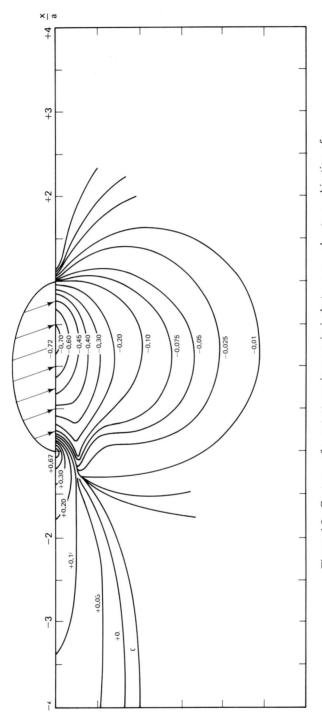

Figure 4.9 Contours of constant maximum principal stress σ_{max} due to combination of elliptical distribution of tangential and normal loads. $Q = P/3$; that is, coefficient of friction is assumed as $\frac{1}{3}$. σ_{max} is obtained by multiplying coefficient on curve by p_0. (From C. K. Liu, 1950.)

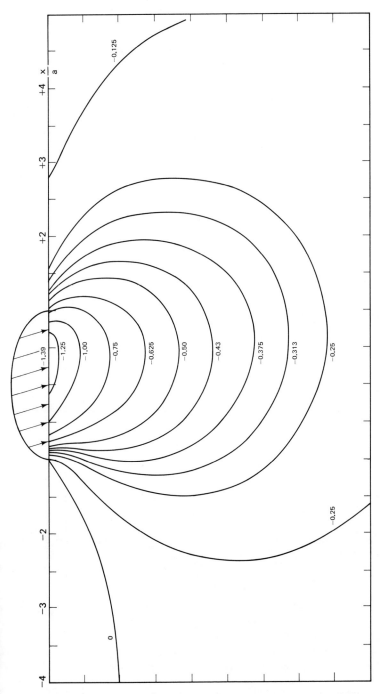

Figure 4.10 Contours of constant minimum principal stresses σ_{min} due to combination of elliptical distribution of tangential and normal loads. $Q = P/3$; that is, the coefficient of friction is assumed as $\frac{1}{3}$. σ_{min} is obtained by multiplying coefficient on curve by P_o. (From C. K. Liu, 1950.)

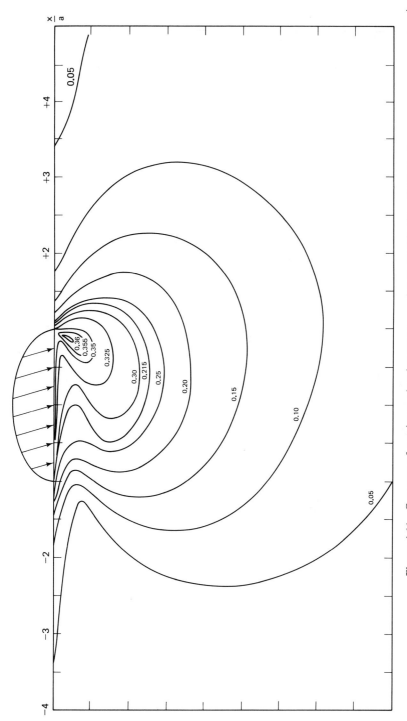

Figure 4.11 Contours of maximum shearing stress τ_{max} due to combination of elliptical distribution of tangential and normal loads. $Q = P/3$; that is, the coefficient of friction is assumed as $\tfrac{1}{3}$. τ_{max} is obtained by multiplying coefficient on curve by P_o. (From C. K. Liu, 1950.)

Sec. 4.2 Response of Elastic Solids to the Cyclic Load

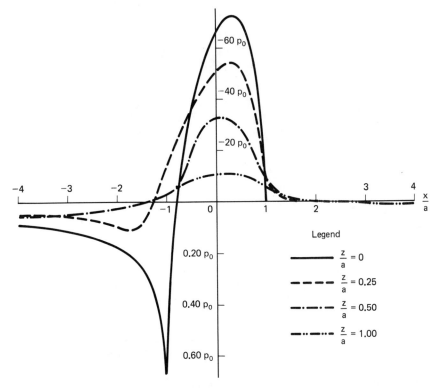

Figure 4.12 Variation of maximim principal stress σ_{max} due to a combination of elliptical distribution of tangential and normal loads with respect to x/a, for several values of z/a. $Q = P/3$; that is, the coefficient of friction is assumed as $\frac{1}{3}$. (From C. K. Liu, 1950.)

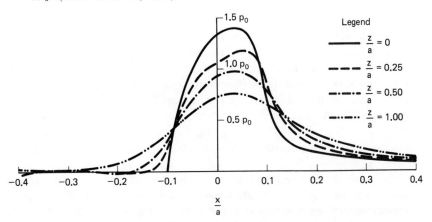

Figure 4.13 Variation of minimum principle stress σ_{min} due to combination of elliptical distribution of tangential and normal loads with respect to x/a, for several values of z/a. $Q = P/3$; that is, the coefficient of friction is assumed as $\frac{1}{3}$. (From C. K. Liu, 1950.)

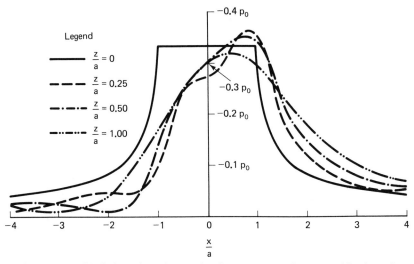

Figure 4.14 Variation of maximum shearing stress τ_{max} due to combination of elliptical distribution of tangential and normal loads with respect to x, for several values of z/a. $Q = P/3$; that is, the coefficient of friction is assumed as $\frac{1}{3}$. (From C. K. Liu, 1950.)

4.2.2 Elastic Stress Fields Under Sharp and Blunt Indenters and Crack Formation in Glassy Solids

The analyses given in this section provide information on the stress fields created under indenters of various shapes, and on the location and the magnitude of the maximum tensile stresses. The information is important in tribology since the wear of glassy solids is caused by fracture which occurs when the maximum tensile stress exceeds the fracture strength of the solid (see Appendix 4.A). However, the fracture of glass may also occur due to shear and after a long time delay. In some glasses the tensile stress at the crack tip is created as a consequence of shear slip of the glass under loading. Furthermore, the cohesive strength of the glass is weakened when moisture is present due to the replacement of strong silanol bridging bonds by hydrogen bonds. Glasses with residual stress fail after a long time delay, since the diffusion and absorption of moisture is a time-dependent phenomenon. Notwithstanding these failure modes of glass, the most important failure of glass occurs when the tensile stress exceeds the cohesive strength.

The analysis showed that under a sharp (point force) indenter, the maximum tensile stress is created right at and underneath the indenter. The trajectories of the principal stresses shown in Fig. 4.2 and the magnitudes of the principal stresses plotted in Fig. 4.4 indicate that cracks may propagate along the z axis from the point of indentation and also along the circumferential direction, both due to the σ_{11} component. They also indicate that radial

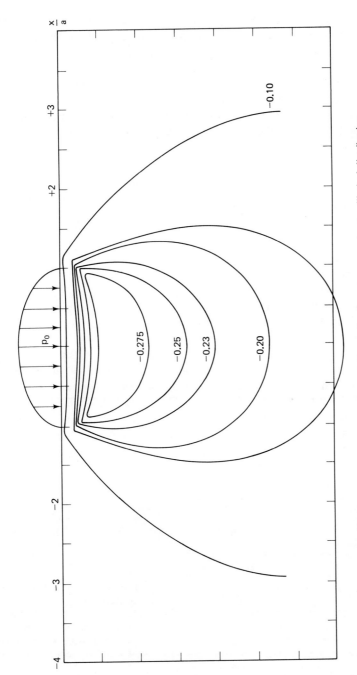

Figure 4.15 Contours of constant maximum shearing stress τ_{max} due to elliptical distribution of normal force. τ_{max} is obtained by multiplying coefficient on cure by p_o. (From C. K. Liu, 1950.)

cracks along the r axis may propagate due to the σ_{22} stress component, possibly starting from the subsurface crack generated along the z axis by the σ_{11} stress component. However when a blunt indenter is used, the maximum tensile stress is radial and is located right at or outside the periphery of the contact area. Since the magnitude of this stress component is inversely proportional to the square of the radius of contact, the magnitude of this tensile stress is much less than the stresses under a sharp indenter.

Figure 4.16 shows the sequence of the ring crack formation in soda-lime glass when indented by a stationary spherical indenter. The cracks develop at the surface and propagate perpendicular to the surface for a short distance and then propagate obliquely, forming a conical crack surface. When a spherical indenter slides over a soda-lime glass surface, exerting a tangential load as well as the normal load, semicircular rings are created at the trailing edge of the circular contact zone, as shown in Fig. 4.17. The shape of the ring crack depends on isotropy of the solid. Figure 4.18 shows the surface traces of ring cracks in three densely packed surfaces of silicon. The rings are not perfect circles because of the anisotropic structure of single-crystal silicon.

When a sharp conical indenter is used, "star"-like radial cracks are formed, following the prediction of the analysis. Figure 4.19 shows radial cracks formed in soda-lime glass when indented by a conical indenter with an included angle of 120°. However, the formation of cracks under sharp indenters is very complicated due to the fact that none of the indenters have infinitely sharp tips and also due to the fact that even normally brittle solids can undergo plastic deformation when stress is applied to a very small microscopic region where flaws are absent. Therefore, even a spherical indenter may act as a sharp indenter when its radius is smaller than a critical radius, at which the transition from the ring crack formation to the radial crack formation occurs.

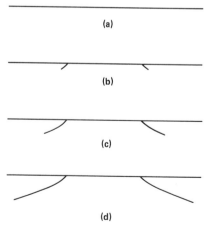

Figure 4.16 Section-and-etch sequence of Hertzian cone crack growth in soda-lime glass in air. Load durations: (a) 0.5 second; (b) 1.4 seconds; (c) 1.7 seconds; (d) 100 seconds. Note growth of "embryo" crack in (a) and (b) prior to sudden full development in (c) and (d). Width of surface ring crack 0.86 mm. (After Mikosza and Lawn, 1971.)

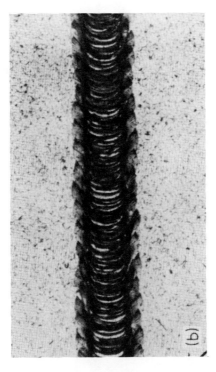
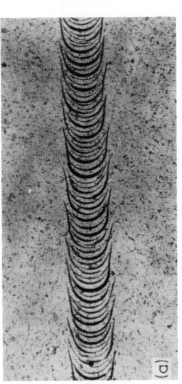

Figure 4.17 Crack patterns on soda-lime glass produced by sliding tungsten carbide sphere, $r = 1.5$ mm. $P = 10$ N, across surface (from left to right): (a) in n-decanol, $f \approx 0.12$; (b) in water, $f \approx 0.44$. Etched surfaces, viewed in transmitted light. Note surface flaws. Width of field 1.25 mm. (After Lawn and Wilshaw, 1975.)

125

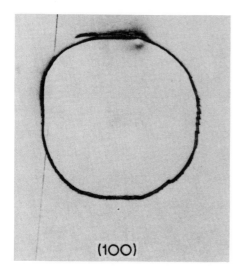
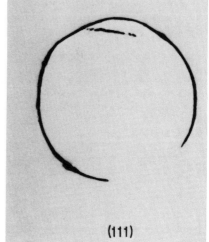
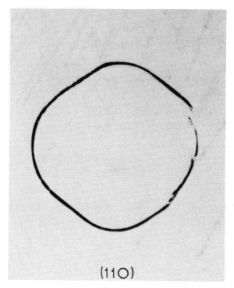

Figure 4.18 Surface traces of Hertzian cracks on three low-index surfaces of silicon. Specimens indented, etched. and viewed in reflected light. Diameter surface cracks 1.0 mm. (After Lawn, 1968.)

The sequence of crack formation under a very small spherical indenter was studied by Evans and Wilshaw (1976). They chose spherical indenters whose diameters were so small that ring cracks could not be produced in silicon nitride and sapphire. They showed that when the indenter radii are smaller than the critical radius, the cracks shown in Fig. 4.20 are generated rather than ring cracks. The figure shows the sequence of crack formation during a loading cycle. Upon commencement of loading with a small spherical indenter, a small plastic indentation is made with no discernible bulging at

Sec. 4.2   Response of Elastic Solids to the Cyclic Load   127

Figure 4.19 "Star" crack system in soda-lime glass, produced by conical indenter (included angle 120°). Note chipping where median and lateral vent cracks intersect. Note also dislodged fragment. Reflected light. Indenter load 160 N. Width of field, 2 mm. (After Lawn and Wilshaw, 1975.)

the adjacent surface. Then shallow radial cracks form at the surface, emanating from the periphery of the crater. As the load is increased, the average crack length increases and new cracks form. Then the subsurface median crack is formed below the indenter, which is circular in shape and symmetric about the z axis. These median cracks join with the radial cracks when they are coplanar or continue to grow with the increase in loading. Then lateral subsurface cracks are formed and propagate approximately parallel to the surface, between the radial cracks. Upon unloading, all of these cracks extend. The tendency for crack extension on unloading decreases as the peak load and the duration of loading at peak load increases. Under large loads, the subsurface lateral cracks extend to the surface, creating loose wear particles. Even after unloading, the radial cracks extend due to the residual strain. In many cases a large number of microcracks form underneath the indenter.

The work of Evans and Wilshaw (1976) suggests that the wear of brittle solids is likely to depend sensitively on the contact size. Therefore, when abrasive particles apply loads at the contact area, the particle size should have a profound effect on the wear mechanisms and rate, since the crack nucleation and propagation phenomena differ depending on the size of the contact. These findings are incorporated in Chapter 7 in analyzing the erosive wear due to solid particle impingement.

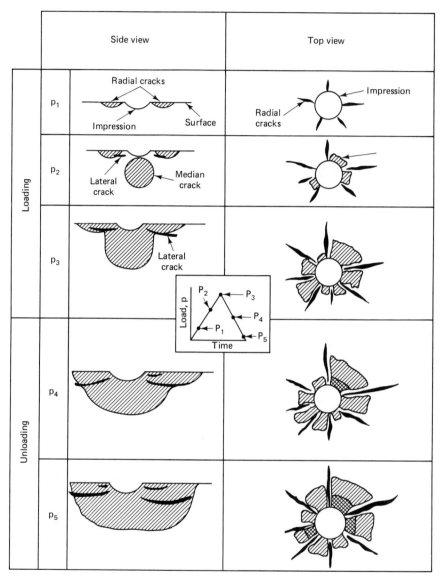

Figure 4.20 Schematic showing the sequence of crack formation and growth events during loading and unloading. (From Evans and Wilshaw, 1976.)

4.3 PLASTIC DEFORMATION OF THE SURFACE LAYER

Many materials used in tribological applications are elastoplastic rather than purely elastic. They are different from elastic solids in that they can undergo large permanent deformation. When these materials are subjected to cyclic loading during sliding, plastic strain accumulates with the increase in the

Sec. 4.3   Plastic Deformation of the Surface Layer   129

loading cycle. Unlike the elastic solids discussed in the preceding section, unstable cracks cannot propagate from the surface of these plastic solids since the surface layer is repeatedly sheared by the sliding motion, changing the crack shape and orientation. Also, it takes much larger energy to sustain the crack growth in elastoplastic solids. However, plastic deformation induces subsurface crack nucleation, which propagate due to the cyclic loading of the surface layer. The subsurface damage processes have dominant effects on the tribological behavior of these materials.

In this section we consider two types of plastic deformation: the formation of surface ridges by plowing and the gross plastic deformation of the surface layer. The former leads to the formation of small thin wear particles, while the latter leads to the generation of large wear sheets by delamination wear. The mechanisms of crack nucleation and propagation are treated in later sections of this chapter.

4.3.1 Formation of Surface Ridges

In Chapter 3 it was shown that plowing occurs as a consequence of sliding action. When plowing occurs, ridges form along the plowed grooves regardless of whether or not chips are formed by the plowing action. These ridges become flattened as shown in Fig. 4.21, which eventually fracture upon repeated loading. When wear occurs primarily by this mechanism, the wear rate is low. This process of wear particle formation has not been analyzed, although many experimental results support the qualitative view.

The plowing process also causes subsurface plastic deformation and may also contribute to the generation and propagation of subsurface cracks. When subsurface cracks propagate rapidly, the measured wear rate will be

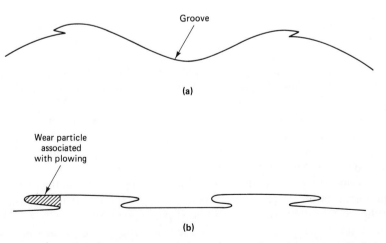

Figure 4.21 Schematic illustration of wear particle formation due to plowing: (a) ridges formed along the sides of the plowed groove; (b) flattened ridges.

due primarily to the large wear particles created when these subsurface cracks link up. However, when subsurface cracks do not propagate, the removal of the surface ridges may be the primary mechanism for wear in normal sliding situations, with the consequent low wear rate.

4.3.2 Deformation of a Semi-infinite Elastoplastic Solid

As the asperities move over the surface, applying a large force at the contact area, plastic deformation of the surface and subsurface occurs. Each time an asperity slides over the surface, small incremental plastic strains accumulate and residual stresses develop in the deformed layer. Since asperities are about 1000 μm or so apart and the asperity contact size is only about 10 μm, the material experiences cyclic loading as it moves over the surface. Understanding this cyclic deformation process is important since anisotropy in the surface layer and subsurface crack generation are direct consequences of this deformation process.

The deformation process can be analyzed using either the finite element method (FEM) or an approximate analytical/numerical technique (Merwin and Johnson, 1963; Johnson and Jefferis, 1963). The finite element method is a powerful tool that can deal with various constitutive relationships, but the computation time is long and costly. The approximate analytical/numerical technique provides a good physical insight and the computation scheme is very simple.

In this section the deformation solution will be obtained using the Merwin–Johnson method (Jahanmir and Suh, 1976) and the results will be compared with the solution obtained using the finite element method (Sin and Suh, 1984).

Determination of stress and strain in elastoplastic solids by the Merwin–Johnson method. Figure 4.22 shows the load exerted on a semi-infinite solid surface by an asperity. For convenience it will be assumed that the asperity is stationary and the semi-infinite solid is moving with velocity U. The contact length is $2a$. We will assume a plane strain condition (i.e., $\sigma_{yy} = 0$ and $\partial/\partial y = 0$) where the y axis is perpendicular to the plane of the paper. The stress distribution at the asperity contact will be assumed to be elliptic over the contact area as given by Eqs. (4.11), (4.12), and (4.20), although the pressure at the contact area may become more uniform as the asperities deform plastically.

To find the stresses in the plastically deforming region and the cumulative plastic deformation after every passage of the asperity, it is necessary to trace the loading cycle of each point as it passes under the asperity. During plastic deformation it is initially assumed that the plastic strains are equal to the elastic strains found from Eq. (4.14) by applying Hooke's law. (The reasons behind this assumption will be discussed later.) The stresses at

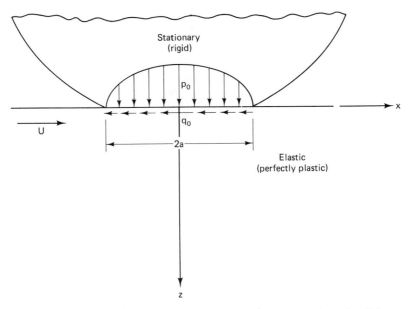

Figure 4.22 Model of a contact between a stationary rigid asperity and a sliding elastic–perfectly plastic plane. (From Jahanmir, S., and Suh, N. P., "Mechanics of Subsurface Void Nucleation in Delamination Wear," *Wear*, Vol. 44, 1977, pp. 17–38; also Jahanmir, S., Ph.D. Thesis, Massachusetts Institute of Technology, 1976.)

each point in the plastic region are then found by employing the Prandtl–Reuss equations for an elastic–perfectly plastic material (see Appendix 2.A).

During plastic deformation the von Mises flow rule (i.e., the rule for determining the locus of yield points) requires that the second invariant of the stress deviators, J'_2, remain constant and equal to k^2, where k is the yield strength in shear. J'_2 is defined as

$$-J'_2 = \frac{1}{2}\sum_{i=1}^{3}\sum_{j=1}^{3}(\sigma'_{ij}\sigma'_{ij}) = \frac{1}{3}\bar{\sigma}^2 = k^2 \tag{4.21}$$

We may now proceed to find the stress–strain relations in the plastic region. The total strain rate, $\dot{\varepsilon}_{ij}$, is the summation of the elastic strain rate, $\dot{\varepsilon}^e_{ij}$, and the plastic strain rate, $\dot{\varepsilon}^p_{ij}$, or

$$\dot{\varepsilon}_{ij} = \dot{\varepsilon}^e_{ij} + \dot{\varepsilon}^p_{ij} \tag{2.A12}$$

The plastic strain rate can be related to the applied stress using the Prandtl–Reuss equations [see Eqs. (2.A14) and (2.A17)]:

$$\dot{\varepsilon}^p_{ij} = \frac{\dot{W}^p}{2k^2}\sigma'_{ij} = \frac{3}{2}\frac{d\bar{\varepsilon}^p}{\bar{\sigma}}\sigma'_{ij} \tag{4.22}$$

where $\dot{W}^p$ is the rate of plastic work done, which may be expressed as

$$\dot{W}^p = \sum_{i=1}^{3}\sum_{j=1}^{3} \dot{\varepsilon}_{ij}^p \sigma'_{ij} = \bar{\sigma}\dot{\bar{\varepsilon}}^p \qquad (2.A17)$$

By substituting Eqs. (2.A17) and (4.22) in Eq. (2A.12), we obtain

$$\dot{\varepsilon}_{ij} = \frac{\dot{\sigma}'_{ij}}{2G} + \frac{\dot{W}_p}{2k^2}\sigma'_{ij} \qquad (4.23)$$

which can be rewritten as

$$\dot{\sigma}'_{ij} = 2G\left(\dot{\varepsilon}'_{ij} - \frac{\dot{W}_p}{2k^2}\sigma'_{ij}\right) \qquad (4.24)$$

Since the rate of work done to deform the material plastically is much larger than the rate of elastic energy stored,

$$\dot{W}^p \simeq \dot{W} \qquad (4.25)$$

where $\dot{W}$ is the rate of total work done. Therefore,

$$\dot{\sigma}'_{ij} = 2G\left(\dot{\varepsilon}'_{ij} - \frac{\dot{W}}{2k^2}\sigma'_{ij}\right) \qquad (4.26)$$

where $\dot{\varepsilon}_{ij}$ is the total strain rate. The basic assumption of the Merwin–Johnson method is that the total strain is identical to the elastic strains obtained by solving Eqs. (4.10) through (4.15), even when the materials yield plastically. Plastic deformation occurs as long as J'_2 equals $-k^2$ and $\dot{W}$ is positive. Otherwise, the solid is in an elastic state.

The stress field of the plastically deforming solid can be determined by integrating Eq. (4.26). To integrate Eq. (4.26), it is convenient if the time rates of change can be transformed to gradients with respect to x as follows:

$$\frac{d}{dt}(\sigma'_{ij};\, \varepsilon'_{ij};\, W) = U\frac{\partial}{\partial x}(\sigma'_{ij};\, \varepsilon'_{ij};\, W) \qquad (4.27)$$

where U is the steady sliding speed of the lower plane. Therefore, by substituting Eq. (4.27) into (4.26), the time rates of change will be replaced with derivatives of x. U is canceled out of the equation.

Since the strains are assumed to be known (i.e., equal to the elastic strains), Eq. (4.26) can be solved for the stress deviators, using the Runge–Kutta–Gill method (Gill, 1951).[3] The step-by-step numerical analysis was

---

[3] The Runge–Kutta–Gill method is a step-by-step numerical/analytical method of obtaining an approximation of $A_K + 1$ from A_K. Canned programs are available at many computer centers.

Sec. 4.3 Plastic Deformation of the Surface Layer

used by Jahanmir and Suh (1977) to determine the stress at any given point z using the following procedure (see Appendix 4.C). The stresses are first determined assuming that the material behaved as an elastic solid. Then the location x_0 of the solid where the state of stress first satisfied the yield condition [Eq. (4.21)] is identified. Starting from this elastic stress state at x_0, the rate of change of stress with x is found from Eq. (4.26). These stress gradients are then used to predict the values of the stress components when the loading point has moved a small distance dx. In this manner the state of stress along a constant depth is obtained, as the asperity contact moves over the surface.

Although the assumption that the total strain is identical to the elastic strain satisfies the geometric compatibility condition and also the stress boundary conditions for the traction-free surface outside the contact, the solution obtained using the procedure described above is only an approximation to the exact solution.

The stresses that are found by the foregoing technique do not satisfy the equilibrium condition, since no attempt was made to maintain equilibrium during the loading cycle. Therefore, it is necessary to restore the condition of equilibrium at the end of the loading cycle by considering elastic unloading phenomena.

Had the loading been entirely elastic the stresses would approach zero when the asperity contact point moved away from the point of interest. However, as a result of plastic deformation, each point must have a state of residual stress because the unloading path is different from the loading path. Since every point at a given depth experiences exactly the same loading history, residual stresses, $(\sigma_{ij})_r$, and strains must be independent of x; that is,

$$(\sigma_{xx})_r = f_1(z)$$
$$(\sigma_{zz})_r = f_2(z) \tag{4.28}$$
$$(\sigma_{yy})_r = \nu(f_1 + f_2)$$
$$(\sigma_{xz})_r = f_3(z)$$

However, the equilibrium condition requires that these residual stresses satisfy the equilibrium condition,

$$\frac{\partial(\sigma_{xx})_r}{\partial x} + \frac{\partial(\sigma_{xz})_r}{\partial z} = 0$$
$$\frac{\partial(\sigma_{xz})_r}{\partial x} + \frac{\partial(\sigma_{zz})_r}{\partial z} = 0 \tag{4.29}$$

Substituting Eq. (4.28) into Eq. (4.29), we obtain

$$\frac{df_3}{dz} = 0 \quad \text{or} \quad f_3 = c_1 \tag{4.30}$$

$$\frac{df_2}{dz} = 0 \quad \text{or} \quad f_2 = c_2$$

The boundary conditions $\sigma_{zz} = 0$ and $\sigma_{xz} = 0$ at $x = \infty$ and $z = 0$ yield $c_1 = c_2 = 0$. Therefore, the only permissible state of residual stress is

$$(\sigma_{xx})_r = f(z)$$
$$(\sigma_{yy})_r = \nu f(z) \tag{4.31}$$
$$(\sigma_{zz})_r = 0$$
$$(\sigma_{xz})_r = 0$$

The stress field obtained by solving Eq. (4.26) does not satisfy the condition specified by Eq. (4.31); that is, σ_{zz} and σ_{xz} are not zero.

To satisfy Eq. (4.31) we apply opposing stresses $-(\sigma_{zz})$ and $-(\sigma_{xz})$ to cancel out $(\sigma_{zz})_r$ and $(\sigma_{xz})_r$, which is equivalent to unloading the material elastically. Then the resulting state of stress and strain is the residual stress and strain. The residual strains at the end of a loading and unloading cycle are

$$(\varepsilon_{zz})_r = -\frac{1-2\nu}{2(1-\nu)G}(\sigma_{zz})_r' \tag{4.32}$$

$$(\gamma_{xz})_r = -\frac{(\sigma_{xz})_r'}{G} \tag{4.33}$$

where $(\sigma_{zz})_r'$ and $(\sigma_{xz})_r'$ are the pseudoresidual stresses and $(\gamma_{xz})_r$ is the residual engineering shear strain. Furthermore, the residual stresses $(\sigma_{xx})_r$ and $(\sigma_{yy})_r$ become

$$(\sigma_{xx})_r = (\sigma_{xx})_r' - \frac{\nu}{1-\nu}(\sigma_{zz})_r' \tag{4.34}$$

$$(\sigma_{yy})_r = (\sigma_{yy})_r' - \frac{\nu}{1-\nu}(\sigma_{zz})_r' \tag{4.35}$$

Since many asperities move over the surface, the residual state of stress and strain changes with each cyclic loading until there can be no further relaxation of σ_{zz} and σ_{xz} components.

Using the residual stresses $(\sigma_{xx})_r$ and $(\sigma_{zz})_r$ as initial conditions, new values of residual stresses are calculated by repeating the numerical integration scheme outlined in the preceding paragraph. This procedure is repeated until there is no further change in $(\sigma_{xx})_r$ and $(\sigma_{yy})_r$ [which corresponds to $(\sigma_{zz})_r'$ approaching zero]. This condition is typically satisfied after 5 to 10 integration cycles. The residual shear strain per pass $(\gamma_{xz})_r$ also approaches a constant value and the total shear strain accumulates with cyclic loading by this amount of incremental loading. Therefore, the residual stress observed on typical sliding surfaces is invariant, although many asperities move over the surface once it achieves a steady state, whereas the residual shear strain continues to increase by steady-state increments of shear strain for each passage of asperities over the surface.

Discussion of the numerical solutions obtained using the Merwin–Johnson method. The Merwin and Johnson method was used to find the size of the plastically deforming zone under a contact, the state of stress, the residual stresses, and the residual strains during steady-state sliding. The result for the applied normal stress $p_0 = 4k$ and different tangential stresses ranging from $q_0 = 0$ to $q_0 = 4k$ is given in Fig. 4.23. It should be noted that for zero friction a state of shakedown is reached and the steady-state deformation is purely elastic. The size of the plastic region increases with increasing friction coefficient. For friction coefficients smaller than a critical value (0.25 for the case shown in Fig. 4.23), the plastic region is below the surface, whereas at larger friction coefficients the plastic region extends to the surface. The large size of the plastic zone in front of the slider for large friction coefficient is surprising. Perhaps this is due to the fact that plowing was assumed for the stress boundary condition, but the displacements of the raised material in front of the slider were not considered. If the solution would allow a raised surface in front of the contact, stresses would be relieved below the surface in front of the contact, and the size of the plastic zone would decrease.

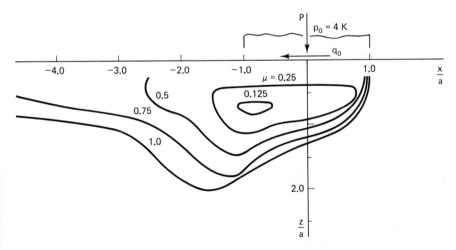

Figure 4.23 Steady-state plastic deformation regions in an elastic–perfectly plastic material under a sliding contact, for a maximum applied normal stress $p_o = 4k$ and different friction coefficients. (From Jahanmir, S., and Suh, N. P., "Mechanics of Subsurface Void Nucleation in Delamination Wear," *Wear*, Vol. 44, 1977, pp. 17–38; also Jahanmir, S., Ph.D. Thesis, Massachusetts Institute of Technology, 1976.)

The steady-state σ_{xx}, σ_{zz}, and σ_{xz} components of stress at various depths are given in Figs. 4.24 through 4.26. (The coordinates are described

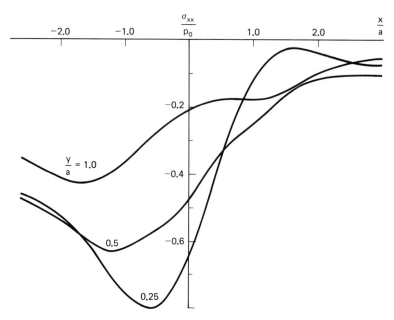

Figure 4.24 Steady state σ_{xx} component of the state of stress at different depths. normalized with respect to the maximum applied normal stress $p_o = 4k$; for friction coefficient $\mu = 0.5$. (From Jahanmir, S., and Suh, N. P., "Mechanics of Subsurface Void Nucleation in Delamination Wear," *Wear*, Vol. 44, 1977, pp. 17–38; also Jahanmir, S., Ph.D. Thesis, Massachusetts Institute of Technology, 1976.)

in Fig. 4.22). It is noted that σ_{xx} is always compressive, whereas σ_{zz} becomes tensile behind the contact and close to the surface. This tensile zone has also been found in obtaining a similar solution by a finite element method. The distribution of steady-state residual stress $(\sigma_{xx})_r$ is given in Fig. 4.27 for different friction coefficients. (As discussed earlier, the only possible residual stresses are σ_{xx} and σ_{yy}.) Figure 4.27 shows that the residual stress is compressive and its magnitude, for large friction coefficients, is largest near the surface.

The steady-state increment of shear strain, $(\gamma_{xz})_r$, for each passage of the asperity, is given in Fig. 4.28. It is observed that the increment of shear strain is largest at the surface for the large friction coefficients, and its magnitude increases with the friction coefficient. It should be noted that during steady-state sliding, the only nonzero plastic strain is the shear strain, which accumulates with an amount $(\gamma_{xz})_r$ after each passage of the slider.

FEM analysis of the deformation of elastoplastic solids. The deformation of an elastoplastic solid was also modeled and solved by using a finite element method (Sin, 1981). The deformation of a semi-infinite

Sec. 4.3 Plastic Deformation of the Surface Layer

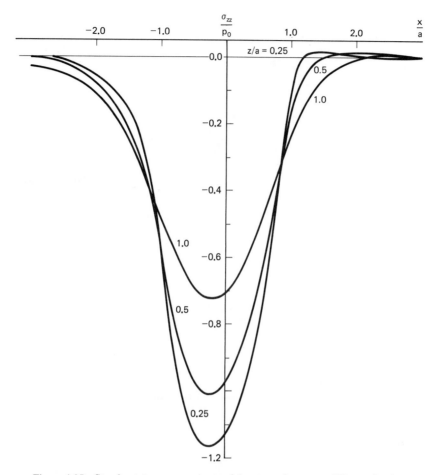

Figure 4.25 Steady-state σ_{zz} component of the state of stress at different depths, normalized with respect to the maximum applied normal stress $p_o = 4k$; for friction coefficient $\mu = 0.5$. (From Jahanmir, S., and Suh, N. P., "Mechanics of Subsurface Void Nucleation in Delamination Wear," *Wear*, Vol. 44, 1977, pp. 17–38; also Jahanmir, S., Ph.D. Thesis, Massachusetts Institute of Technology, 1976.)

slightly work-hardening elastoplastic solid was loaded by a moving asperity. The normal load was assumed to be $4k$ and the coefficient of friction 0.25. The material properties used were as follows: isotropic, slightly work hardening (slope $d\sigma/d\varepsilon$ of the work-hardening region $= 10^{-4}E$, where E is Young's modulus), $E = 1.96 \times 10^5$ MPa $= 2 \times 10^4$ kg/mm^2, $\nu = 0.28$, and $\sigma_y = \sqrt{3}\, k = 424$ MPa $= 43.3$ kg/mm^2.

Figure 4.29 shows the size of the plastically deformed zone under a stationary asperity contact, whereas Fig. 4.30(a) and (b) show the plastically deformed regions under a moving asperity during the first and fourth cycles

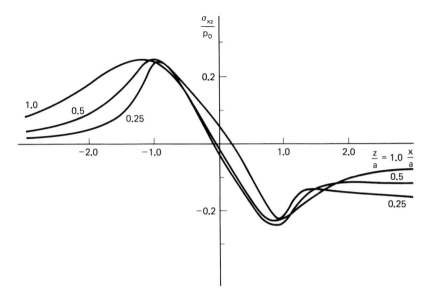

Figure 4.26 Steady-state σ_{xz} component of the state of stress at different depths, normalized with respect to the maximum applied normal stress $p_o = 4k$; for friction coefficient $\mu = 0.5$. (From Jahanmir, S., and Suh, N. P., "Mechanics of Subsurface Void Nucleation in Delamination Wear," *Wear*, Vol. 44, 1977, pp. 17–38; also Jahanmir, S., Ph.D. Thesis, Massachusetts Institute of Technology, 1976.)

of loading, respectively. Under the moving load the material just in front of the load deforms plastically. The size of the plastically deformed zone is much smaller than that under static load given in Fig. 4.29. Furthermore, the repeated cyclic loading makes the plastic region smaller, as shown in Fig. 4.30. This is very consistent with the results obtained by the Merwin–Johnson method. With further repeated loading, the plastically deformed zone may become smaller due to the residual stresses remaining in the solid. According to the numerical investigation of a rolling contact by Anand (1977), the steady-state deformation, after a few revolutions of a disk, would eventually reach a purely elastic state for a strain-hardening material, whereas elastic–perfectly plastic solids have elastoplastic deformation even at high loads. Therefore, the case shown in the figure may eventually reach a purely elastic state since it was assumed that the material is work hardening.

The residual stress $(\sigma_{xx})_r$ is given in Fig. 4.31 as a function of depth from the surface. It can be seen that the difference in $(\sigma_{xx})_r$ is almost negligible between the third and fourth cycles. Also, the dominant residual stress is compressive in a direction parallel to the surface. The other

Sec. 4.3   Plastic Deformation of the Surface Layer   139

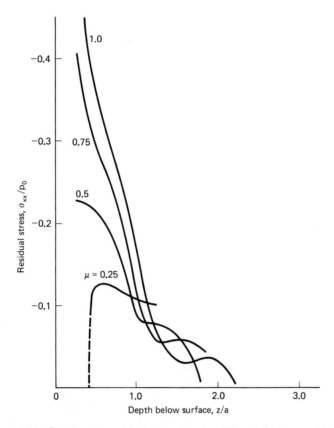

Figure 4.27 Steady-state residual stress σ_{xx} for different friction coefficients, normalized with respect to the maximum applied normal stress $p_o = 4k$. (From Jahanmir, S., and Suh, N. P., "Mechanics of Subsurface Void Nucleation in Delamination Wear," *Wear*, Vol. 44, 1977, pp. 17–38; also Jahanmir, S., Ph.D. Thesis, Massachusetts Institute of Technology, 1976.)

component of residual stresses, $(\sigma_{zz})_r$, is very small. These two are the only components that can affect crack opening under sliding conditions. Therefore, it can be concluded that residual stresses do not affect crack propagation.

These FEM results are in good agreement with the results obtained using the Merwin–Johnson method (see Fig. 4.23). These theoretical results are also in reasonable agreement with the experimentally determined plastic strain fields which were obtained by measuring the grain deformation (Augustsson, 1974; see Appendix 2.B for detailed analysis). The plastic strain at the surface can be extremely large, especially on highly ductile copper (see Fig. 4.32).

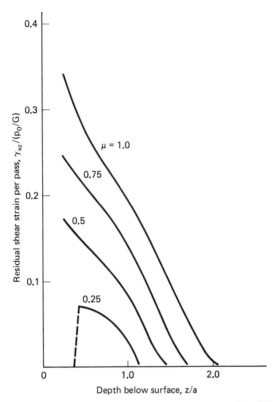

Figure 4.28 Steady-state residual shear strain per pass, for different friction coefficients, normalized with respect to the yield strain in pure shear and the maximum applied normal stress $P_o = 4k$. (From Jahanmir, S., and Suh, N. P., "Mechanics of Subsurface Void Nucleation in Delamination Wear," *Wear*, Vol. 44, 1977, pp. 17–38; also Jahanmir, S., Ph.D. Thesis, Massachusetts Institute of Technology, 1976.)

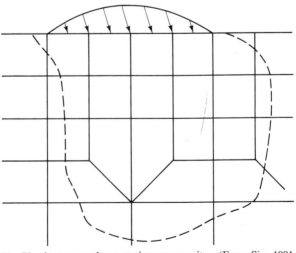

Figure 4.29 Plastic zone under a stationary asperity. (From Sin, 1981.)

Sec. 4.4 Void and Crack Nucleation at the Subsurface 141

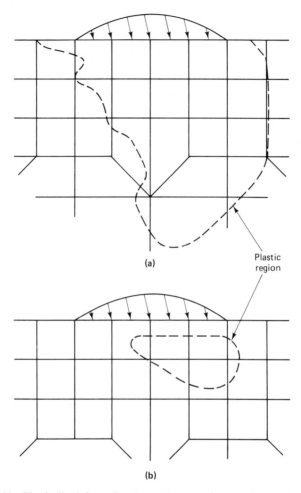

Figure 4.30 Plastically deformed region under a moving asperity: (a) during the first cycle; (b) during the fourth cycle. (From Sin, 1981.)

4.4 VOID AND CRACK NUCLEATION AT THE SUBSURFACE

The failure of an initially flaw-free material proceeds in two sequential steps: (1) the initiation and nucleation of microcracks or microvoids, and (2) crack propagation or void growth that leads to catastrophic failure (Suh et al., 1977). The first step of the failure process may not be an important process if flaws are initially present in the material. In this case, the crack propagation may control the failure process. However, in some materials crack nucleation may determine the rate of wear rather than crack propagation, if the former occurs at a much slower rate than the latter. In this section the mechanism of crack nucleation in a material without any flaw will be analyzed in order to understand the origin of deformation-induced failure of elastoplastic solids.

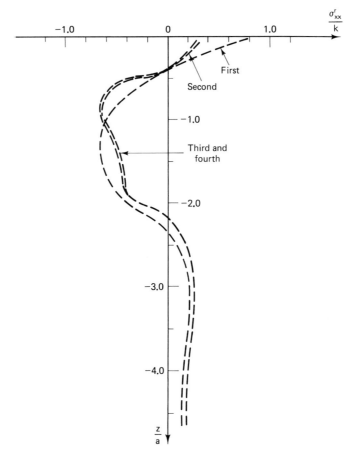

Figure 4.31 Variation of residual stress σ_{xx}^r as a function of depth under a moving asperity. (From Sin, 1981.)

When the surface layer undergoes plastic deformation, voids and cracks can nucleate. Cracks in two phase metals are nucleated around the hard particles due to the displacement incompatibility between the particle and the matrix, which occurs when the matrix deforms plastically near the surface with repeated loading, but the hard particles cannot deform. Micrographs of crack nucleation around hard particles are shown in Figs. 4.33 and 4.34. Even in single-phase metals cracks are present, as shown in Fig. 4.35. Although the exact mechanism for crack and void nucleation in single-phase metals is not definitively known, it has been suggested that interactions of dislocations and the formation of dislocation cells may be responsible (Suh, 1977; Argon, 1976; Hirth and Rigney, 1976). It has also been suspected that even in single crystals crack nucleation may be a result of displacement incompatibility between impurity inclusions and the matrix (Suh, 1977).

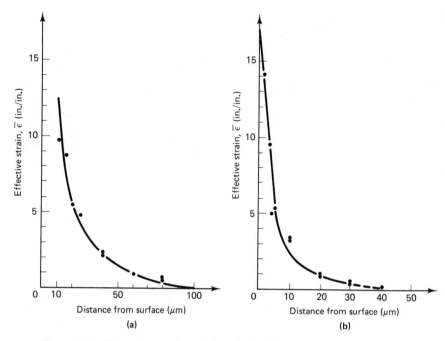

Figure 4.32 Subsurface strain variation obtained by grain shape measurements: (a) copper specimen tested under a normal load of 2.1 kg and after a sliding distance of 68 m in argon; (b) AISI 1020 steel tested under a normal load of 2.4 kg and a sliding distance of 180 m in argon. (From Augustsson, 1974.)

Intersection of twins may also create cracks (McClintock and Argon, 1966).

The fact that the crack nucleation in two-phase materials is due to the displacement incompatibility between spherical inclusions and the matrix can be illustrated as follows. Consider a hard rigid spherical particle surrounded by an elastoplastic matrix which undergoes shear deformation. In the absence of the hard inclusion the matrix will deform in such a manner that an imaginary spherical boundary will assume an ellipsoidal shape, as shown in Fig. 4.36. However, in the presence of the rigid sphere which is well bonded to the matrix, the interface cannot assume the ellipsoidal shape because of the geometric constraint imposed on the matrix by the sphere. Consequently, normal stresses are developed at the sphere–matrix interface. At location A the normal stress is compressive, while at location B it is tensile. When the tensile normal stress at B exceeds the cohesive (or adhesive) strength of the bonding at the particle–matrix interface, a crack may nucleate if the energy criterion is also satisfied.

The strength criterion may be expressed as

$$(\sigma_{kk})_{\max} \geq \sigma_i \qquad (4.36)$$

where $(\sigma_{kk})_{\max}$ is the maximum principal normal stress in tension and σ_i is

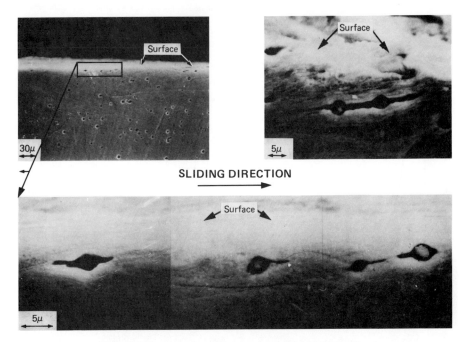

Figure 4.33 Void formation around inclusions and crack propagation from these voids near the surface in annealed Fe-1.3% Mo. (From Jahanmir, S., and Suh, N. P., "Mechanics of Subsurface Void Nucleation in Delamination Wear," *Wear*, Vol. 44, 1977, pp. 17–38; also Jahanmir, S., Ph.D. Thesis, Massachusetts Institute of Technology, 1976.)

the ideal cohesive strength at the interface. This is strictly a local criterion. When applied to an inclusion-filled material, the criterion itself is invariant to the size or shape of the inclusion. The shape of the inclusion will have to be considered to find the location of $(\sigma_{kk})_{\max}$ and the relationship between the applied stress $(\sigma_{ij})_A$ and the interfacial stress $(\sigma_{kk})_{\max}$. Once the maximum stress concentration factor K around a particle is found, the stress criterion can be expressed in terms of the applied stress. In an elastoplastic solid the interfacial stress concentration is limited by the flow strength of material $\sigma_y(\varepsilon^p)$. For strain-hardening materials, the strain concentration around inclusions of different geometric shape in an inhomogeneous deformation field are generally bound by two limiting idealizations of the plastic behavior of the material: a nonhardening rigid–plastic behavior and a linear hardening behavior with zero yield stress (Rhee and McClintock, 1962). These results indicate that the interfacial stress $(\sigma_{kk})_{\max}$ at the surface of a cylindrical inclusion after yielding is bound by

$$\frac{3}{2} k \leq (\sigma_{kk})_{\max} \leq 2k \qquad (4.37)$$

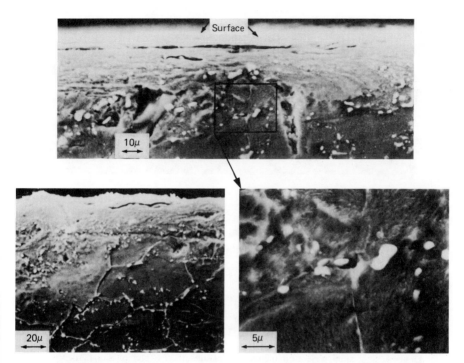

Figure 4.34 Subsurface cracks and deformation in annealed AISI 1020 steel. (From Jahanmir, S., and Suh, N. P., "Mechanics of Subsurface Void Nucleation in Delamination Wear," *Wear*, Vol. 44, 1977, pp. 17–38; also Jahanmir, S., Ph.D. Thesis, Massachusetts Institute of Technology, 1976.)

where $(\sigma_{kk})_{\max} = \sigma_{rr} - \sigma$

k = flow stress in shear

σ = hydrostatic tensile component of the applied stress

Since the limits on $(\sigma_{kk})_{\max}$ are very close to each other and nearly equal to $\sigma_y(\bar{\varepsilon}^p)$, the interfacial radial stress on the cylindrical inclusion may be taken as

$$\sigma_{rr} \simeq \sigma_y(\bar{\varepsilon}^p) + \sigma \qquad (4.38)$$

The energy criterion for void nucleation must be satisfied in addition to Eq. (4.36). When the inclusion-filled material is subject to an external load, strain energy is stored in the elastic field within and around the inclusion. This strain energy will change as the elastic field changes during void nucleation. The energy in the matrix–inclusion system, E^s, should be sufficient to provide for the surface energy created by the void nucleation process. This may be expressed as

$$E^s_{\text{before}} - E^s_{\text{after}} \geq \Delta \gamma \, A \qquad (4.39)$$

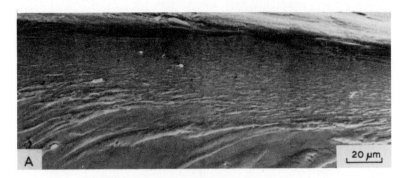

Figure 4.35 Scanning electron micrographs of subsurface of precipitation hardened Cu-0.81 at. pct Cr alloy for aging time of (a) 5 min and (b) 10,000 min. Other details are the same as in Figure 9. (From Pamies-Teixeira, J. J.; Saka, N.; and Suh, N. P., "Wear of Copper-based Solid Solutions," *Wear*, Vol. 44, 1977, pp. 65–75.)

where A is the surface area of the nucleated void (or crack) and $\Delta\gamma$ is the surface energy change during void nucleation. If the void nucleates at the matrix–inclusion interface,

$$\Delta\gamma = -\gamma_{M-I} + (\gamma_M + \gamma_I) \quad (4.40)$$

where the subscripts M and I denote the matrix and inclusion, respectively, and γ the surface energy. Equation (4.39) is inclusion-size dependent, since the strain energy is proportional to the volume and the surface energy is proportional only to the surface area. This is in contrast to the strength criterion, which is independent of the size. Figure 4.37 shows schematically both the strength and the energy criteria expressed in units of applied elastic strain as a function of the inclusion size. For inclusion size larger than d^*, the energy criterion is always satisfied whenever the strength criterion is reached. However, for inclusion sizes smaller than d^*, satisfying the strength criterion does not necessarily guarantee the satisfaction of the energy requirement. In elastoplastic solids the elastic strain that can be imposed at the particle–matrix inclusion is limited to the elastic yield strain corresponding to its strain-hardened state. Therefore, when the particle is smaller than

Sec. 4.4   Void and Crack Nucleation at the Subsurface

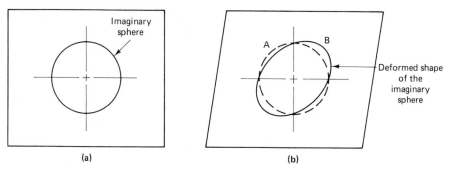

Figure 4.36 Illustration of the displacement incompatibility between the matrix and the inclusion: (a) before deformation; (b) after deformation of the matrix. (From Jahanmir, S., and Suh, N. P., "Mechanics of Subsurface Void Nucleation in Delamination Wear," *Wear*, Vol. 44, 1977, pp. 17–38; also Jahanmir, S., Ph.D. Thesis, Massachusetts Institute of Technology, 1976.)

a critical size, the energy criterion cannot be satisfied. In metals this critical size is in the range of a few hundred angstroms.

To find σ_{rr} due to the accumulation of plastic shear strain around a hard spherical particle, Eq. (4.38) may be used. Consider a rigid cylindrical particle which is inserted at (x, z) of the center of a small volume element below the surface. If the volume element is distorted by applying a shear strain γ_{xz}, an interfacial normal stress σ_{rr} will develop around the particle (see Fig. 4.38). σ_{rr} is a function of γ_{xz}, which may be stated as

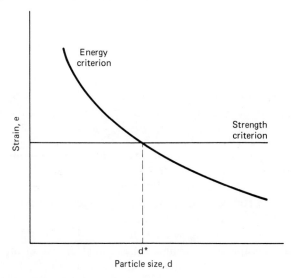

Figure 4.37 Schematics of the energy and stress criterion for void nucleation. (From Su, 1980.)

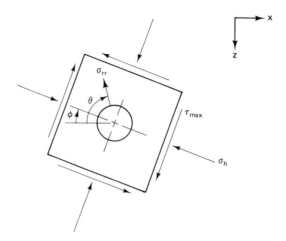

Figure 4.38 Rigid cylindrical inclusion under a general state of stress. (From Jahanmir, S., and Suh, N. P., "Mechanics of Subsurface Void Nucleation in Delamination Wear," *Wear*, Vol. 44, 1977, pp. 17–38; also Jahanmir, S., Ph.D. Thesis, Massachusetts Institute of Technology, 1976.)

$$\sigma_{rr} = f(\gamma_{xz}) \tag{4.41}$$

The shear strain γ_{xz} is the total accumulated strain and it is increased by increments of $(\gamma_{xz})_r$ after each asperity pass. In an elastoplastic solid a maximum σ_{rr} is developed at yield, that is,

$$\gamma_{xz} = \gamma_y = \frac{k}{G} \tag{4.42}$$

where γ_y is yield strain in pure shear and G is the elastic shear modulus. The number of slider passages required to reach the maximum σ_{rr} can be found from the shear strain increment per pass $(\gamma_{xz})_r$, which was determined in the preceding section. For an elastic–perfectly plastic solid, the maximum σ_{rr} due to accumulated plastic strain is equal to $\sqrt{3}\,k$ once the yield strength is exceeded. However, σ_{rr} depends on the position around the particle, or

$$(\sigma_{rr})_1 = \sqrt{3}\,k \sin 2\theta \tag{4.43}$$

where $(\sigma_{rr})_1$ is the normal interfacial stress due to the cumulative plastic deformation around a particle located at a depth z below the surface.

In order to find the maximum interfacial normal stress σ_{rr} due to the applied stress at the contact, Eq. (4.38) may be used. However, the stresses at each point must first be transformed to a state of maximum shear stress and hydrostatic stress by using Mohr's circle. Following the procedure of the preceding paragraph, it is assumed that the stresses act on an element and a particle inserted at the center of the element (Fig. 4.38). Therefore, using Eq. (4.38),

$$(\sigma_{rr})_2 = \sqrt{3}\,\tau_{max} \sin 2(\theta - \phi) + \sigma \tag{4.44}$$

Sec. 4.4 Void and Crack Nucleation at the Subsurface

where τ_{max} is the maximum shear stress at the point (x, z), θ is the angle from the x axis to the axis of the maximum positive shear stress, and σ is the hydrostatic stress at (x, z); $\sqrt{3}$ appears in the equation since the von Mises yield criterion is used.

The total σ_{rr} due to both the state of stress and the accumulation of plastic shear strain is found by adding Eq. (4.43) and Eq. (4.44), and the maximum σ_{rr} which occurs at some angle $\theta = \theta_0$ can be obtained for each point (x, z):

$$(\sigma_{rr})_{max} = \sqrt{3}\, k \sin 2\theta_0 + \sqrt{3}\, \tau_{max} \sin 2(\theta_0 - \phi) + \sigma \qquad (4.45)$$

The number of asperity passages before $(\sigma_{rr})_{max}$ can be developed at a given depth can also be determined once the shear strain increment per pass is known at that depth.

Equation (4.45) was solved numerically using the Merwin–Johnson method to calculate the state of stress and the residual stresses and strains. The contours of constant $(\sigma_{rr})_{max}$, found from Eq. (4.45), are plotted in Figs. 4.39 through 4.41 for an applied normal stress at each asperity contact of $4k$ and friction coefficients of 1.0, 0.5, and 0.25.[4] It should be noted that $(\sigma_{rr})_{max}$ is normalized with respect to k, the yield strength in shear, and all distances are normalized with respect to a, the half-contact length. The figures show that $(\sigma_{rr})_{max}$ is compressive below the contact and attains its largest values in front of the slider, well below the surface.

If the particle–matrix bond strength is equal to $2k$, Figs. 4.39 through 4.41 show that the size of the region in which void nucleation is possible increases with friction coefficient. The size of the void nucleation region is smaller for a stronger particle–matrix bond. The fact that voids can nucleate only well below the surface ($\sigma_{rr} \geq 2k$) is consistent with the subsurface observations of worn samples. As indicated in Chapter 3, experimental results show that voids nucleate below the surface.

The minimum and maximum depth of the void nucleation region is plotted in Fig. 4.42 as a function of friction coefficient for the applied normal contact stresses, p_0, of $4k$ and $6k$. The range of depth for void nucleation increases with friction coefficient for both applied normal stresses, but the voids nucleate deeper below the surface and the number of cycles required for void nucleation increases with the increase in applied normal stress. It is interesting to observe that at $p_0 = 6k$ and zero friction (i.e., a case of pure rolling), void nucleation is possible in a small region if voids nucleate at $(\sigma_{rr})_{max} = 2k$. However, at $p_0 = 4k$ and zero friction coefficient, voids do not nucleate since the stresses shake down to an elastic state during steady-state condition.

---

[4] Note that these solutions are not valid very near the asperity contact since the yield condition is violated. This problem comes about because the effect of plowing, asperity deformation, and adhesion at the asperity contacts are simply summed in specifying the resultant surface tractions at an asperity contact.

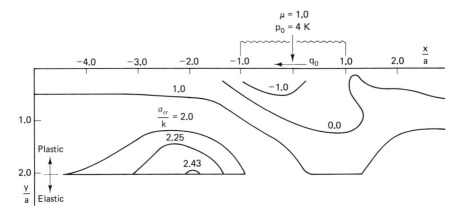

Figure 4.39 Contours for σ_{rr} under a sliding contact, normalized with respect to k, the shear yield strength, for $p_o = 4k$ and $\mu = 1.0$. (From Jahanmir, S., and Suh, N. P., "Mechanics of Subsurface Void Nucleation in Delamination Wear," *Wear*, Vol. 44, 1977, pp. 17–38; also Jahanmir, S., Ph.D. Thesis, Massachusetts Institute of Technology, 1976.)

The depth of the void nucleation region ($\sigma_{rr} > 2k$) is plotted in Figs. 4.43 and 4.44 as a function of the number of passes required for void nucleation at different friction coefficients and p_0 of $4k$ and $6k$. At a given depth, the number of passes required for void nucleation decreases with increase in the friction coefficient. It should be noted that voids begin to nucleate after only about 10 passes; this implies that in many materials with hard inclusions, crack propagation may be the wear-rate-determining

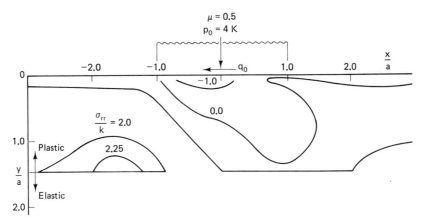

Figure 4.40 Contours for σ_{rr} under a sliding contact, with respect to k, the shear yield strength, for $p_o = 4k$ and $\mu = 0.50$. (From Jahanmir, S., and Suh, N. P., "Mechanics of Subsurface Void Nucleation in Delamination Wear," *Wear*, Vol. 44, 1977, pp. 17–38; also Jahanmir, S., Ph.D. Thesis, Massachusetts Institute of Technology, 1976.)

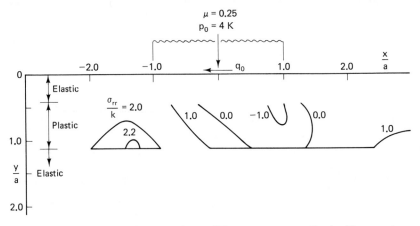

Figure 4.41 Contours for σ_{rr} under a sliding contact, normalized with respect to k, the shear yield strength, for $p_o = 4k$ and $\mu = 0.25$. (From Jahanmir, S., and Suh, N. P., "Mechanics of Subsurface Void Nucleation in Delamination Wear," *Wear*, Vol. 44, 1977, pp. 17–38; also Jahanmir, S., Ph.D. Thesis, Massachusetts Institute of Technology, 1976.)

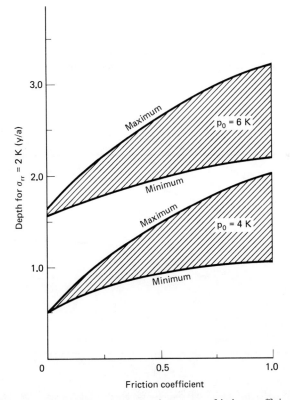

Figure 4.42 Depth of void nucleation regions versus friction coefficient, for two different maximum applied normal stress $p_o = 4k$ and $6k$. (From Jahanmir, S., and Suh, N. P., "Mechanics of Subsurface Void Nucleation in Delamination Wear," *Wear*, Vol. 44, 1977, pp. 17–38; also Jahanmir, S., Ph.D. Thesis, Massachusetts Institute of Technology, 1976.)

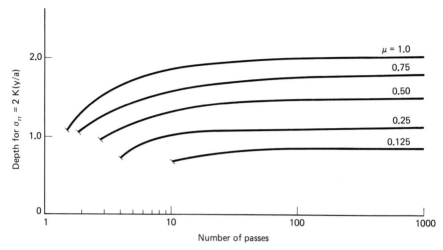

Figure 4.43 Depth of void nucleation regions for different friction coefficients versus the number of passes required for void nucleation, for a maximum applied normal stress, $p_o = 4k$. (From Jahanmir, S., and Suh, N. P., "Mechanics of Subsurface Void Nucleation in Delamination Wear," *Wear*, Vol. 44, 1977, pp. 17–38; also Jahanmir, S., Ph.D. Thesis, Massachusetts Institute of Technology, 1976.)

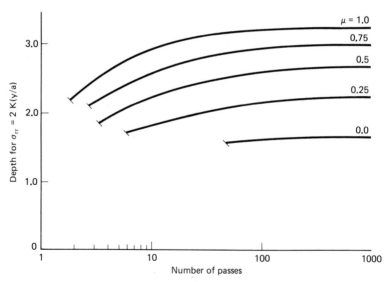

Figure 4.44 Depth of void nucleation regions for different friction coefficients versus the number of passes required for void nucleation, for maximum applied normal stress $p_o = 6k$. (From Jahanmir, S., and Suh, N. P., "Mechanics of Subsurface Void Nucleation in Delamination Wear," *Wear*, Vol. 44, 1977, pp. 17–38; also Jahanmir, S., Ph.D. Thesis, Massachusetts Institute of Technology, 1976.)

process. It was estimated that about 10^5 asperity passes are required to create a wear sheet in AISI 1020 steel (Jahanmir and Suh, 1977).

As the number of passes (i.e., cyclic loading) increases, the region of void nucleation becomes larger. Cracks begin to propagate as soon as they are nucleated and as cracks get larger, they propagate faster since the stress and strain concentrations at the crack tips become larger with crack length until they approach asymptotic values. Therefore, in normal sliding situations where the relative sliding distance along a given direction is larger than the typical crack length in the material, the cracks that nucleated first propagate fastest. The cracks that nucleate later cannot propagate as much, since the stress and strain concentration is less for the cracks that are farther away from the surface and since these cracks are subjected to less stress and strain due to the cracks already existing above them. However, in fretting, where the relative displacement is very small and the direction of sliding reverses, cracks at all depths can propagate at nearly the same rate. Experimental results show that this is the case (Sproles et al., 1980).

In the analysis of void nucleation it was assumed that the hard particles are rigid. However, in real materials the particles have some elasticity, which would result in values of σ_{rr} smaller than the ones calculated by the foregoing procedure. An exact analysis of the interfacial stress for elastic inclusions has been done for an elastic solid which may be adopted for an elastoplastic solid (Eshelby, 1957; Su, 1980).

The correct criterion for void nucleation from large particles may be a combination of a local shear strain and a local interfacial tensile stress criterion. In the analysis above it was assumed that the local stress criterion was sufficient. This assumption, however, may be a good approximation for equiaxed particles, but not for elongated particles. In the analysis of elongated particles the local strain concentrations are large and void nucleation generally occurs by particle fracture. Therefore, as the particles become more elongated, the local shear strain criterion for void nucleation may become the dominating criterion.

4.5 CRACK PROPAGATION DUE TO SURFACE TRACTION

In the preceding section crack nucleation in an elastoplastic solid was discussed in great detail. It was shown that cracks can nucleate at the subsurface when large plastic deformation occurs, although the applied load by the asperity creates compressive and shear stresses. It was also stated in Section 4.2 that cracks nucleate at the surface of brittle solids, especially under rolling contacts. These cracks are perpendicular to the surface and propagate into the subsurface since the maximum normal stress at the surface is parallel to the surface, as shown in Fig. 4.45. These cracks eventually change direction due to the changes in the maximum normal stress field. The important point is that the crack location and the direction of crack propagation depend very sensitively on material properties and

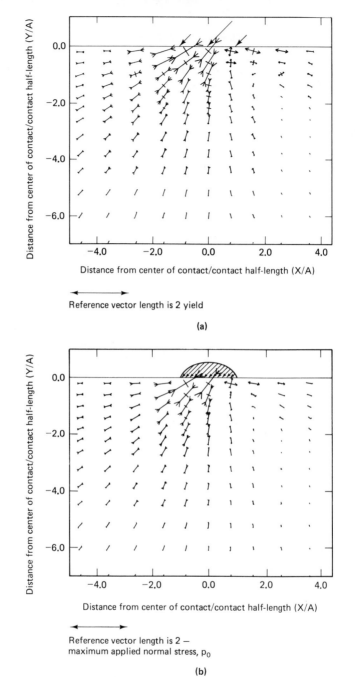

Figure 4.45 Stress field under an asperity contact (coefficient of friction = 1; lines show the directions and magnitudes of principal stresses, arrowheads indicate signs and "p" indicates plasticity): (a) an elastoplastic solid (finite element solution, applied load vectors not to scale); (b) an elastic solid (solutions for an elliptically distributed load, vectors not to scale). (From Fleming, J. F., and Suh, N. P., "Mechanics of Crack Propagation in Delamination Wear," *Wear*, Vol. 44, 1977, pp. 39–56.)

Sec. 4.5 Crack Propagation Due to Surface Traction

the nature of the surface traction. Various possible cases of crack propagation were illustrated by Suh (1980).

In this section the crack propagation in an elastoplastic solid under sliding conditions will be analyzed. The propagation of these cracks can be quite different from that in an elastic solid, although there are also many similarities. As shown earlier, cracks in an elastoplastic solid under sliding contact loads are nucleated within a narrow zone below the surface due to the compressive and tangential load applied by an asperity. These cracks may propagate due to the repeated loading applied by moving asperities at the surface. These subsurface cracks will propagate at different rates depending on where the cracks are located. The crack that propagates fastest will control the wear process. Although the growth rate of these cracks increases with the crack length and eventually reaches a steady-state rate, the analysis will be done for cracks that have already grown to a finite size for steady-state growth.

When stresses are applied at the surface of a solid with a subsurface crack as shown in Fig. 4.46, part of the crack is in a tensile region. These cracks always experience combined loading of normal and shear stresses. The crack tip in the tensile region experiences a crack opening–closing mode of loading (i.e., mode I) as well as shear loading, whereas the crack in the compression region experiences only compressive and shear loading, which cause crack tip sliding displacement parallel to the crack.

In the fields of fracture mechanics and fatigue it is well established that when cracks are loaded cyclically in mode I and when the plastic zone surrounding the crack tip is much smaller than the crack and the dimensions of the part, the crack propagation rate per cycle dC/dN correlates with the cyclic change in the stress intensity factor ΔK_I[5] as shown in Fig. 4.47. The direction of the propagation in mode I correlates well with the direction of the maximum tensile stress and also with the direction of minimum strain energy (Sih, 1974). When the change in the stress intensity factor per cycle is less than the threshold value ΔK_{th}, cracks cannot propagate under mode I loading.

Except in the case of surface crack propagation in elastic solids, mode I crack propagation is not often encountered in sliding wear of many elastoplastic solids. As will be shown in this section, when the loading is both compressive and shearing, and when the crack tip is fully plastic, extending to the surface, such as encountered in sliding wear of many elastoplastic solids, the crack propagation direction is parallel to the maximum shear stress (τ_{max}) direction and the crack propagation rate is best correlated with

[5] The stress intensity factor K_I is a measure of the strength of the stress singularity at an infinitely sharp crack tip of an elastic solid and is proportional to the product of the applied stress σ_∞ and the square root of the crack length [i.e., $K_I = $ (constant) $\cdot \sigma_\infty \sqrt{c}$]. When the plastically deformed zone at the crack tip is small in comparison to the dimension of the part and the crack length, the fracture toughness of a given elastoplastic solid correlates well with the critical stress intensity factor of the material.

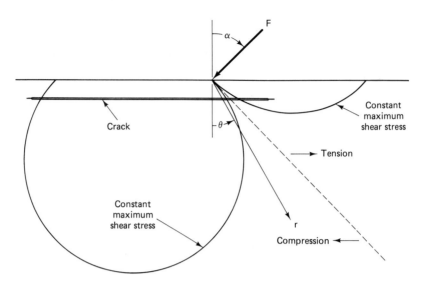

Figure 4.46 Line load applying normal and shear forces on the surface of a semi-infinite solid with a subsurface crack. (From Fleming, J. F., and Suh, N. P., "Mechanics of Crack Propagation in Delamination Wear," *Wear*, Vol. 44, 1977, pp. 39–56.)

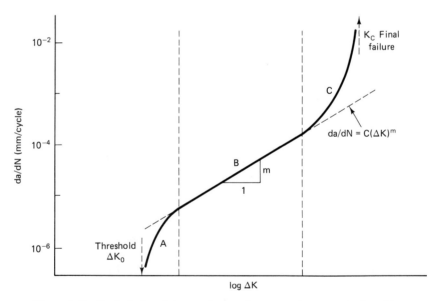

Figure 4.47 Typical plot of the crack extension per cycle versus the logarithm of the change in the stress intensity factor. The primary mechanisms may be divided into three regimes: regime A, noncontinuum mechanism showing a large influence of (1) microstructure, (2) mean stress, and (3) environment; regime B, continuum mechanisms (striation growth), showing little influence of (1) microstructure, (2) means stress, (3) dilute environment, and (4) thickness; regime C, "static mode" mechanisms (cleavage, intergranular, and fibrous), showing a large influence of (1) microstructure and (2) thickness but little influence of the environment. (From Ritchie, 1977.)

Sec. 4.5 Crack Propagation Due to Surface Traction   157

the crack tip sliding displacement. In this case the concept of stress intensity factor is not useful in correlating with the crack propagation rate.

In investigating the crack propagation phenomena in sliding wear, Suh and his coworkers first studied the problem using an elastic approximation because of the mathematical simplicity (Fleming and Suh, 1977). It was found later, however, that according to the finite element method, the elastic analysis is a gross approximation of the phenomenon and may not be valid due to the extremely large plastic zone around the crack tip. However, the elastic analysis has been useful in exploring the crack propagation direction analytically. Therefore, in this section the crack propagation problem in sliding wear will be discussed following the historical development, starting from the analysis of crack propagation in elastic solids and ending with the FEM analysis of crack propagation in elastoplastic solids.

4.5.1 Subsurface Crack Propagation in Linear Elastic Solids and Its Implications

For the case shown in Fig. 4.46, Fleming and Suh (1977) analyzed the propagation of a subsurface crack parallel to the surface using a linear elastic fracture mechanics (LEFM) model. This stress intensity factor due to an elliptically distributed load at the asperity contact was calculated using an approximate method based on weighting factors. The stress intensity factor for mode I computed in this manner was very small (being in the neighborhood of the threshold stress intensity factor ΔK_{th}) for most metals, while the stress intensity factor for mode II was about an order of magnitude larger than that for mode I. Essentially similar results were obtained by Hills and Ashelby (1979), Rosenfield (1980), Keer et al. (1980), and Hearle and Johnson (1983).

The stress intensity factor determined using the finite element method for an elastic solid shows that (Sin and Suh, 1983):

1. The stress intensity factors for these subsurface horizontal cracks are small.
2. The maximum values of K_{II} occur when the crack tips lie right below the asperity contact.
3. The closer the crack is to the surface, the larger is the maximum value of K_{II}.
4. K_{II} increases with crack length.

Since the largest stress component due to the surface traction exerted by the asperity contact is the normal stress at and parallel to the surface, the largest stress intensity factor for an elastic solid is expected to be associated with cracks perpendicular to and emanating from the surface.

Whether the linear elastic fracture mechanics (LEFM) approach is appropriate for elastoplastic solids depends on the plastic zone size at the crack tip. If this size is too big or comparable with such dimensions as the distance from the crack tip to the free surface and/or the crack size, LEFM is no longer valid and plasticity plays a significant role. In this case the local stress and strain history become important in determining the fracture criterion.

The plastic zone size for mode II loading can be estimated using

$$r_p = \frac{1}{2\pi}\left(\frac{\Delta K_{II}}{k}\right)^2 \qquad (4.46)$$

where ΔK_{II} is the change in the stress intensity factor in mode II and k is the shear yield strength of the material.

Tables 4.1 and 4.2 show the changes in stress intensity factor and estimated plastic zone sizes as a function of the depth of crack location for a small crack during one cycle of loading. Values for all parameters are the same as given before: k = 245 MPa = 25 kg/mm^2 and a = 10 μm. From these two tables it can be seen that the plastic zone sizes estimated are comparable with the depth of crack, and the stress intensity factor ranges calculated are close to or less than the threshold intensity ΔK_{th}. Although no fracture toughness data on mode II are available, an approximate analysis (Sin, 1981) indicates that the threshold intensity factor in mode II is very close to that in mode I.

The linear elastic fracture mechanics solutions cannot be used to predict the behavior of the subsurface cracks in an elastoplastic solid for two reasons. The cracks very near the surface can have large values of ΔK_{II} (if the material is elastic), but the solution is meaningless in the case of elastoplastic solids due to the large plastic zone that surrounds the crack tip. On the other hand, cracks far below the surface cannot grow since the stress intensity factor for these is much smaller than the threshold values for crack propagation, although the plastic zones at the crack tips are small. Therefore, a different approach that includes the effect of plastic

TABLE 4.1 Stress Intensity Factor Range, K_{II} (MNm$^{3/2}$), at Left and Right Tips for a Small Crack ($c = \frac{1}{4}a$)[a]

d/a		μ		
		0.25	0.5	1.0
0.3	Left	1.05	1.07	1.44
	Right	1.28	1.34	1.54
0.5	Left	1.08	1.11	1.15
	Right	1.35	1.40	1.49
1.0	Left	0.67	0.68	0.70
	Right	0.89	0.94	1.06

[a] a, half-length of the asperity contact; c, length of the crack; d, depth of the crack; μ, friction coefficient.

Sec. 4.5 Crack Propagation Due to Surface Traction

TABLE 4.2 Ratio of Plastic Zone Size to Depth of Crack at Left and Right Tips for a Small Crack $(c = \frac{1}{4}a)$[a]

d/a		μ		
		0.25	0.5	1.0
0.3	Left	0.96	1.00	1.48
	Right	1.44	1.56	2.08
0.5	Left	0.76	0.80	0.88
	Right	1.20	1.28	1.48
1.0	Left	0.12	0.12	0.12
	Right	0.20	0.24	0.32

[a] Nomenclature defined in Table 4.1.

deformation and the combined compressive and shear loading on the subsurface crack behavior must be used to analyze the subsurface crack propagation in elastoplastic solids under the moving asperity. One such an approach is discussed in Section 4.5.3.

4.5.2 Crack Trajectory: Fracture Criteria in Mixed Mode

One of the major issues in studying crack propagation is the prediction of the crack propagation direction. Discussion of this topic must precede the analysis of the crack propagation rate and the plasticity effects at the crack tip.

Until recently, no suitable failure criterion was available for the mixed-mode fracture under combined compression and shear loading. Much of the attention has been given only to the problem of predicting the direction of crack extension when a body with cracks is simply loaded. There are two major criteria for this case: maximum hoop stress (Erdogan and Sih, 1963) and minimum strain energy density (Sih, 1974). The maximum hoop stress criterion states that crack growth will occur in a direction perpendicular to the maximum principal stress. On the other hand, the minimum strain–energy–density factor criterion postulates (1) that the initial crack growth takes place in the direction along which the strain–energy–density factor possesses a stationary value, and (2) that crack initiation occurs when the factor reaches a critical value. Comparison of the two criteria by several authors has shown that the differences between them are small for tensile loading. For compressive loading, however, not only do these two not agree well, but neither criterion correlates well with physically observed behavior (Swedlow, 1976).

When these criteria are applied to the subsurface cracks in a compressive zone, they predict crack extension direction to be about 110° at the left (trailing) tip and about 70° at the right (leading) tip from the direction parallel to the surface, implying that crack extension occurs toward the surface at both tips (Sin, 1981). However, experimental results given in Chapter 5 show that the subsurface cracks grow parallel to the surface most of the time before they become loose. McClintock (1978) has suggested that cracks in a compressive field are most likely to grow in shear. In fact, Forsyth (1961) has observed that fatigue cracks have two growth regimes. In stage I, cracks formed on the slip planes of the persistent slip bands grow when they are most closely aligned with the maximum shear stress direction. In sliding wear, since the slip planes tend to line up parallel to the surface, the maximum shear stress direction is likely to be the crack propagation direction.

In a two-dimensional deformation field in an elastic solid (or elastoplastic solid with a very small plastic zone at the crack tip), the stresses at the crack tip are expressed as (Sih, 1974)

$$\sigma_{rr} = \frac{1}{2\sqrt{2\pi r}}\left[K_I \cos\frac{\theta}{2}(3 - \cos\theta) + K_{II}\sin\frac{\theta}{2}(3\cos\theta - 1)\right]$$

$$\sigma_{\theta\theta} = \frac{1}{\sqrt{2\pi r}}\cos\frac{\theta}{2}\left(K_I \cos^2\frac{\theta}{2} - \frac{3}{2}K_{II}\sin\theta\right) \tag{4.47}$$

$$\sigma_{r\theta} = \frac{1}{2\sqrt{2\pi r}}\cos\frac{\theta}{2}[K_I \sin\theta + K_{II}(3\cos\theta - 1)]$$

and the maximum shear stress τ_{max} is given by

$$\tau_{max} = \frac{1}{2\sqrt{2\pi r}}[K_I^2 \sin^2\theta + 2K_I K_{II} \sin 2\theta + K_{II}^2(4 - 3\sin^2\theta)]^{1/2} \tag{4.48}$$

τ_{max} will have maximum values when the conditions $\partial\tau_{max}/\partial\theta = 0$ and $\partial^2\tau_{max}/\partial\theta^2 > 0$ are satisfied. If τ_{max} occurs at $\theta = \theta_{max}$, $[\sigma_{rr}]_{\theta_m}$, $[\sigma_{\theta\theta}]_{\theta_m}$, and $[\sigma_{r\theta}]_{\theta_m}$ can be computed using Eq (4.47). Using the Mohr's circle transformation, the maximum of τ_{max} and the angle between the θ_m direction and the τ_{max} direction can be determined. The direction at $r \to 0$ ultimately becomes the direction of crack propagation.

If the foregoing criterion is applied to the results obtained in the preceding section for subsurface cracks, the predicted angle is between -5 and $5°$. These values are very small, being nearly equal to zero. This implies that cracks propagate in a plane collinear with the original cracks which are parallel to the surface.

The preceding analytical results for an elastic solid compare favorably with the numerically determined maximum shear stress direction near the crack tip of an elastoplastic solid. One has to be cautious, however, in generalizing the elastic results, since the work-hardening rate affects the direction of the maximum shear stress direction. Furthermore, the crack propagation direction in an elastoplastic solid is likely to be affected by the texture that develops owing to the large plastic deformation that occurs in

4.5.3 The Crack Propagation Mechanism in Elastoplastic Solids

Linear elastic fracture mechanics is found to be useful in assessing the order of the crack tip stress concentration and in determining the crack trajectory using mixed-mode fracture criteria. However, the exact stress concentration and the crack propagation rate for elastoplastic solids cannot be determined using LEFM since the plastic zone extends all the way to the surface. Therefore, a plastic fracture mechanics approach is required. Sin and Suh (1984) investigated this problem using the finite element method.

The solutions were obtained using the ADINA[6] Program (Bathe, 1975). The material model used in the ADINA Program was for infinitesimal deformation and allowed material nonlinearity only (Bathe, 1976). The model used to calculate the elastoplastic response under the moving-load condition was the same as that for the elastic case. No dynamic effect was considered in the analysis. The material was assumed to be slightly work-hardening ($E_T = 10^{-4}E$). The material properties used were as follows: isotropic, slightly work-hardening, $E_T = 1.96 \times 10^5$ MPa $= 2 \times 10^4$ kg/mm^2, $E_T = 19.6$ MPa $= 2$ kg/mm^2, $\nu = 0.28$, $\sigma_y = \sqrt{3}\, k = 424$ MPa $= 43.3$ kg/mm^2.

For the investigation of crack propagation only a short crack ($c = \frac{1}{4}a$) was used. Due to the prohibitively expensive computer cost, only a limited parameter study was conducted for the case of $a = 10$ μm, $p_0 = 4k$, and $\mu = 0.25$.

The problem was solved incrementally by moving the load step by step. For an accurate solution, the increments per step must be sufficiently small. However, such a load step requires a large number of calculations, which make the analysis very expensive. Therefore, larger load steps were used with iteration to obtain efficient and accurate solutions. The use of iteration can introduce some difficulties. The convergence process may be slow, requiring a large number of iterations, which can be expensive. Also, some iterative methods do not converge for certain types of problems or large load increments. Figure 4.48 shows the development of plastic zone at each step loading. At the beginning the shape of the plastically deformed zone is more or less the same as for the case of no crack except right around the crack tips. However, as the load moves over the crack, the stress field changes substantially due to the presence of the crack. It is also interesting to note that there are some areas inside the plastic region

---

[6] Automatic Dynamic Incremental Nonlinear Analysis, a finite element computer program for the static and dynamic displacement and stress analysis of solids, fluid structure systems, and structures.

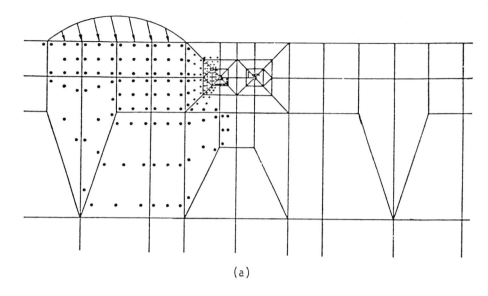

(a)

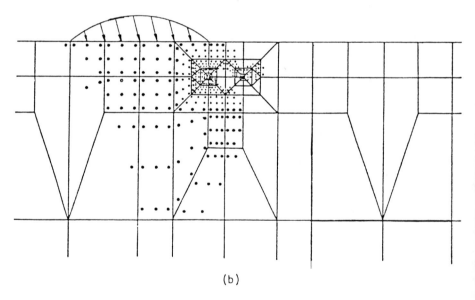

(b)

Figure 4.48 Variation of plastically deformed zone around a crack under a moving asperity: $a = 10\mu m$, $d = 0.5a$, $\mu = 0.25$, $p_o = 4k = 980$ MPa. Dots indicate the integration points. (From Sin and Suh, 1984.)

Sec. 4.5 Crack Propagation Due to Surface Traction 163

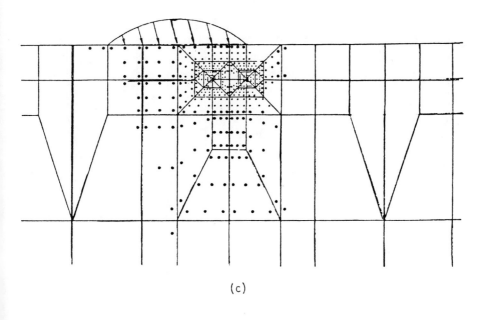

(c)

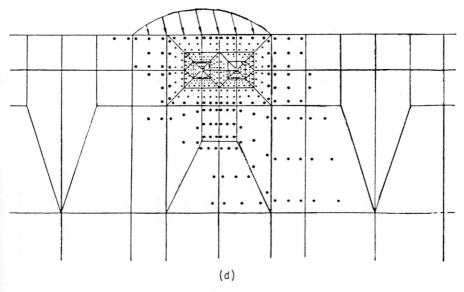

(d)

Figure 4.48 (continued)

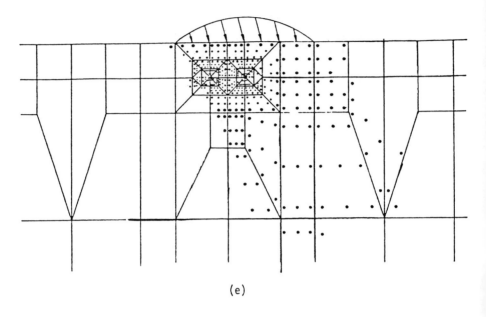

(e)

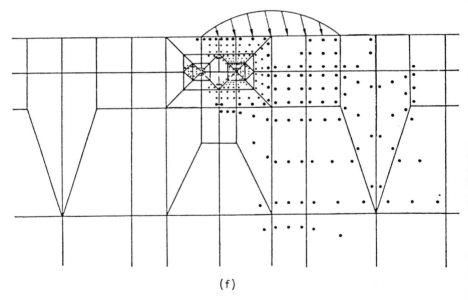

(f)

Figure 4.48 (continued)

Sec. 4.5 Crack Propagation Due to Surface Traction

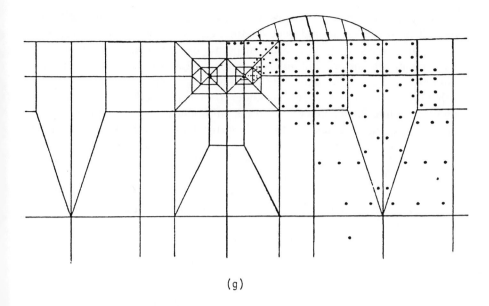

(g)

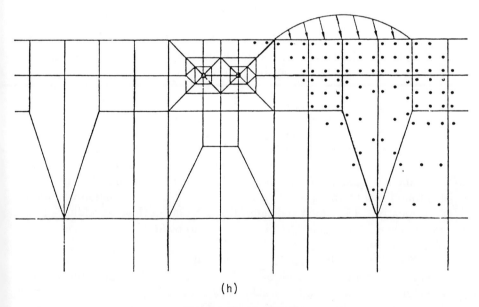

(h)

Figure 4.48 (continued)

where unloading has taken place. With repeated loading–unloading the overall plastic zone should become smaller, as in the case of no crack.

The displacement of the nodal points on the crack surfaces is such that the upper surface initially slides forward and then slides backward. Similar results were obtained by McClintock (1978), who investigated the behavior of a subsurface crack under rolling contact. In Table 4.3 the relative sliding displacements of crack tip nodal points are shown for two different depths of crack location. It shows that the relative displacement increases with sliding, then decreases, and finally changes the sign. Furthermore, the crack closer to the surface and the left tip have large values of relative sliding displacements.

In Figs. 4.49 and 4.50 the shear strains at crack tips are plotted as a function of the distance from the crack tip at different stages of the loading as the asperity contact moves from left to right. The shear strain increases to a maximum value and then decreases. After it attains a minimum, it increases again. Very near the tip the shear strain changes from positive to negative, and then to positive again. The shear strains at the crack tip can be very large.

To determine the crack growth rate under cyclic loading the crack-opening displacement concept may be applied even to mode II cracks. In this mode the relative sliding displacement of the crack tip occurs as a result of deformation-induced slip. If the maximum displacement at the crack tip is employed as a crack tip sliding displacement (CTSD), ΔS, the crack growth length, ΔC, may be expressed as

$$\Delta C = \Delta S - \Delta C_W \tag{4.49}$$

where ΔC_W is the length of the crack that has rewelded. There is no theoretical basis for determining ΔC_W. Kikukawa et al. (1979) found that rewelding of cracks under mode I loading was affected by environmental conditions, whereas the length of rewelding and the crack growth length under mode II loading were nearly the same both in air and vacuum. They also found that the ratio of $\Delta C/\Delta S$ for mode II type deformation was found to be nearly 0.16, which indicates that a great deal of rewelding of the cracks occurs under shear loading. The ratio of $\Delta C/\text{COD}^7$ for mode I was found to be 0.55.

The FEM results and Kikukawa et al.'s study show that the crack propagation occurs as a result of the large strain field that develops in front of the crack tip. That is, the crack propagates to the point where the shear strain exceeds a certain critical value. This condition may be addressed in terms of fracture shear strain, γ_f. If the material element at a distance r from the crack tip can be written as

$$\gamma_{xy} = \frac{K_\varepsilon}{(r/r_0)^m} \tag{4.50}$$

[7] COD stands for crack tip opening displacement.

TABLE 4.3 Relative Crack Tip Sliding Displacement[a] (μm)

(a) $d = 0.5a$

	a	b	c	d	e	f	g	h
Right tip	0.0009	0.0023	0.0033	0.0027	0.0015	−0.0009	−0.0008	−0.0001
Left tip	0.0013	0.0030	0.0035	0.0028	0.0004	−0.0012	−0.0005	−0.00003

(b) $d = 0.25a$

	a	b	c	d
Right tip	0.0003	0.0012	0.0077	0.0093
Left tip	0.0002	0.0031	0.0141	0.0138

[a] Each step corresponds to the relative position shown in Figure 4.48.

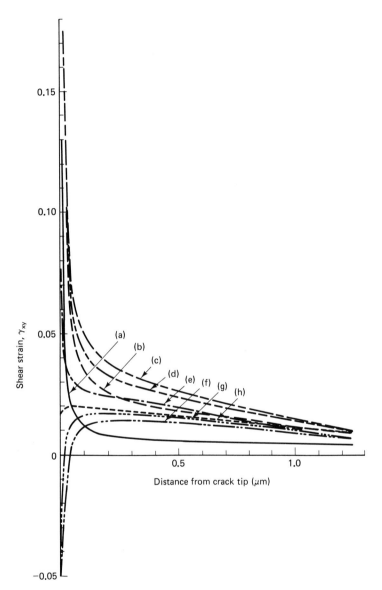

Figure 4.49 Shear strain versus distance from the left crack tip when plastic elements are used. (a)–(h) correspond to the position of the asperity contact relative to the crack tip shown in Fig. 4.48. (From Sin, 1981.)

where K_ε is the strain intensity factor, r_0 is a characteristic distance used for nondimensionalization, and m is a parameter that depends on the strain-hardening exponent. For a nonhardening material and for the HRR field, m is unity. Therefore, r_f may be obtained by substituting γ_f for γ_{xy} in Eq. (4.50).

Sec. 4.5 Crack Propagation Due to Surface Traction

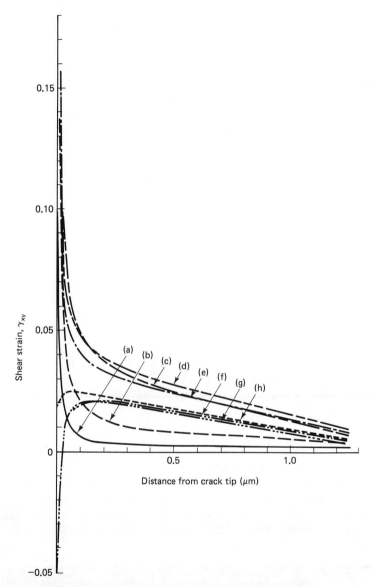

Figure 4.50 Shear strain versus distance from the right crack tip when plastic elements are used. (a)–(h) correspond to the position of the asperity contact relative to the crack tip shown in Fig. 4.48. (From Sin, 1981.)

$$r_f = r_0 \left(\frac{K_\varepsilon}{\gamma_f}\right)^{-m} \tag{4.51}$$

Figure 4.51 shows the shear strain distribution near the crack tip of cracks located at $d/a = 0.25$ and $d/a = 0.5$. If the contact length, 2a, is

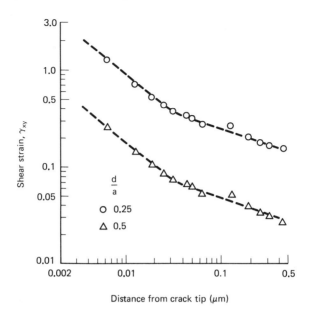

Figure 4.51 Shear strain as a function of distance from the left tip for different depths of crack location. (From Sin, 1981.)

20 μm and if the fracture strain γ_f is assumed to be 0.4, the material within 0.0030 μm of the crack tip of the crack located at $d = 5$ μm and the material within 0.02 μm of the crack tip of the crack located at $d = 2.5$ μm will fracture.

4.5.4 Experimental Results

Figures 4.52 and 4.53 show a typical micrograph of subsurface cracks in OFHC copper and iron solid solution, respectively. Figure 4.33 shows the early stages of crack propagation in annealed Fe–1.3% Mo. Cracks are not exactly parallel to the surface. This could be due to the loading condition, anisotropy of materials, and unique texture around second phases.

Figure 4.52 Void and crack formation in annealed OFHC copper.

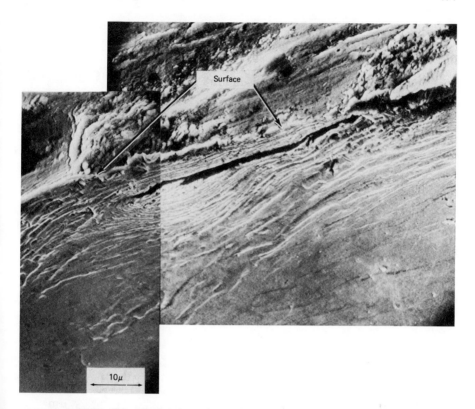

Figure 4.53 Subsurface deformation and crack formation in iron solid solution.

Cracks are more prevalent in two-phase metals. In high-purity single-phase metals cracks are not a common occurrence and consequently the wear rate is low. However, cracks have been observed even in high-purity metals. Whether these cracks are due to dislocation interactions or due to small impurities we do not know. Further work is needed to understand crack nucleation mechanism in single-phase metals.

Cracks have been also observed in polymers, composites, and ceramics. Clerico (1980) showed that chopped glass-fiber-reinforced polymers also develop subsurface cracks. Swain (1975) also reported formation of subsurface cracks in alumina. All these cracks lead to failure of the surface layer.

4.6 CONCLUDING REMARKS

In this chapter it was shown that materials respond to the forces exerted at the asperity contacts by deformation, crack nucleation, and crack propagation. It is shown that in the case of glassy elastic solids loaded by blunt asperities the crack is likely to propagate from the surface because the tensile stress parallel to the surface is largest right behind the moving asperity. The direction of these cracks change below the surface due to the changes in the tensile stress direction. When the glassy solid is loaded by a sharp asperity, cracks propagate from the asperity tip into and perpendicular to the surface—also radially from the contact area along the surface, and laterally at the subsurface.

In the case of elastoplastic solids the plastic strain of the surface layer accumulates under the cyclic loading exerted by moving asperities. With the plastic strain accumulation the stress at the particle–matrix interface grows. When the interfacial stress around a hard particle larger than about 500 Å reaches a critical stress for fracture, cracks nucleate. Analysis shows that cracks nucleate readily, requiring less than one hundred cycles of asperity loading. Once these cracks are present, they can elongate and propagate due to the large plastic deformation at the crack tip. These cracks propagate along the direction parallel to the surface, because this is the maximum shear stress direction.

Depending on the material and the loading condition, either crack nucleation or propagation can be the wear-rate-controlling process. In many two-phase metals crack propagation controls the wear rate, whereas in extremely pure single-phase metals the crack nucleation rate controls the wear.

The crack propagation rate can be computed by determining the crack tip sliding displacement under the combined loading of compression and shear. These cracks propagate along the maximum shear stress direction because of the compressive stress superimposed on the shear stress at the crack tip. This is in contrast to the case when the crack tips are under negative hydrostatic pressure. In this case, the crack propagates along the direction perpendicular to the maximum tensile stress.

REFERENCES

ANAND, S. C., "Numerical Investigation of Stresses in the Inelastic Range in a Rolling Contact," *Proceedings of the Second International Conference on Vehicle Structural Mechanics,* Society of Automotive Engineers, 1977, pp. 121–127.

ARGON, A. S., "Formation of Cavities from Non-deformable Second-Phase Particles in Low Temperature Ductile Fracture," *Journal of Engineering Materials and Technology,* Transactions of the ASME, 1976, pp. 60–68.

AUGUSTSSON, G., "Strain Field near the Surface due to Surface Friction," S.M. thesis, MIT, 1974.

BATHE, K.-J., "ADINA—A Finite Element Program for Automatic Dynamic Incremental Non-linear Analysis," Report 82448-1, Acoustics and Vibration Laboratory, MIT, Cambridge, Mass., 1975.

BATHE, K.-J., "Static and Dynamic Geometric and Material Non-linear Analysis Using ADINA," Report 82448-2, Acoustics and Vibration Laboratory, MIT, Cambridge, Mass., 1976.

BOUSSINESQ, J., *Application des potentiels à l'étude de l'équilibre et du mouvement des solides elastiques,* Gauthier-Villars, Paris, 1885 (also discussed in Timoshenko and Goodier, 1970).

CLERICO, M., "Sliding Wear Mechanisms of Polymers," in *Fundamentals of Tribology,* N. P. Suh and N. Saka, eds., MIT Press, Cambridge, Mass., 1980.

ERDOGAN, F., and SIH, G. C., "On the Crack Extension in Plates under Plane Loading and Transverse Shear," *Journal of Basic Engineering,* Vol. 85, 1963, pp. 519–527.

ESHELBY, J. D., "The Determination of the Elastic Field of an Ellipsoidal Inclusion, and Related Problems," *Proceedings of the Royal Society of London,* Series A, Vol. 241, 1957, pp. 376–396.

EVANS, A. G., and WILSHAW, T. R., "Quasi-static Solid Particle Damage in Brittle Solids: I. Observations, Analysis, and Implications," *Acta Metallurgica,* Vol. 24, 1976, pp. 939–956.

FLEMING, J. F., and SUH, N. P., "Mechanics of Crack Propagation in Delamination Wear," *Wear,* Vol. 44, 1977, pp. 39–56.

FORSYTH, P. J. E., "A Two-Stage Process of Fatigue Crack Growth," *Proceedings of the Crack Propagation Symposium,* Cranfield, England, Vol. 1, 1961, pp. 76–94.

FUCHS, S., "Hauptspannungstrajekorien bei der Beruhrung einer Kugel mit einer Platte," *Physikalische Zeitschrift,* Vol. 14, 1913, pp. 1282–1285.

GILL, S., "A Process for the Step by Step Integration of Differential Equations in an Automatic Digital Computing Machine," *Proceedings of the Cambridge Philosophical Society,* Vol. 47, 1951, pp. 90–108; and IBM, "Scientific Subroutine Package," Subroutine RKGS, 1975.

HARDY, C., BARONET, C. N., and TORDION, G. V., "The Elasto-plastic Indentation of a Half-Space by a Rigid Sphere," *International Journal for Numerical Methods in Engineering,* Vol. 3, 1971, pp. 451–462.

HEARLE, A. D., and JOHNSON, K. L., "Mode II Stress Intensity Factors for a Crack Parallel to the Surface of an Elastic Half-Space Subjected to a Moving Point Load," Cambridge University Report, CUED/C-Mech/TR26, 1983.

HERTZ, H., *Gesamte Werke,* Vol. 1, Leipzig, Germany, 1895. English translation in *Miscellaneous Papers,* Macmillan, New York, 1896.

HILLS, D. A., and ASHELBY, D. W., "On the Application of Fracture Mechanics to Wear," *Wear,* Vol. 54, 1979, pp. 321–330.

HIRTH, J. P., and RIGNEY, D. A., "Crystal Plasticity and the Delamination Theory of Wear," *Wear,* Vol. 39, 1976, pp. 133–141.

HUBER, M. T., "Zur Theorie der Beruhrung fester elastischer Korper," *Annalen der Physik,* Vol. 14, 1904, pp. 153–163.

HUBER, M. T., and FUCHS, S., "Spannungsverteilung bei der Beruhrung zweier elastischer Zylinder," *Physikalische Zeitschrift,* Vol. 15, 1914, pp. 298–303.

JAHANMIR, S., and SUH, N. P., "Mechanics of Subsurface Void Nucleation in Delamination Wear," *Wear,* Vol. 44, 1977, pp. 17–38.

JOHNSON, K. L., "An Experimental Determination of the Contact Stresses between Plastically Deformed Cylinders and Spheres," *Engineering Plasticity,* J. Heyman and F. A. Leckie, eds., Cambridge University Press, London, 1968.

JOHNSON, K. L., and JEFFERIS, J. A., "Plastic Flow and Residual Stresses in Rolling and Sliding Contact," *Proceedings of the Symposium on Fatigue in Rolling Contact*, Institution of Mechanical Engineers, London, 1963, pp. 54–65.

JOHNSON, K. L., O'CONNOR, J. J., and WOODWARD, A. C., "The Effect of the Indenter Elasticity on the Hertzian Fracture of Brittle Materials," *Proceedings of the Royal Society of London*, Series A, Vol. 334, 1973, pp. 95–117.

KEER, L. M., BRYANT, M. D., and HARITOS, G. K., "Subsurface Cracking and Delamination," *Solid Contact and Lubrication*, ASME Publication, AMD-Vol. 39, 1980, pp. 79–95.

KIKUKAWA, M., JONO, M., and ADACHI, M., "Direct Observation and Mechanism of Fatigue Crack Propagation," *Fatigue Mechanisms*, ASTM STP 675, 1979, pp. 234–253.

KRAUSE, H., and DEMIRCI, A. H., "Texture Changes in the Running Surface of f.c.c. Metals as a Result of Frictional Stress," *Wear*, Vol. 61, 1980, pp. 325–332.

LAWN, B. R., "Hertzian Fracture in Single Crystals with the Diamond Structure," *Journal of Applied Physics*, Vol. 39, 1968, pp. 4828–4836.

LAWN, B., and SWAIN, M. V., "Microfracture beneath Point Indentations in Brittle Solids," *Journal of Materials Science*, Vol. 10, 1975, pp. 113–122.

LAWN, B., and WILSHAW, R., "Review—Indentation Fracture: Principles and Applications," *Journal of Materials Science*, Vol. 10, 1975, pp. 1049–1081.

LIU, C. K., "Stresses and Deformations due to Tangential and Normal Loads on an Elastic Solid with Applications to Contact Stresses," Ph.D. thesis, University of Illinois, 1950.

LUNDBERG, G., "Elastische Beruhrung zweier Halbraeume, Forschung auf dem Gebiete des Ingenieurwesens," *Ausgabe B*, Vol. 10, No. 5, September–October 1939, pp. 201–211.

MCCLINTOCK, F. A., "Plastic Flow around a Crack under Friction and Combined Stress," in *Fracture 1977*, D. M. R. Taplin, ed., Vol. 4, Pergamon Press, Oxford, 1978, pp. 49–64.

MCCLINTOCK, F. A., and ARGON, A. S., *Mechanical Behavior of Materials*, Addison-Wesley, Reading, Mass., 1966.

MERWIN, J. E., and JOHNSON, K. L., "An Analysis of Plastic Deformation in Rolling Contact," *Proceedings of the Institution of Mechanical Engineers*, Vol. 177, 1963, pp. 676–690.

MICHELL, J. H., "The Inversion of Plane Stress," *Proceedings of the London Mathematical Society*, Vol. 34, 1902, pp. 134–142.

MIKOSZA, A. G., and LAWN, B. R., "Section-and-Etch Study of Hertzian Fracture Mechanics," *Journal of Applied Physics*, Vol. 42, 1971, pp. 5540–5545.

MORTON, W. B., and CLOSE, L. J., "Notes on Hertz's Theory of the Contact of Elastic Bodies," *Philosophical Magazine*, Vol. 43, 1922, pp. 320–329.

OH, H. L., and FINNIE, I., "The Ring Cracking of Glass by Spherical Indenters," *Journal of the Mechanics and Physics of Solids*, Vol. 15, 1967, pp. 401–411.

OH, H. L., and FINNIE, I., "On the Location of Fracture in Brittle Solids I—due to Static Loading," *International Journal of Fracture Mechanics,* Vol. 6, 1970, pp. 287–300.

RHEE, S. S., and McCLINTOCK, F. A., "On the Effects of Strain Hardening on Strain Concentrations," *Proceedings of the 4th National Congress of Applied Mechanics,* ASME, New York, 1962, pp. 1007–1013.

RITCHIE, R. O., "Influence of Microstructure on Near-Threshold Fatigue Crack Propagation in Ultra-high Strength Steel," presented at the Metal Society Conference, "Fatigue 1977," Cambridge, England, March 28–30, 1977.

ROSENFIELD, A. R., "A Fracture Mechanics Approach to Wear," *Wear,* Vol. 61, 1980, pp. 125–132.

SIH, G. C., "Strain–Energy–Density Factor Applied to Mixed Mode Crack Problem," *International Journal of Fracture,* Vol. 10, 1974, pp. 305–321.

SIN, H.-C., "Surface Fraction and Crack Propagation in Delamination Wear," Ph.D. thesis, MIT, 1981.

SIN, H.-C., and SUH, N. P., "Subsurface Crack Propagation due to Surface Traction in Sliding Wear," *Journal of Applied Mechanics,* Vol. 51, 1984, pp. 317–323.

SMITH, J. O., and LIU, C. K., "Stresses due to Tangential and Normal Loads on an Elastic Contact with Application to Some Contact Problems," *Journal of Applied Mechanics,* Transactions of the ASME, Vol. 20, 1953, pp. 157–166.

SPROLES, E. S., JR., GAUL, D. J., and DUQUETTE, D. J., "New Interpretation of the Mechanisms of Fretting and Fretting Corrosion Damage," in *Fundamentals of Tribology,* N. P. Suh, and N. Saka, eds., MIT Press, Cambridge, Mass., 1980, pp. 585–596.

SU, K.-Y., "Void Nucleation in Particulate Filled Polymeric Materials and Its Implications on Friction and Wear Properties," Ph.D. thesis, MIT, 1980.

SUH, N. P., "An Overview of the Delamination Theory of Wear," *Wear,* Vol. 44, 1977, pp. 1–16.

SUH, N. P., "Wear Mechanisms: An Assessment of the State of Knowledge," in *Fundamentals of Tribology,* N. P. Suh and N. Saka, eds., MIT Press, Cambridge, Mass., 1980.

SUH, N. P., and co-workers, *The Delamination Theory of Wear,* Elsevier, New York, 1977.

SWAIN, M. V., "Microscopic Observations of Abrasive Wear of Polycrystalline Alumina," *Wear,* Vol. 35, 1975, pp. 185–189.

SWEDLOW, J. L., "Criteria for Growth of the Angled Crack," *Cracks and Fracture,* ASTM STP 601, 1976, pp. 506–521.

THOMAS, H. R., and HOERSCH, V. A., "Stresses due to the Pressure of One Elastic Solid on Another," *University of Illinois Bulletin 212,* 1930.

TIMOSHENKO, S. P., and GOODIER, J. N., *Theory of Elasticity,* 3rd ed., McGraw-Hill, New York, 1970.

WHEELER, D. R., and BUCKLEY, D. H., "Texturing in Metals as a Result of Sliding," *Wear,* Vol. 33, 1975, pp. 65–74.

APPENDIX 4.A

FRACTURE AND FATIGUE[8]

Brittle Fracture

The eventual result of nearly any tensile deformation experiment when carried far enough is the separation of the test sample into two or more pieces. Except for the case of rupture, where ultimate failure occurs when the cross-sectional area is reduced to zero, the ultimate failure of materials is the result of some type of fracture. This broad heading of fracture encompasses a large number of phenomena. When fracture occurs at the end of the elastic extension range without extensive preceding plastic deformation, the process is called *brittle fracture*. Inorganic glasses, glassy polymers at low temperatures, and ceramics are subject to brittle fracture. When substantial plastic deformation precedes fracture, the process is called *ductile fracture*. Most metals at room temperature will undergo plastic deformation and ultimate failure by ductile fracture. Many metals also fracture after many cycles of repeated loading, which is commonly known as *fatigue fracture*. More structures fail due to fatigue fracture than any other form of failure. When subsurface cracks under the sliding surface propagate parallel to the crack surface due to compressive and shear loading, it is also a specific form of fatigue fracture.

The simplest case of fracture is the brittle fracture of glassy materials. For brittle fracture under tension to occur, the tensile stress must exceed the theoretical tensile strength of a material, σ_T, which can be related to Young's modulus, E, of the material. To within the uncertainty related to the lack of knowledge of the exact interatomic potential, the theoretical tensile strength is estimated to be

$$\sigma_T = \frac{E}{10} \text{ to } \frac{E}{20} \qquad (4.\text{A}1)$$

For typical glasses, the tensile strength is on the order of 10^6 psi, which is much greater than the observed values of the fracture strength of commercial glasses, which fall in the range of 5×10^3 to 10^5 psi.

This large discrepancy was explained by the proposal of Griffith that glassy materials contained cracklike defects which act as stress raisers. Griffith argued that for the case of uniaxial tensile loading of a material containing a crack in the plane perpendicular to the tensile axis, the crack would begin to grow and cause ultimate failure at stresses below the theoretical strength. Griffith's criterion says that if the rate of increase in surface energy resulting from the creation of a free surface by the extension of the crack

---

[8] Adapted from N. P. Suh and A. P. L. Turner, *Elements of the Mechanical Behavior of Solids*, McGraw-Hill, New York/Scripta, Technica, Washington, D.C., 1975.

is equal to the sum of the rate of the decrease in elastic strain energy around the crack and the rate of the work done by the applied constant loads, fracture can occur without requiring further increase in the external loads. From this development the breaking strength in tension, σ_c, can be related to the specific surface energy of the material, α, and the half-length of the preexisting cracks, c, by the expression

$$\sigma_c = \sqrt{\frac{2\alpha E}{\pi c}} \qquad (4.A2)$$

In an alternative development, Orowan argued that in the absence of plastic deformation the radius of curvature at the tip of the crack must be nearly equal to the atomic radius a. If this value is assumed, one can calculate the local stress at the tip of the crack by using the theory of stress concentration factors. Then the microscopic fracture stress σ_c is the stress required to make the local stress at the crack tip equal to the theoretical strength σ_T given in Eq. (4.A1). This can be expressed as

$$\sigma_c = \frac{\sigma_T}{\text{SCF}} \qquad (4.A3)$$

For a long sharp crack, the stress concentration factor is given by

$$\text{SCF} \simeq 1 + 2\sqrt{\frac{c}{a}} \simeq 2\sqrt{\frac{c}{a}} \qquad (4.A4)$$

If σ_T is taken to be $E/10$, the fracture criterion for tension becomes

$$\sigma_c \simeq \frac{E}{20}\sqrt{\frac{a}{c}} \qquad (4.A5)$$

This expression predicts results which are similar to those of the original Griffith formula, Eq. (4.A2).

Using Eq. (4.A5) and the measured fracture strengths of glasses, one may calculate the size of cracks necessary to account for the observed reduction in strength. Since breaking strengths of the order of 0.01 to $0.1\sigma_T$ are commonly observed, the required stress concentration factors are of the order of 10 to 100. This requires that the material contain cracks with lengths on the order of 25 to 2500 atomic distances, or roughly 100 to 10,000 Å. Most severe cracks in commercial glass result from mechanical damage of the surface. Such damage occurs even during careful handling as a result of the very high contact stresses which can occur when the glass touches other hard objects.

Fracture of brittle materials depends very sensitively on the volume of the material at a given stress level. This is caused by the fact that the probability of finding cracks of a given length increases with an increase in the size of the specimen.

Brittle fracture involves pulling the atoms apart. A reasonable local fracture criterion is that fracture occurs when the maximum principle stress at a point in the body reaches the theoretical strength regardless of the magnitude of other principal stresses. If σ_I^L, σ_{II}^L, and σ_{III}^L are the local principal stresses at a point such that $\sigma_I^L > \sigma_{II}^L > \sigma_{III}^L$, fracture occurs when

$$\sigma_I^L = \sigma_T \simeq \frac{E}{10} \qquad (4.\text{A}6)$$

In a perfect material, this would imply that fracture could occur only when a positive tensile stress existed in the material. However, in a material that contains an array of randomly oriented microcracks, one finds that the fracture criterion given in Eq. (4.A6) can be satisfied near the cracks even if the nominal stress is compressive.

To determine the brittle fracture conditions for a material containing microcracks of every possible orientation and subjected to a complex state of stress, one must first determine what crack orientation produces the most severe stress concentration under this state of stress. Then the magnitude of the nominal principal stresses $\sigma_I^N > \sigma_{II}^N > \sigma_{III}^N$, which will cause the local stress to reach the theoretical strength, can be found. The results of such a calculation depend on the assumptions made about the shape of the microcracks. One set of assumptions that allows the use of known solutions for the stress distribution is that the cracks have elliptical cross section, that conditions of plane strain exist near the crack, and that the cracks cannot close up under applied stress (i.e., the surfaces of the cracks are stress-free even for compressive loading). Since the stress distribution around elliptical cracks in plane strain is known, this calculation can be carried out even though it is algebraically complicated. The results expressed in terms of the tensile breaking strength σ_c, defined in Eq. (4.A5), are

If $-3\sigma_c < \sigma_{III}^N$, fracture occurs when

$$\sigma_I^N = \sigma_c \qquad (4.\text{A}7)$$

If $\sigma_{III}^N < -3\sigma_c$, fracture occurs when

$$(\sigma_I^N - \sigma_{III}^N)^2 + 8\sigma_c(\sigma_I^N + \sigma_{III}^N) = 0$$

The resulting fracture locus for plane stress is shown in Fig. 4.A1.

An important result of the derivation above is that the fracture strength of glass in compression should be approximately eight times the tensile strength.

Fracture in Materials with Limited Ductility

Unlike glasses and ceramics, which nearly always fail in a brittle manner, at room temperature, metals are generally considered to be ductile. In fact, nearly all metals can be made to deform plastically at room temperature

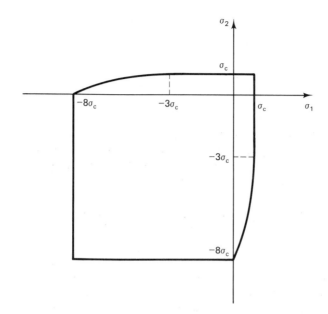

Figure 4.A1 Brittle-fracture locus for plane stress in terms of the two in-plane principal stresses σ_1 and σ_2, when the third principal stress perpendicular to the page $\sigma_3 = 0$. (From Suh and Turner, 1975.)

when tested in the form of a smooth tensile bar. Some metals, particularly those with high plastic yield stresses, however, can be made to fracture without undergoing general yield when they contain notches, slots, or other severe stress concentrators.

The Griffith fracture theory, which works well for glasses and other brittle materials such as ceramics, is meant to apply only to materials that do not undergo plastic deformation during the fracture process. Both the Griffith and the Orowan developments include assumptions that cannot be satisfied if plastic deformation occurs around the crack. Griffith made the assumption that all the work done during fracture goes into the creation of a new free surface, which does not allow for any dissipation of energy by plastic deformation. Orowan's stress concentration analysis assumes purely elastic behavior. These requirements are reasonably satisfied in the case of glasses and ceramics, where shear deformation at room temperature requires exceedingly high stresses, but they are almost certainly not satisfied by metals. It is well documented that even the most brittle fracture in a metal is accompanied by considerable plastic deformation in a small region at the tip of the crack.

In spite of the fact that the physical arguments on which the Griffith theory is based do not carry over to fracture of metals, the important functional relationships predicted by the theory are often found to work.

Both the Griffith formula (4.A2) and the Orowan formula (4.A5) predict that the fracture criterion can be expressed in the form

$$\sigma_0 \sqrt{c} = \text{constant} \tag{4.A8}$$

or that the tensile fracture stress σ_c is inversely proportional to the square root of the crack length c.

The concept of a stress intensity factor begins with the elastic stress distribution about a sharp crack in an infinite elastic body under conditions of plane strain. Figure 4.A2 shows a crack that is loaded by tension perpendicular to the crack. This is known as mode I loading. There are two other modes of loading which may cause crack growth. These other loading modes are shear on the plane of the crack perpendicular to the crack front, mode II, and shear on the plane of the crack parallel to the crack front, mode III. In terms of the coordinates in Fig. 4.A2, the stresses for mode I, mode II, and mode III loadings are σ_{22}, σ_{21}, and σ_{23}, respectively. In tribology involving sliding interfaces, the development of cracks behind the slider in glassy solids is caused by mode I, whereas the subsurface crack propagation in metals is often caused by combined loads involving compression and shear, and in some cases, tension and shear.

For the case of tensile loading, mode I, the elastic stress distribution near the crack tip is

$$\sigma_{11} = \frac{k_1}{\sqrt{2r}} \cos\frac{\theta}{2}\left(1 - \sin\frac{\theta}{2}\sin\frac{3\theta}{2}\right)$$

$$\sigma_{22} = \frac{k_1}{\sqrt{2r}} \cos\frac{\theta}{2}\left(1 + \sin\frac{\theta}{2}\sin\frac{3\theta}{2}\right) \tag{4.A9}$$

$$\sigma_{12} = \frac{k_1}{\sqrt{2r}} \sin\frac{\theta}{2}\cos\frac{\theta}{2}\cos\frac{3\theta}{2}$$

where in the infinite body,[9] $k_1 = \sigma_{22\infty}\sqrt{c}$, and is called the *stress intensity factor*. The parameter $\sigma_{22\infty}$ is the tensile stress perpendicular to the crack applied far away from the crack, and c is the crack half-length. This solution is valid in a region where r is much less than the crack length c. The $1/\sqrt{r}$ singularity in the stresses is characteristic of the elastic stress distribution around a sharp crack. The stress intensity factor is therefore a parameter which describes the strength of the elastic crack tip singularity. The stress intensity factor is a parameter that combines information about the applied loads $\sigma_{22\infty}$ and the geometry of the body, in this case the crack length c. Since the stress distribution in the region is dominated by the $1/\sqrt{r}$ crack tip singularity, the stress intensity factor completely describes the local stress distribution at the tip of a crack in an elastic body. Any combination of applied stress σ_{22} and crack length c that gives the same value of k_1 will

---

[9] Some authors use different symbol and definition for the stress intensity factor, $K_1 = k_1\sqrt{\pi}$. This definition is used, for example, in ASTM standards and publications.

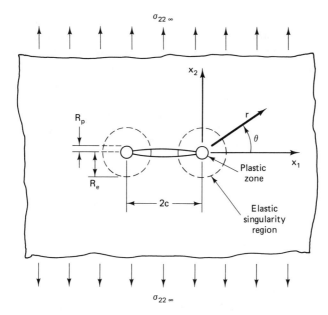

Figure 4.A2 Schematic of the state of stress and strain around the tip of an initially sharp crack. The stress distribution inside a radius R_e is essentially given by the crack tip singularity solution. Plastic deformation has occurred in a region of radius R_p which is much smaller than R_e. (From Suh and Turner, 1975.)

therefore give the same stress distribution near the crack tip. Since the fracture process should depend only on the local stresses at or near the tip of the crack, the fracture criterion for elastic bodies should be as expressible as

$$k_1 = k_{1c} \qquad (4.\text{A}10)$$

at fracture where k_{1c} is the critical stress intensity factor. The critical stress intensity factor is a material constant which in theory can be derived from the microscopic details of the fracture process.

The concept of the stress intensity factor can be generalized to situations other than the crack in an infinite body, such as edge cracks in a semi-infinite body and cracks in finite bodies. Since the $1/\sqrt{r}$ singularity caused by the presence of the crack tip itself dominates the stress distribution near the crack tip, the expressions for the stress distributions in other geometric situations can also be put in the form given by Eq. (4.A9) plus additional terms which become negligible relative to the $1/\sqrt{r}$ term, when r is sufficiently small. Each geometric situation therefore has associated with it an expression for the stress intensity factor k_1 which is a function of the applied loads and the geometric parameters of the body. Once this expression is known, the stress distribution around the crack tip is given by the expressions in Eq. (4.A9). The functional form of k_1 has been determined for a number of different geometries. Since even for geometries other than a crack in an infinite body, the stress distribution near the crack tip is uniquely determined by the stress intensity factor, the fracture criterion for these cases should also be expressible in terms of k_1 as given in Eq. (4.A10).

Up to this point, the discussion has been limited to cases where the deformation of the body is completely elastic. As might be expected, the fracture criterion given by Eq. (4.A10) is completely equivalent to the Griffith fracture criterion. If this were all that could be done with the concept of the stress intensity factor, it would have little relevance to fracture in metals. Fortunately, it can be shown that in many cases, in metals, when the plastic deformation is limited to a small area near the crack tip, the argument justifying the fracture criterion is still valid.

Consider the case of a metal sample containing a sharp crack and subjected to mode I stresses which increase slowly with time. When the stress is still very small, the deformation is elastic, and the stress distribution near the crack is given by Eq. (4.A9), which is valid in a region of radius R_e such that $R_e \ll c$ and $\sqrt{c/r} \gg 1$ (see Fig. 4.A2). Because of the singularity in the stresses, the yield condition is quickly reached at the crack tip. Assume that the plastic deformation is limited to a region of radius R_p which is smaller than R_e. The plastic deformation disturbs the stress distribution inside R_p and also in the elastic region outside R_p. However, since the plastic deformation does not change the net traction on the boundary of the plastic region, St. Venant's principle should be applicable to this situation. Thus the difference between the stress distribution in the real elastic–plastic problem and an idealized case where the plastic flow is not allowed should decrease as the distance from the plastic zone increases. When $R_p \ll R_e$, St. Venant's principle would indicate that a region must still exist outside R_p, where the stress distribution is still given by Eq. (4.A9) even after plastic flow has occurred. The stress distribution in this elastic region is uniquely specified by the value of the stress intensity factor k_1.

As the applied stresses continue to increase, the size of the plastic zone, R_p, continues to increase, but as long as R_p remains small compared to R_e, the plastic region is always surrounded by an elastic region with a stress distribution given by Eq. (4.A9). Since the tractions of the plastic region at any time are uniquely specified by the value of k_1, the stress and strain distribution inside the plastic region must also be uniquely determined by the value of k_1. Thus two different bodies of the same material containing cracks which have been loaded from zero to the same value of k_1 will experience the same stress–strain history at the crack tips even when limited plastic deformation occurs.

As the applied stresses are increased further, the plastic strain at the crack tip increases, and if general plastic yield of the body does not occur first, the microscopic fracture condition will eventually be satisfied at the crack tip. What this microscopic fracture criterion is cannot be specifically stated, but it should be possible to express it in terms of the local stress and strain history at the crack tip, even when the strain is partially plastic. When the plastic zone size R_p remains small compared to the singularity region bounded by R_e, all the way to fracture, the stress and strain distributions in the plastic zone are at all times in monotonic loading uniquely determined by the current value of k_1. In this case the microscopic conditions for crack advance should be produced at a specific value of k_1, independent of the stress distribution outside the singularity region. This would mean that a fracture criterion of the type $k_1 = k_{1c}$ at fracture should still be applicable even for elastic–plastic fracture. The critical stress intensity factor k_{1c} should be a material constant which is independent of the details of the shape of the body in question, provided that the plastic zone is sufficiently small.

Although the argument above, which justifies the validity of Eq. (4.A10) in cases where plastic deformation occurs, has avoided the need to specify the details of the plastic deformation, it does require that a number of assumptions about the size of the plastic zone, R_p, be satisfied. The argument above will not be valid unless the plastic zone is limited to a suitably small region so that the $1/\sqrt{r}$ behavior of the stresses is still valid in the surrounding region. This will be true only if the plastic zone radius R_p is small compared to the crack half-length c and small compared to the distance from the crack tip to any other boundaries of the body. If these requirements are not met, the argument that a region still exists in which the crack tip stress distribution, Eq. (4.A9), still describes the stress distribution is no longer valid. For example, if the plastic zone gets to be nearly the same size as the crack length, the plastic zones at the opposite ends of the crack will interact with each other and link up to form a plastic zone completely surrounding the crack. In this case the $1/\sqrt{r}$ singularities in the stress will no longer exist. If the plastic zone extends too near a free surface, the plastic flow may extend through from the crack tip to the free surface, and a plastic failure could result.

There are two additional conditions that must be satisfied for the development proposed to be completely valid. One is that conditions of plane strain must be present at the crack tip, and the other is that the crack tip must be sufficiently sharp. The elastic solution is based on a crack with zero radius of curvature at the tip. In practice, this requirement can be relaxed when plastic deformation occurs at the crack tip. During the plastic deformation, the crack tip opens up and blunts out. If the initial crack tip radius is small compared to the displacement that occurs at the crack tip during loading, the initial shape of the crack will have only a secondary effect on the shape of the crack tip which results from the deformation.

The plane strain requirement can be related to the size of the plastic zone. If the plastic zone radius R_p is small compared to the thickness of the part, the plastic zone has the shape of a long slender cylinder. In this case the constraints imposed by the surrounding elastic material keep the material inside the plastic zone from moving parallel to the crack tip so that the strain in the thickness of direction must be zero. Because of the constraint on the strains parallel to the crack tip, a tensile stress develops through the thickness of the part. If the plastic zone radius is large compared to the part thickness, the plastic zone has the shape of a thin disk, and there is no effective restraint on a contraction in the thickness direction. In this case the stress through the thickness should be nearly zero, so that conditions of plane stress would prevail at the crack tip. When the plastic zone is an intermediate size such that R_p is nearly equal to the thickness of the part, the constraint of the surrounding material is insufficient to enforce conditions of plane strain, but the thickness is too great to give conditions of plane stress, and a more complex state of stress is present. Since the tension through the thickness in plane strain makes the stress more hydrostatic, thus suppressing plastic flow and increasing the tendency to fracture, it is expected that the measured value of the critical stress intensity will be greater for plane stress than for plane strain.

To determine whether the size restrictions stated above are satisfied in any given situation, it is necessary to estimate the size of the plastic zone, R_p. Although an exact value for the plastic zone size can be determined only by solving the elastic–plastic problem, an estimated value can be determined from the undisturbed elastic stress distribution given in Eq. (4.A9). From the stress distribution it is determined that the equivalent stress along the maximum stress directions has a magnitude of approximately $k_1/\sqrt{2r}$. This implies that the material is at a stress above yield, according to the elastic solution, out to a radius R_p given by

$$R_p \approx \frac{1}{2}\left(\frac{k_1^2}{\sigma_y^2}\right) \tag{4.A11}$$

where σ_y is the tensile yield strength of the material. The maximum value of k_1 is k_{1c}, which occurs at fracture. Therefore, the maximum plastic zone size R_c is the one that is present just before the fracture is reached.

$$R_c \approx \frac{1}{2}\left(\frac{k_{1c}^2}{\sigma_y^2}\right) \tag{4.A12}$$

The crack propagation of mode I discussed so far in this appendix assumes that the state of loading is uniaxial tension. In that case the crack tip is opened when a tensile stress is applied across the crack. Therefore, a criterion based on the magnitude of tensile stress at the crack tip was a sufficient criterion. Even under a combined loading involving both tensile and shear stresses, the cracks will open and the crack propagation direction will be along a direction perpendicular to the maximum principle stress. However, when the combined loading consists of compression and shear of the crack, the crack propagation criteria based on the stress intensity factor or an energy criteria are no longer valid. Therefore, for sliding wear

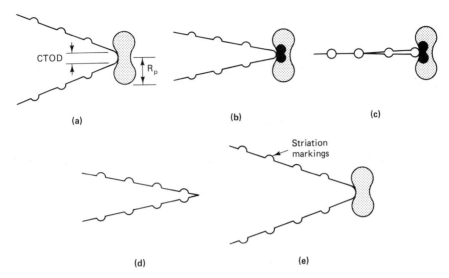

Figure 4.A3 Schematic illustration of the deformation at the crack tip during a fatigue cycle. (a) Maximum tension. The plastic zone is indicated by the shaded area and the plastic crack tip opening displacement (CTOD) is indicated. (b) Partially closed. A reverse plastic zone, solid region has begun to form. (c) Fully closed. The reverse plastic zone is at its maximum extent. (d) Partially reloaded. The crack advances into new material which has been strained plastically on the previous cycle. (e) Fully opened. The cycle begins to repeat (From Suh and Turner, 1975.)

of metals where the state of loading consists of compression and shear, another criterion must be used.

Fatigue Failure under Cyclic Loading

The foregoing discussion of fracture was based on the assumption that the failure occurs due to unstable crack growth under a monotonic loading. Cracks also grow incrementally under cyclic loading. The crack tip extension under a tension–compression type of cyclic loading is illustrated schematically in Fig. 4.A3. Under this type of loading, when the plastic zone near the crack tip is small, the rate of crack growth, dC/dN, of many metals is a function of the change in the stress intensity factor[10] Δk_1, as shown in Fig. 4.47. For cracks to grow, Δk_1 has to be larger than the threshold value. When the maximum value of k_1 reaches the critical stress intensity factor, unstable crack propagation occurs.

The relationship between dC/dN and Δk is a useful concept when the plastic zone is confined to a very small region near the crack tip. This kind of situation may exist in brittle solids or even with *deep* subsurface

---

[10] Δk_1 is defined as $(k_{1\max} - k_{1\min})$, where $k_{1\min}$ is set equal to zero when it has a negative value.

cracks that have been seen in rails that were subjected to pure rolling contact loading.

Crack Growth in Sliding Wear of Elastoplastic Solids

When the plastic zone size near the crack tip of an elastoplastic solid is comparable to the distance between the free surface and the crack tip, the stress intensity factor is not a useful concept in determining the crack growth rate. In many sliding wear situations, the subsurface cracks are very near the surface and the plastic zone size at the crack tip is often of the same size as the depth of the crack. Furthermore, in sliding wear, the crack is under a combined loading of compression and shear. In this case the crack propagation rate is best analyzed in terms of the crack tip sliding displacement (CTSD), which occurs due to the extensive plastic deformation at the crack tip. Crack extension occurs as a result of cyclic shearing of the crack tip along a direction parallel to the maximum shear stress direction, which is nearly parallel to the surface.

Direction of Crack Propagation

One of the important issues in fracture and fatigue is the direction of crack propagation. In most glass and metals with limited ductility, the cracks under mode I loading propagate along the direction perpendicular to the maximum principle stress [i.e., $(\sigma_{\theta\theta})_{max}$ in Fig. 4.A2]. Even under mode II and mode III loading without a superimposed compressive load across the crack tip, the crack propagation direction may be perpendicular to the maximum principle direction, although there are exceptions; cracks in some material propagate parallel to the maximum shear stress direction.

The statement above is true even when the cracks experience a combined state of loading (i.e., normal stress as well as shear stress) if the normal stress acting at the crack tip is tensile. On the other hand, when the crack is under the combined loading of compression and shear, such as that experienced by the subsurface cracks under many sliding wear situations, the crack propagation direction is parallel to the maximum shear stress direction rather than being perpendicular to the maximum principal stress direction.

APPENDIX 4.B

THE BOUSSINESQ SOLUTION AND HERTZIAN STRESS

The contact stress that exists between two elastic spherical bodies under the normal load is commonly known as the *Hertzian stress*, after H. Hertz (1895). Hertz's original analysis was improved by Huber (1904), Fuchs

(1913), Huber and Fuchs (1914), Thomas and Hoersch (1930), and Morton and Close (1922). In 1939, Lundberg developed a general theory of elastic contacts between two semi-infinite bodies with tangential and normal forces at the contact. A detailed analysis of the axisymmetric loading problem can be found in Timoshenko and Goodier (1970), which is reviewed briefly here.

Consider two spherical bodies with radii R_1 and R_2 in contact, as shown in Fig. 4.B1. The axes z_1 and z_2 pass through the contact point 0, which is the only point of contact when the normal force applied along the z_1 axis is equal to zero. The distance z_1 from the plane tangent to the contact point 0 to a point M on the meridian section of the sphere can be related to the radial distance r from the z_1 axis to the point M with sufficient accuracy as

$$z_1 = \frac{r^2}{2R_1} \tag{4.B1a}$$

Similarly, on the lower sphere, z_2 is related to r as

$$z_2 = \frac{r^2}{2R_2} \tag{4.B1b}$$

The distance between these two points is

$$z_1 + z_2 = r^2 \left(\frac{1}{2R_1} + \frac{1}{2R_2} \right) = \beta r^2 \tag{4.B2}$$

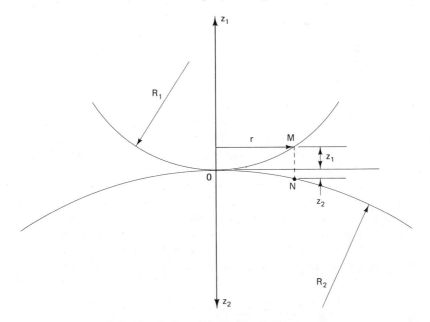

Figure 4.B1 Two spherical bodies in contact.

When the bodies are pressed together by a concentrated load P, the spheres deform, the deformation being maximum near the point of contact. The contact area will increase with the load, distributing the load over the entire area. When the radius of the contact area is much smaller than the radii of the spheres R_1 and R_2, the problem can be approximated as that of semi-infinite solid loaded by a concentrated load over a small area.

Denoting the displacements of the points M and N (see Fig. 4.B1) along the z_1 and z_2 axes as w_1 and w_2, respectively, and the relative displacement of any two points far away from the contact area as α, the original distance z_1 and z_2 can be related to the displacements as

$$z_1 + z_2 = \alpha - (w_1 + w_2) = \beta r^2 \qquad (4.\text{B3})$$

Then the displacement of any point on the surface of contact is given by

$$w_1 + w_2 = \alpha - \beta r^2 \qquad (4.\text{B4})$$

These geometric relationships can now be related to the stresses existing in the body by considering the equilibrium of forces of an axisymmetric element in cylindrical coordinates (r, θ, z). The equilibrium equations in cylindrical coordinates are

$$\frac{\partial \sigma_{rr}}{\partial r} + \frac{\partial \sigma_{rz}}{\partial z} + \frac{\sigma_{rr} - \sigma_{\theta\theta}}{r} = 0$$

$$\frac{\partial \sigma_{rz}}{\partial r} + \frac{\partial \sigma_{zz}}{\partial z} + \frac{\sigma_{rz}}{r} = 0 \qquad (4.\text{B5})$$

Equation (4.B5) and the compatibility condition can be satisfied if we choose a stress function ϕ that satisfies the biharmonic equation

$$\nabla^2 \nabla^2 \phi = \left(\frac{\partial^2}{\partial r^2} + \frac{1}{r}\frac{\partial}{\partial r} + \frac{\partial^2}{\partial z^2}\right)\left(\frac{\partial^2 \phi}{\partial r^2} + \frac{1}{r}\frac{\partial \phi}{\partial r} + \frac{\partial^2 \phi}{\partial z^2}\right) = 0 \qquad (4.\text{B6})$$

The displacements may be expressed as

$$u = -\frac{1}{2G}\frac{\partial^2 \phi}{\partial r \, \partial z}$$

$$w = \frac{1}{2G}\left[2(1 - \nu)\nabla^2 \phi - \frac{\partial^2 \phi}{\partial z^2}\right] \qquad (4.\text{B7})$$

In terms of polear coordinates R and ψ the biharmonic equation (4.B6) may be expressed as

$$\left(\frac{\partial^2}{\partial R^2} + \frac{2}{R}\frac{\partial}{\partial R} + \frac{1}{R^2}\cot\psi\frac{\partial}{\partial \psi} + \frac{1}{R^2}\frac{\partial^2}{\partial \psi^2}\right) \cdot$$
$$\left(\frac{\partial^2 \phi}{\partial R^2} + \frac{2}{R}\frac{\partial \phi}{\partial R} + \frac{1}{R^2}\cot\psi\frac{\partial \phi}{\partial \psi} + \frac{1}{R^2}\frac{\partial^2 \phi}{\partial \psi^2}\right) = 0 \qquad (4.\text{B8})$$

The solution to $\nabla^2\phi = 0$ is of the form

$$\phi_n = R^n \psi_n \tag{4.B9}$$

Solutions to a variety of different axisymmetric problems can be obtained using Eq. (4.B9), including the stress distribution in semi-infinite solids due to concentrated and distributed loads.

The stress distribution in a semi-infinite solid due to a concentrated normal force P acting at $z = 0$ and $r = 0$ can be determined from Eq. (4.B9), which was first solved by Boussinesq (1885). The Boussinesq solution for the stress distribution may be expressed as

$$\sigma_{rr} = \frac{P}{2\pi}\left\{(1-2\nu)\left[\frac{1}{r^2} - \frac{z}{r^2(r^2+z^2)^{1/2}}\right] - \frac{3r^2z}{(r^2+z^2)^{5/2}}\right\}$$

$$\sigma_{zz} = -\frac{3P}{2\pi}\frac{z^3}{(r^2+z^2)^{5/2}} \tag{4.B10}$$

$$\sigma_{\theta\theta} = \frac{P}{2\pi}(1-2\nu)\left[-\frac{1}{r^2} + \frac{z}{r^2(r^2+z^2)^{1/2}} + \frac{z}{(r^2+z^2)^{3/2}}\right]$$

$$\sigma_{rz} = -\frac{3P}{2\pi}\frac{rz^2}{(r^2+z^2)^{5/2}}$$

where r, z, and θ are the cylindrical coordinates. The displacements are given as

$$u = \frac{(1-2\nu)(1+\nu)P}{2\pi E r}\left[\frac{z}{(r^2+z^2)^{1/2}} - 1 + \frac{1}{1-2\nu}\frac{r^2z}{(r^2+z^2)^{3/2}}\right] \tag{4.B11}$$

$$w = \frac{P}{2\pi E}\left[\frac{(1+\nu)z^2}{(r^2+z^2)^{3/2}} + \frac{2(1-\nu^2)}{(r^2+z^2)^{1/2}}\right]$$

The displacements at $z = 0$ are

$$u = -\frac{(1-2\nu)(1+\nu)P}{2\pi E r}$$

$$w = \frac{(1-\nu^2)P}{\pi E r} \tag{4.B12}$$

where u and w are the displacements along the r and z axes, respectively, of a point distance r away from the concentrated load P.

The stress distribution and the displacements due to the concentrated load P can be used to obtain the solution for the case of the load distributed over a small area by superposing solutions. Consider the problem of determining the Hertzian stress distribution at the contact area of two elastic spheres under load. The displacement of a point M of the upper ball (see

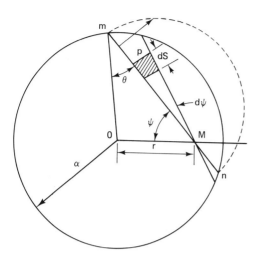

Figure 4.B2 Contact area between two spheres. The load is applied at the shaded area and its effect on the deflection of a point M is of interest. (Adapted with permission from Timoshenko, S. P., and Goodier, J. N., *Theory of Elasticity*, 3rd Ed., McGraw-Hill, New York, 1970. Copyright 1970 by McGraw-Hill.)

Fig. 4.B2) at $z = 0$ may be written, using the results given by Eq. (4.B12), as

$$w_1 = k_1 \int_A \frac{p \, dA}{s} \tag{4.B13}$$

where

$$k_1 = \frac{1 - \nu^2}{\pi E_1}$$

s is the distance from the load applied at dA to M, and p is the distributed load (i.e., local pressure). The total load is given by

$$P = \int p \, dA \tag{4.B14}$$

Substituting Eq. (4.B13) and a similar expression for w_2 into Eq. (4.B4), we obtain

$$(k_1 + k_2) \int p \frac{dA}{s} = \alpha - \beta r^2$$

or

$$(k_1 + k_2) \int p \, ds \, d\psi = \alpha - \beta r^2 \tag{4.B15}$$

In order to satisfy the equilibrium condition and the geometric compatibility, we have to find an expression for p that satisfies Eq. (4.B15). This requirement is satisfied if we assume that the pressure distribution is hemispherical with its maximum pressure p_0 acting at the center of the contact area and zero pressure along the periphery [i.e., $p = p_0(1 - r^2/a^2)^{1/2}$]. To integrate Eq. (4.B15), p acting on the shaded area in Fig.

4.B2 must be expressed in terms of its coordinate s and ψ. The integration is straightforward; $\int p \, ds$ is simply equal to the area under the dashed semicircle of Fig. 4.B2, since p is the ordinate of the semicircle and ds is an infinitesimal element along the chord mn. Since the maximum pressure is proportional to the radius of contact area a, p_0 may be expressed as $p_0 = ka$, where k is a constant representing the scale of the pressure distribution. Then

$$\int p \, dA = kA = \frac{p_0}{a} A \tag{4.B16}$$

where A is the area under the dashed semicircle, which is equal to $\frac{1}{2}\pi(a^2 - r^2 \sin^2\psi)$. Then, substituting Eq. (4.B16) into Eq. (4.B15) and integrating with respect to ψ, one obtains

$$(k_1 + k_2)\frac{p_0 \pi^2}{4a}(2a^2 - r^2) = \alpha - \beta r^2 \tag{4.B17}$$

which is satisfied when

$$\alpha = (k_1 + k_2)\frac{\pi^2 a}{2} p_0 \tag{4.B18}$$

$$a = (k_1 + k_2)\frac{\pi^2 p_0}{4\beta}$$

The total contact force P is

$$P = \int p \, dA = \int_0^a p_0 \left(1 - \frac{r^2}{a^2}\right)^{1/2}(2\pi r)\, dr = \frac{2\pi a^2}{3} p_0 \tag{4.B19}$$

Therefore,

$$p_0 = \frac{3}{2}\frac{P}{\pi a^2} = 1.5 \text{ times the average pressure} \tag{4.B20}$$

The stress distribution in the semi-infinite solid due to the distributed load can be obtained by the superposition of the stresses due to concentrated load given by Eq. (4.B10). For example, the stress σ_{zz} at a point on the z axis can be determined by substituting $p_0(1 - r^2/a^2)^{1/2}(2\pi r \, dr)$ for P in Eq. (4.B10) and integrating the expression with respect to r from 0 to a. The determination of stresses away from the axis of symmetry is complicated by the fact that the applied load is no longer symmetrical about the point of interest. Nevertheless, by using the superposition principle the stresses can be computed (Huber, 1904).

APPENDIX 4.C

CALCULATION OF ELASTIC–PLASTIC SUBSURFACE STRESSES IN SLIDING CONTACTS BY THE MERWIN–JOHNSON METHOD[11]

This appendix describes the method that was used to calculate the state of stress and the residual stresses and strains during cyclic sliding. The procedure was first developed by Merwin and Johnson (1963) for rolling contacts and was later modified (Johnson and Jefferis, 1963) to be used for sliding contacts.

The general problem is simplified by assuming that an elastic–perfectly plastic plane slides with a speed U past a rigid stationary surface which makes a contact of length $2a$. The analysis is carried out for the plane strain condition.

The analysis is performed by the following steps:

1. The steady-state residual stresses and strains in the plane xz are initially assumed to be zero.
2. The state of stress at a fixed point in the plane (initially at $x = -\infty$ and $z = z_0$) is computed using the elastic solution, Eq. (4.14). The state of stress is calculated by superimposing the elastic and residual stresses, which were obtained from the preceding cycle. The stresses are recalculated at this point as it moves toward the contact with a speed U and by increments of dx.
3. When the state of stress at the point satisfies the von Mises yield criterion, the Prandtl–Ruess equations are used to calculate the state of stress for successive movements of the point, assuming that the total strains are the same as those given by elastic relations.
4. The Prandtl–Ruess equations are integrated using Gill's modification of the Runge–Kutta method to fourth order.
5. When the von Mises yield criterion is not satisfied any more or when the rate of plastic work becomes negative as the asperity contact moves away from this point, plastic deformation terminates. For successive movements of the point, the state of stress is found from elastic equations.
6. Since no attempt is made to satisfy the equilibrium condition during the cycle, the final state of stress violates the equilibrium condition. Therefore, at the end of the cycle, the stresses are relaxed elastically to satisfy the equilibrium condition. This procedure may give nonzero residual stresses and strains.

[11] From Jahanmir and Suh, 1977.

7. Steps 2 to 6 are repeated for the same point using the residual stresses from the previous cycle. The procedure is repeated until a steady-state condition in the state of stress, the residual stresses, and the residual strains is reached. (This is when the residual stresses and

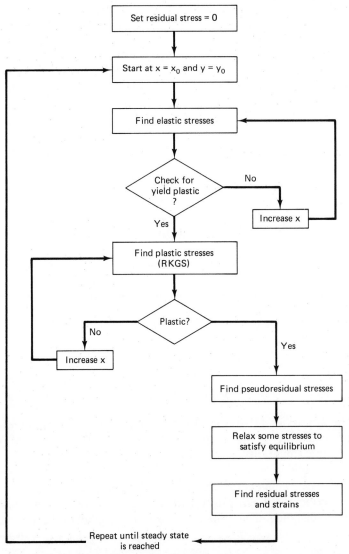

Figure 4.C1 Flowchart for the numerical calculation of stress and strain under a sliding contact. (From Jahanmir, S., and Suh, N. P., "Mechanics of Subsurface Void Nucleation in Delamination Wear," *Wear*, Vol. 44, 1977, pp. 17–38; also Jahanmir, S., Ph.D. Thesis, Massachusetts Institute of Technology, 1976.)

strains found in step 6 are not significantly different from the residual stresses and strains at the beginning of the cycle.) The final result is an approximate solution of the steady-state stresses in a sliding contact at a given depth.
8. The same procedure is followed for points at different depths to approximate the total state of stress.

The preceding method was programmed in DGC FORTRAN IV langauge, and a Nova 2 minicomputer (which uses 32 bits for real numbers) was used to perform the computations. The procedure is shown by the flowchart in Fig. 4.C1. It should be noted that the DGC FORTRAN IV language used differs from ANSI FORTRAN IV (especially in its handling of DATA statements), so that the program may require modification to run on another computer.

The final state of stress and the residual stresses and strains calculated by the foregoing procedure is an approximate solution for the case of steady-state sliding (i.e., when the residual stresses become constant and the pseudoresidual stress in the z direction becomes zero). Merwin and Johnson considered the integration cycles before a steady state is reached to be the actual transient solution, but this is not necessarily correct since it is only after the last integration cycle that the pseudoresidual stress in the z direction approaches zero.

The assumption of allowing the total strains during plastic deformation to be identical with the elastic strains is reasonable for low-friction coefficients (i.e., less than 0.5) since the region that continuously deforms plastically is contained by an elastic region around it. Therefore, the total strains cannot be much different from the elastic strains. For larger friction coefficients one boundary of the plastic region is the free surface, which may allow larger strains near the surface. Therefore, the solution becomes less exact at larger friction coefficients and near the free surface. However, it should still be reasonable deep below the surface near the elastic–plastic boundary.

The approximate solution found from the preceding method has two types of instabilities. The plastic stresses at the surface cannot be obtained, owing to the singularity of the elastic strain gradients at $x = \pm a$. The solution also becomes unstable in that the residual stresses do not converge to steady-state values very near the steady-state elastic–plastic boundaries for low friction coefficients (lower than 0.5). However, there is no problem at a small distance from the elastic–plastic boundary inside the plastic region.

5

SLIDING WEAR OF METALS

5.1 INTRODUCTION

In the preceding two chapters we have examined the generation and transmission of forces at the sliding interface and how these cyclic forces deform or fracture the materials at or below the surface. The ultimate consequence of the interfacial rubbing process is the gradual removal of materials, which is called *wear*. This is the subject of this chapter and the next three chapters. This chapter deals with the wear of elastoplastic metals under sliding conditions, whereas Chapter 6 analyzes the sliding wear of polymers and composites, and Chapter 7 is concerned with abrasive and erosive wear. The wear processes considered in these chapters are controlled by the mechanical behavior of materials, in contrast to the wear process controlled by the chemical behavior of materials, which is treated in Chapter 8.

Sliding wear of materials depends on the constitutive relationship of materials, contact geometry, and the ambient conditions. When a blunt asperity contact is sliding over an elastic glassy material, semicircular ring cracks can form behind the moving asperity, whereas when the contact geometry is very sharp, cracks can propagate from the asperity tip radially and laterally as well as axially, as discussed in Chapter 4. Since there are many asperity contact points in a given contact area, there are many sources for cracks. When these cracks emanating from various different locations intersect, loose wear particles are generated. In contrast to the brittle case, the wear of elastoplastic materials is largely caused by the cumulation of deformation-induced damages.

Wear particles in elastoplastic solids are generated during sliding by the following three mechanisms:

1. *Asperity deformation and fracture:* Wear particles can be generated by the removal of original surface asperities and the asperities generated during the delamination wear. Asperities may be removed in a single asperity interaction or in multiple, repeated interactions.
2. *Plowing:* Wear particles are also generated as a consequence of plowing. Surfaces can be plowed by wear particles, hard particles entrapped from the environment, and by hard asperities of the counterface. Plowing can generate chips by cutting or through the formation of ridges which deform and fracture due to subsequent asperity interactions.
3. *Delamination:* Large wear particles are removed by the process of plastic deformation of the surface layer, subsurface crack nucleation, and crack propagation.

The wear particles generated by these processes can either transfer to the counterface by mechanical interlocking and by adhesion or leave the interface in the form of loose wear particles. The latter is predominant in most sliding situations.

This chapter briefly describes the asperity removal process first and then extensively treats the delamination theory of wear. (Wear due to plowing is analyzed in detail in Chapter 6.) At first, the qualitative aspects of the delamination theory are discussed based on the basic mechanisms of subsurface damage discussed in Chapter 4. Following the qualitative discussion of the theory, the wear coefficient for sliding wear is derived from first principles. The theoretically predicted values of wear coefficients are then compared with the experimentally determined values. The theoretically postulated wear mechanisms are also supported with micrographic evidence and other experimental results. For example, it is shown that the critical experiments designed to confirm the microstructural effects on wear yield results that are consistent with the theoretically predicted wear behavior of single-phase and two-phase metals.

5.2 ASPERITY REMOVAL DURING SLIDING

The surface roughness and waviness affect the wear characteristics of unlubricated and boundary lubricated surfaces. In some cases the initial surface roughness generated by the machining process is removed by "running-in" the surface. Once these wear particles generated during the run-in process are filtered out, the wear rate decreases dramatically and a long wear life is obtained between the sliding surfaces. A systematic study of the asperity removal process gives a great deal of insight as to how asperity removal is related to geometry and applied load. It will be shown in this section that the roughness of machined surfaces affects only the initial wear behavior, not the steady-state wear process. This initial wear rate is a function of the distance slid and the applied load in addition to the roughness of machined surfaces.

Figure 5.1 shows the effect of the original surface roughness on wear as a function of the sliding distance. The results were obtained using the pin-on-disk geometry. AISI 52100 steel pin and AISI 1020 steel disk were tested in an argon gas environment at a sliding speed of 4.7 m/min. Under the large load of 0.3 kg, the rougher surfaces resulted in a larger initial weight loss. However, once the original machining marks were removed and the steady-state delamination wear was begun, the wear rates of all surfaces were equal. When the normal load is much lower (0.075 kg), the reverse was true; that is, the rough surfaces had lower initial wear than smoother surfaces (Fig. 5.2). This is due to the fact that the rough surfaces under light loads delay the commencement of the delamination process until the original machining marks are removed. The smooth surface initiates the delamination process soon after sliding takes place. During steady-state delamination wear, the wear rate is independent of the initial surface finish. Since the rate of the removal of asperities was much lower than the delamination rate under light loads, the rough surfaces had smaller weight losses initially than those of smooth surfaces. At larger loads, the initial machined asperities were removed rapidly and the delamination wear process was initiated early.

Figure 5.3 shows the micrographs of the surface asperities undergoing deformation and fracture. The machined surfaces of AISI 1018 steel cylinders were subjected to cylinder-on-cylinder sliding against AISI 52100 steel rod.

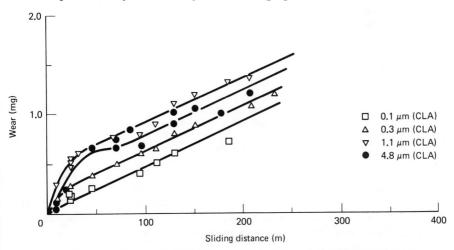

Figure 5.1 Weight loss of AISI 1020 steel disk sliding against AISI 52100 steel pin as a function of sliding distance (normal load 0.30 kg). (After Abrahamson et. al., 1975.)

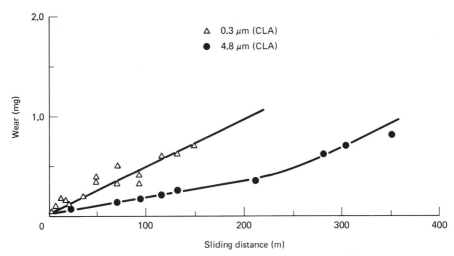

Figure 5.2 Weight loss of AISI 1020 steel disk sliding against AISI 52100 steel pin as a function of sliding distance (normal load 0.075 kg). (After Abrahamson et. al., 1975.)

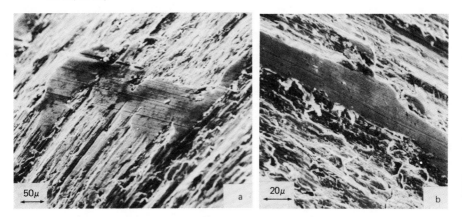

Figure 5.3 Plastic deformation of original asperities for AISI 1018 steel (cylinder on cylinder): (a) sliding perpendicular to the machining marks, 2.0 μm (CLA) surface finish, under a normal load of 0.91 kg for 10 passes; (b) sliding parallel to the machining marks, 3.3 μm (CLA) surface finish, uner a normal load of 0.35 kg after 0.25 m of sliding. Testing done in argon at 1.8 m/min. (After Suh et al., 1975.)

The sliding was perpendicular to the machining marks in Fig. 5.3(a) and parallel in Fig. 5.3(b). It is clear that the rough surface [i.e., 3.3 μm (CLA)] has become smooth at the wear track, and there is no evidence of adhesive or transferred particles, in spite of the fact that the tests were performed in an argon atmosphere.

As the asperities are gradually deformed and removed, the subsurface undergoes plastic deformation, leading to delamination wear. When wear sheets come off the surface as a result of the delamination process, new surface asperities are generated. These newly generated surface asperities constitute a new surface. With further sliding, these new asperities are also deformed and eliminated. Once the asperities are removed, the surface remains smooth until the surface lifts off again, when the subsurface damage process is completed. This process exposes a new rough surface. This process of wear goes on as long as sliding takes place.

5.3 DELAMINATION THEORY OF WEAR

5.3.1 Description of the Theory

The delamination theory of wear was introduced in 1972 to explain the wear of metals and other solid materials (Suh, 1973). The theory has been supported later by a large number of experimental results (Jahanmir et al., 1974; Pamies-Teixeira et al., 1977; Saka et al., 1977; Clerico, 1980). The significance and implications of the mechanisms postulated by the delamination theory of wear can best be understood by considering the following questions:

1. Where does all the energy supplied by the external agent go?
2. Why and how does the coefficient of friction affect the wear rate?
3. Why do some hard metals wear faster than softer metals?
4. Why do most wear particles have an aspect ratio greatly different from unity?
5. Why does seizure occur?
6. How does the microstructure of metals affect the wear rate?
7. How do initial surface roughness and waviness influence the wear phenomenon?

The delamination theory of wear describes the following sequential (or independent, if there are preexisting subsurface cracks) events which lead to loose wear sheet formation:

1. When two surfaces come into contact, normal and tangential loads are transmitted through the contact points. Asperities of the softer surface are easily deformed and fractured by the repeated loading action, forming small wear particles. Hard asperities are also removed but at slower rates. A relatively smooth surface is generated initially, either when these asperities are deformed or when they are removed.
2. The surface traction exerted by the harder asperities at the contact points induces incremental plastic deformation per cycle of loading, accumulating with repeated loading. The increment of permanent de-

formation remaining after given cyclic loading is small compared with the total plastic deformation that occurs in that cycle, since the direction of shear deformation reverses during a given cycle and the magnitude of elastic unloading is comparable to plastic strain.

3. As the subsurface deformation continues, cracks are nucleated below the surface. Crack nucleation very near the surface cannot occur, due to the triaxial state of compressive loading which exists just below the contact region.
4. Once cracks are present (either by crack nucleation or from preexisting voids and cracks), further loading and deformation causes the cracks to extend and propagate, eventually joining neighboring cracks. The cracks tend to propagate parallel to the surface at a depth governed by material properties and the state of loading.
5. When the cracks finally shear to the surface, long and thin wear sheets delaminate. The thickness of the wear sheet is determined by the location of subsurface crack growth, which is affected by the normal and tangential loads at the surface. The wear rate is controlled by the crack nucleation rate or the crack propagation rate, whichever is the slower.

A series of experimental studies conducted at MIT (Jahanmir et al., 1974) has substantiated the theory, showing that the delamination process initiates when the subsurface plastic deformation causes the nucleation of voids (Fig. 5.4). With further deformation these voids elongate and link up to form long cracks in a direction nearly parallel to the wear surface (Figs. 4.52, 4.53, and 5.5). At a critical length, these cracks shear to the surface, yielding a wear particle in the form of a long thin sheet, as shown in Fig. 5.6. The top surface of the wear sheet is generally smooth, while the fractured surface is rough, sometimes showing dimples (Fig. 5.7) and in some cases, shear failure surfaces. These experimental observations have been explained in terms of the theoretical models presented in the preceding chapters.

5.3.2 Model for Delamination Wear

A wear equation based on the delamination theory can be readily developed (Fleming and Suh, 1977; Suh and Sin, 1983). For this purpose consider a subsurface crack lying below the surface, as shown in Fig. 5.8. An asperity is moving over the surface from left to right. The wear rate will be dictated by the crack propagation at both ends, L and R. The crack propagation for the i cycle, ΔC_i, may be expressed as

$$\Delta C_i = f(\mu, d, C, \text{material properties}) \tag{5.1}$$

for both ends. If N is the total number of asperity passes required for

Sec. 5.3 Delamination Theory of Wear

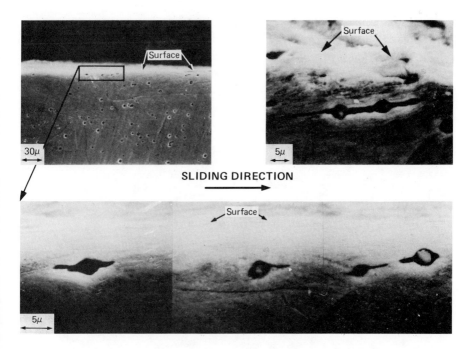

Figure 5.4 Void formation around inclusions and crack propagation from these voids near the surface in annealed Fe–1.3 at % Mo. (From Jahanmir, S.; Suh, N. P.; and Abrahamson, E. P., II, "Microscopic Observations of the Wear Sheet Formation by Delamination," *Wear*, Vol. 28, 1974, pp. 235–249.)

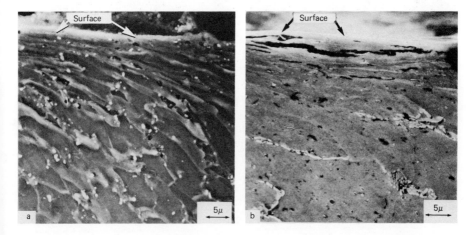

Figure 5.5 Subsurface deformation, void elongation, and crack formation in steel: (a) doped AISI 1020 steel; (b) commercial AISI 1020 steel. (From Jahanmir, S.; Suh, N. P.; and Abrahamson, E. P., II, "Microscopic Observations of the Wear Sheet Formation by Delamination," *Wear*, Vol. 28, 1974, pp. 235–249.)

Figure 5.6 Wear sheet formation in iron solid solution. The sliding direction is always from right to left since the only crack tip behind the slider can be subjected to tensile stress field which propagates the crack along the direction 70° away from the surface. (From Jahanmir et al., 1974.)

removal of one layer, the volume V_1 for one wear sheet of width w lying at a depth d is obtained as

$$V_1 = wd \sum_{i}^{N} (\Delta C_{L_i} + \Delta C_{R_i}) \qquad (5.2)$$

Therefore, the total volume V for one layer may be given by

$$V = N_c N_w wd \sum_{i}^{N} (\Delta C_{L_i} + \Delta C_{R_i}) \qquad (5.3)$$

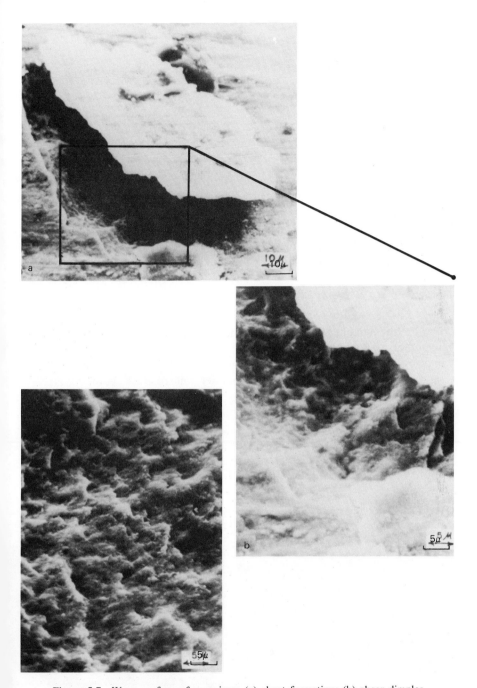

Figure 5.7 Worn surface of pure iron: (a) sheet formation; (b) shear dimples beneath the wear sheet in (a); (c) dimpled appearance of a wear crater. (From Jahanmir, S.; Suh, N. P.; and Abrahamson, E. P., II, "Microscopic Observations of the Wear Sheet Formation by Delamination," *Wear*, Vol. 28, 1974, pp. 235–249.)

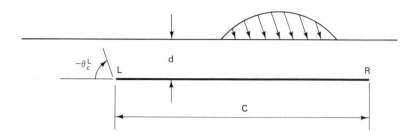

Figure 5.8 Subsurface crack under a moving asperity.

where N_c is the number of cracks along the sliding direction and N_w is the number of wear sheets in the direction of contact width. Since $N_w w$ is of the order of the contact width L_w, the volume V becomes

$$V = N_c L_w d \sum_i^N (\Delta C_{L_i} + \Delta C_{R_i}) \qquad (5.4)$$

Let λ be the spacing of asperity contacts, l_c the crack spacing, D the diameter of a specimen, and ΔL the contact length, as shown in Fig. 5.9. It may be assumed that L_w is equal to ΔL. Using this model the number N_c and N can be determined for a specimen and a slider, respectively.

Since the number of asperities per unit length of contact is $\Delta L/\lambda$, where ΔL is the total contact length, the number of cyclic loading N for

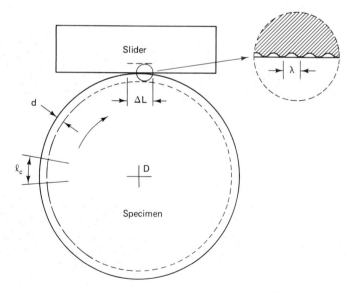

Figure 5.9 Model of wearing specimen and slider.

Sec. 5.3 Delamination Theory of Wear

the specimen is given by

$$N = \frac{S}{\pi D} \frac{\Delta L}{\lambda} \tag{5.5}$$

where S is the sliding distance required for removal of one layer. The number of cracks in the specimen N_c is simply

$$N_c = \frac{\pi D}{l_c} \tag{5.6}$$

Substituting Eqs. (5.5) and (5.6) into Eq. (5.4), the wear rate is obtained as

$$\frac{V}{S} = \frac{\Delta L^2 d(\overline{\Delta C}_L + \overline{\Delta C}_R)}{\lambda l_c} \tag{5.7}$$

where $\overline{\Delta C}_L$ and $\overline{\Delta C}_R$ are the average crack propagation rates during N cycles. The corresponding numbers N and N_c for the slider are

$$N = \frac{S}{\lambda} \tag{5.8}$$

and

$$N_c = \frac{\Delta L}{l_c} \tag{5.9}$$

The substitution of Eqs. (5.8) and (5.9) into Eq. (5.4) also yields Eq. (5.7).

The wear coefficient K can be obtained by substituting Eq. (5.7) into Archard's equation, Eq. (1.12), as

$$K = \frac{3H \Delta L^2 d(\overline{\Delta C}_L + \overline{\Delta C}_R)}{W \lambda l_c} \tag{5.10}$$

where H is the hardness of the material and W is the applied load.

Equation (5.10) states that the wear coefficient K is inversely proportional to the asperity contact spacing and the crack spacing. The crack depth d is determined by the crack nucleation condition. As discussed earlier in detail, cracks cannot nucleate very near the surface because of the large hydrostatic pressure that exists right underneath the asperity contact. In two-phase metals the location of the hard particles and the plastic deformation field of the matrix under a given set of surface traction control the exact crack nucleation site. These voids nucleate in a band of subsurface region. Once nucleated, these voids elongate due to plastic deformation and then propagate during the subsequent loading cycles.

The crack propagation rate is the largest for the crack nearest the surface, since the shear strain at the crack tip increases with decrease in the crack depth from the surface. Therefore, although many cracks may propagate, only the crack nearest the surface propagates the fastest and thus controls the wear rate. This is the case in most unidirectional sliding situations.

When the sliding direction reverses periodically as in fretting wear, all the nucleated cracks may propagate rather than only the crack nearest the surface propagating. When the asperity contact does not move continuously along one direction but reverses its direction of sliding with a very minute displacement amplitude along each direction, the cracks will stop propagating when the crack tip extends outside the *zone of influence* of the specific asperity contact. In this case, other subsurface cracks that have lower propagation rates can catch up with the fastest-propagating crack. Therefore, in fretting wear there can be many layers of cracks present in the subsurface, as shown in Fig. 5.10. The maximum crack propagation distance in fretting wear is limited by the maximum relative displacement amplitude between the slider and the specimen. When the crack propagation distance is larger than the crack spacing l_c, the fretting wear coefficient should be the same as the wear coefficient determined under normal sliding conditions (see Fig. 1.9).

In deriving Eq. (5.10) it was assumed that the cracks propagate with an average growth rate of $\Delta \overline{C}_L$ and $\Delta \overline{C}_R$. This is a reasonable assumption when the crack length is longer than a critical length at which the crack-propagation no longer depends on the critical length. When the crack length is shorter than the critical length, the crack propagation rate is expected to be proportional to the crack length. The critical crack length is just equal to the zone of influence of the surface traction, where both crack tips just extend due to the applied load.

The critical crack length and the crack propagation rate must be functions of the interfacial friction at the crack surfaces. When the crack surfaces completely adhere under the combined compressive and shear loading, there cannot be any strain concentrations at the crack tips. The maximum strain

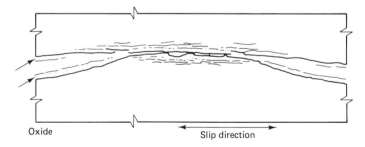

Figure 5.10 Schematic model of fretting phenomenon showing production of metallic platelets, bare surfaces, and surface oxidation. (Reprinted with permission from Sproles, E. S., Jr., Gaul, D. J., and Duquette, D. J., "A New Interpretation of the Mechanism of Fretting and Fretting Corrosion Damage," in *Fundamentals of Tribology,* edited by Suh, N. P., and Saka, N., MIT Press, Cambridge, Mass., 1980. Copyright 1980 by MIT Press.)

concentrations are present when there is no friction between the crack surfaces.

The crack will normally propagate parallel to the surface when the following two conditions are met: (1) the crack experiences combined compressive and shear loading, and (2) if the material behavior is reasonably isotropic or if the cleavage plane is parallel to the surface. When the crack tip enters the tensile region, the cracks will propagate toward the surface. This happens occasionally at the trailing crack tip (i.e., the crack tip behind the asperity contact) since the tensile region exists behind the slider. As shown in Chapter 4, this tensile field forces the crack to propagate perpendicular to the maximum tensile stress direction, which is inclined to the surface by about 70° behind the asperity contact. Figure 5.6 shows these cracks reaching the surface. Experimental results show that wear sheets always lift up right after the asperity moves over the surface, probably because the trailing tip of the crack is the only one that can propagate toward the surface, due to the particular stress field that exists behind the slider.

The crack propagation rate is expected to be affected by material properties such as strain-hardening characteristics, cyclic hardening, cyclic softening, and the Bauschinger effect. Furthermore, the anisotropy associated with the alignment of slip planes with the surface must affect the crack propagation rate since it is often the easiest to deform along the shear plane and since it is easier to cleave along the slip planes. The effect of all these material properties cannot be incorporated in a theoretical model and predict the wear coefficient precisely at this time.

Experimentally, it is well known that the wear behavior of materials depends strongly on the coefficient of friction (Rabinowicz, 1965; Tohkai, 1979). This must be due to the fact that both the crack growth rate and the depth of crack location increase with the frictional force (i.e., tangential load). It was shown in Chapter 4 that a high shear strain field is present ahead of extending cracks when the coefficient of friction is high. Therefore, the crack propagation rate is large when the friction coefficient is large. It was also shown in Fig. 4.42 that the depth of crack nucleation increases with the frictional force.

The effect of friction on wear can be analyzed by using the analysis made on plastic deformation, crack nucleation, and crack propagation of elastoplastic solids, which require costly numerical calculations. Alternatively, the approximate relationship between friction and wear may also be very roughly inferred from the elastic solutions obtained in Chapter 4. In Fig. 5.11 the maximum mode II stress intensity factors for a given d are plotted as a function of the friction coefficient. It can be seen that the relation between K_{II} and μ is approximately linear for a given d. If we assume that Paris's crack growth law is still valid even in this case, the crack growth ΔC may be expressed as

$$\Delta C = (a\mu)^n \qquad (5.11)$$

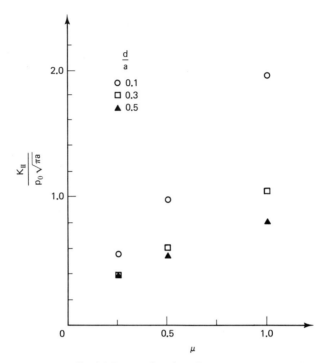

Figure 5.11 Normalized ΔK_{II} as a function of μ : $c = 3a$. (From Sin, 1981.)

where a and n are constants. Since the wear volume is proportional to the total ΔC accumulated, the sliding wear rate is a power function of friction. The exponent n varies from 2 to 4 for most metals in typical fatigue loading situations (stage II of Fig. 4.43), but can be much larger near the threshold value of ΔK_{th} (i.e., stage I) or in the unstable crack growth regime (stage III).

Since the friction coefficient is very high when the entrapped wear particles at the sliding interface plow the surface (see Chapter 3), the wear rate is also very high when those wear particles are present. Therefore, one way of significantly reducing the wear rate is to remove the particles from the interface by using lubricants, blowing the particles away, and by removing the particles through the creation of particle trapping sites on the sliding surfaces. The wear rate may be decreased by two or more orders of magnitude by reducing the friction coefficient by a factor of about 2. The control of friction and wear is further discussed in Chapter 9.

5.3.3 Numerical Examples: Wear Coefficient Prediction

The delamination wear model [i.e., Eq. (5.7)] may be used to predict wear coefficients of metals sliding against a slider. Two examples are given here.

Example 1

The crack tip sliding displacement for cracks at $d = 5$ μm is 0.0035 μm (Table 4.3). If $\Delta \overline{C}_L = \Delta \overline{C}_R = 0.0035$ μm and the apparent contact length $\Delta L = 2$ mm, and both the asperity contact spacing λ and the crack spacing l_c are assumed to be 100 μm, then

$$\frac{V}{S} = \frac{(2 \times 10^{-3})^2 (5 \times 10^{-6})(0.007 \times 10^{-6})}{(100 \times 10^{-6})(100 \times 10^{-6})} = 1.4 \times 10^{-11} \text{ m}^3/\text{m}$$

and for $H = 100$ kg/mm^2 and $W = 1$ kg,

$$K = \frac{3 \times 100 \times 10^6 \times 1.4 \times 10^{-11}}{1} = 4.2 \times 10^{-3}$$

Example 2

For $d = 2.5$ μm, by taking $\Delta \overline{C}_L = \Delta \overline{C}_L = 0.014$ μm again from Table 4.3,

$$\frac{V}{S} = \frac{(2 \times 10^{-3})^2 (2.5 \times 10^{-6})(0.028 \times 10^{-6})}{(100 \times 10^{-6})(100 \times 10^{-6})} = 2.8 \times 10^{-11} \text{ m}^3/\text{m}$$

and

$$K = 8.2 \times 10^{-3}$$

According to the experimental results, K is between 10^{-2} and 10^{-4} (see Table 3.3), which indicates that the model can predict wear behavior fairly well.

5.4 MICROSTRUCTURAL EFFECTS IN DELAMINATION WEAR

The microstructure of metals affects the wear behavior of metals a great deal. Hardness and toughness are generally affected by the microstructure, which in turn affects the wear properties. It has been well established that the hardness and the topography of the surface affect the number of asperity contacts and the size of each individual contact (Gupta and Cook, 1972; Whitehouse, 1980), in addition to the resistance to deformation of the surface layer. As we have seen in Chapter 4, the toughness of metals is closely related to the crack propagation rate, which also affects the rate of wear sheet formation.

Several different aspects of the microstructure of metals affect the wear process: grain size; the property, volume fraction, size, and distribution of second-phase particles; and the texture of the surface layer. Each of these aspects will be discussed in this section.

From the friction and wear point of view, the ideal material that has a low coefficient of friction and a low wear rate is a hard material in which cracks cannot be nucleated, and chemically inert. In an elastoplastic solid, the crack propagation rate is less in a harder material, since the crack tip sliding displacement is smaller. To preclude crack nucleation, such a material must be single phase without any impurities or with extremely small impurities (< 200 Å). If the material is also resistant to crack propagation, the material will be tough and wear resistant. However, many engineering materials cannot be made to be both tough and hard. Hard materials generally have low toughness. Therefore, the choice of materials for tribological applications must be made through compromise.

The wear of single-phase polycrystalline metals was investigated at MIT (Pamies-Teixeira et al., 1977; Saka et al., 1977; Saka, 1980). OFHC copper, copper–chromium (0.58 and 0.81 at % Cr), copper–silicon (2.3 and 8.6 at % Si), and copper–tin (1.4, 3.4, and 5.7 at % Sn) alloys were tested. Table 5.1 gives the heat-treatment conditions, grain size, and chemical composition of these materials. Cylinder-on-cylinder tests were conducted by rotating these single-phase alloys against a stationary AISI 52100 steel rod. Figure 5.12 shows the microstructures of solid solutions, and Fig. 5.13 shows the Vickers hardness of the solid solutions as functions of the solute content. It can be seen that different elements produce different hardening for the same atomic content of solute. The effectiveness of tin in increasing the hardness of copper should be noted.

TABLE 5.1 Experimental Materials: Copper Solid Solutions[a]

Alloy	Composition		Recrystallization Treatment		Grain Size (μm)
	Wt %	At %	Temp. (°C)	Time (min)	
OFCH Cu	—	—	360	250	28
Cu–Sn	2.5	1.4	790	120	35
Cu–Sn	6.0	3.4	790	60	40
Cu–Sn	10.0	5.7	790	30	30
Cu–Si	1.0	2.3	850	300	75
Cu–Si	4.0	8.6	850	300	90
Cu–Cr	0.47	0.58	1070	5	450
Cu–Cr	0.66	0.81	1070	5	485

[a] The purity of the materials used was: OFHC, 99.98%; Sn, 99.89%; Si, 99.9999%.

An immediate consequence of the alloying was the change in the friction coefficient, as shown in Fig. 5.14. The harder Cu–Sn solution seems to have lower coefficients of friction than Cu–Si solutions. Notwithstanding the small difference in friction coefficient, there exists a large difference in wear rates which are shown in Fig. 5.15. Figures 5.16 and 5.17 are SEM micrographs of some of the materials tested. The sliding direction in these micrographs is from the right to the left. These micrographs indicate that cracks tend to propagate parallel to the surface and then extend to the surface. Although the angle of the crack when the crack begins to propagate to the surface is not always 70°, it is interesting to note that most cracks change their direction as they approach the surface, more toward the direction perpendicular to the maximum normal stress direction. The depth at which cracks propagate seems to be different for specimens with different com-

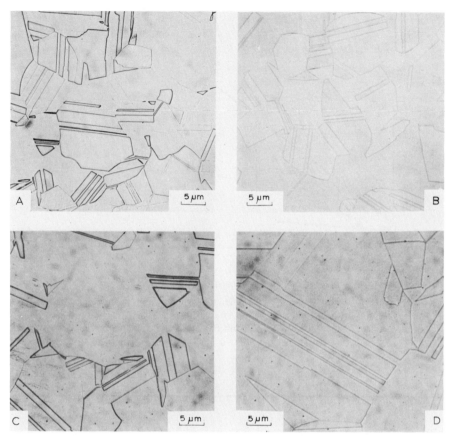

Figure 5.12 Optical micrographs showing the representative microstructures of solid solutions: (a) OFHC copper; (b) Cu–5.7 at % Sn; (c) Cu–8.6 at % Si; (d) Cu–0.81 at % Cr. (From Pamies-Teixeira, J. J.; Saka, N.; and Suh, N. P., "Wear of Copper-based Solid Solutions," *Wear*, Vol. 44, 1977, pp. 65–75.)

positions. This difference is more striking between the micrographs of OFHC copper and Cu–5.7 at % Sn solid solution; OFHC copper shows cracks propagating at a depth of the order of 50 μm, the latter at a depth of the order 15 μm.

The microscopic observations of the worn specimens show four facts: (1) that wear sheets are formed by cracking, (2) that there was extensive plastic deformation, (3) that there are subsurface cracks running parallel to the surface, and (4) that the cracks at the trailing end turn toward the surface. All these observations are in accordance with the delamination theory of wear. These micrographs do not enlighten us as to how the cracks are initiated.

The effects on tribological behavior of having a two-phase structure were investigated at MIT (Saka et al., 1977) using precipitation-hardened

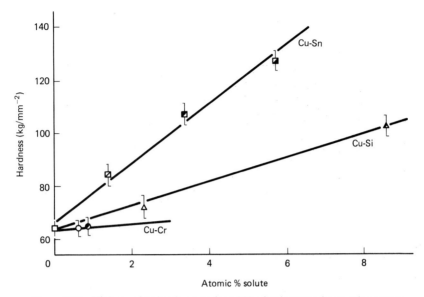

Figure 5.13 Vickers microhardness under a 200-g load versus the atomic percent of solute. Each point represents an average of five measurements and the bars represent the standard deviation. (From Pamies-Teixeira, J. J.; Saka, N.; and Suh, N. P., "Wear of Copper-based Solid Solutions," *Wear*, Vol. 44, 1977, pp. 65–75.)

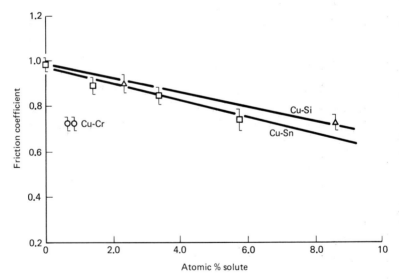

Figure 5.14 Friction coefficient as a function of the atomic content of the solute. The friction coefficient was calculated using the steady-state tangential force. The sliding speed was 2 m/min. (From Pamies-Teixeira, J. J.; Saka, N.; and Suh, N. P., "Wear of Copper-based Solid Solutions," *Wear*, Vol. 44, 1977, pp. 65–75.)

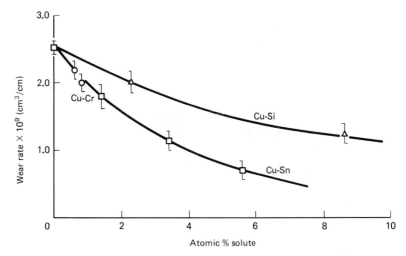

Figure 5.15 Wear rate as a function of the atomic percent of solute. The normal load was 2 kg and the duration of the tests was 100 minutes. (From Pamies-Teixeira, J. J.; Saka, N.; and Suh, N. P., "Wear of Copper-based Solid Solutions," *Wear*, Vol. 44, 1977, pp. 65–75.)

copper–chromium alloys (Cu–0.58 at % Cr and Cu–0.81 at % Cr) aged for different periods of time at 500°C. The characteristics of the materials are given in Table 5.2. Figure 5.18 shows the variation in the hardness of the copper–chromium supersaturated solid solutions as a function of the aging time. The hardness of these materials initially increased with aging time and then decreased. The maximum value was reached after about 100 minutes of aging. The hardness is 65 kg/mm^2 for solid solutions and 140 kg/mm^2 for peak-aged alloys. The aging time for the maximum hardness is about the same. Cylindrical specimens made of these two-phase alloys were rotated against a hard stationary slider made of AISI 52100 steel rod.

Figure 5.19 shows the friction coefficient, the wear rate, and the wear coefficient as functions of the aging time. The friction coefficient is fairly constant for all treatment times for both alloys. The increase in hardness resulting from the aging treatment does not seem to affect the friction coefficient, probably because the hardness of the slider was much greater than the specimens. The wear rate initially decreases by a factor of 3 for both Cu–Cr alloys and then increases approximately linearly; the slope seems to be the same for both alloys. The minimum wear rate does not correspond to the maximum hardness. (In these figures the peak hardness is indicated by arrows A and B for Cu–0.58 at % Cr and Cu–0.81 at % Cr, respectively.) Figure 5.19 also gives a plot of the wear coefficient as a function of aging time, which shows that it increases rapidly after 5 minutes of aging and then levels off asymptotically to a constant value.

As the microstructural differences between the two Cu–Cr alloys are

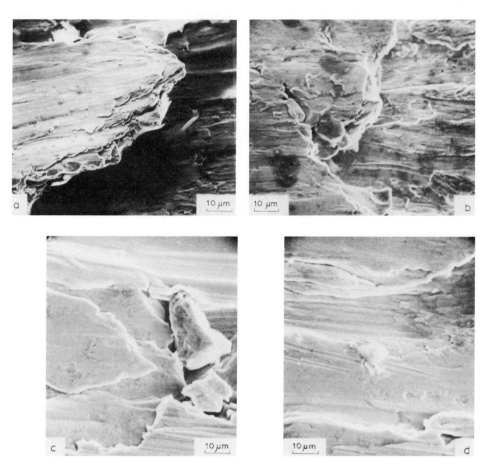

Figure 5.16 Scanning electron micrographs of wear tracks: (a) OFHC copper; (b) Cu–5.7 at % Sn; (c) Cu–8.6 at % Si; (d) Cu–0.81 at % Cr. The normal load was 2 kg at a sliding speed of 2 m/min and the sliding distance was 200 m. (From Pamies-Teixeira, J. J.; Saka, N.; and Suh, N. P., "Wear of Copper-based Solid Solutions," *Wear*, Vol. 44, 1977, pp. 65–75.)

characterized by the volume fraction and the mean free path of particles, the wear resistance (the inverse of wear coefficient) is plotted as a function of the volume fraction and the inverse of mean free path, as shown in Fig. 5.20. The wear resistance decreases with an increase in the volume fraction and with the inverse of the mean free path of particles for the overaged alloys. These results are expected since the crack nucleation sites increase with the number of hard particles.

Figure 5.21 shows the micrographs of wear tracks of the precipitation-hardened alloys aged for 5 and 10,000 minutes. The sliding direction is from left to right. It can be seen that the surface details are similar to

Sec. 5.4 Microstructural Effects in Delamination Wear   215

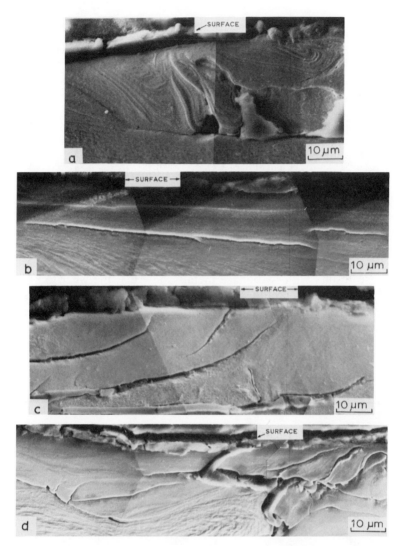

Figure 5.17 Scanning electron micrographs of the subsurface. The materials and test conditions are the same as in Fig. 5.16. (From Pamies-Teixeira, J. J.; Saka, N.; and Suh, N. P., "Wear of Copper-based Solid Solutions," *Wear*, Vol. 44, 1977, pp. 65–75.)

those shown earlier. The subsurface features for the same alloys are shown in Fig. 5.22, where some second-phase particles can be seen in the overaged alloys. However, it is interesting to note that in Fig. 5.22(a) subsurface cracks of the specimen aged for 5 minutes are very close to the surface, while for the specimen aged for 10,000 minutes cracks are formed at a large depth. As an example of the morphology of wear particles, scanning electron

TABLE 5.2 Experimental Materials

Parameter	Alloy	Aging Time (min)		
		100	1000	10,000
Volume fraction,	Cu–0.58 at % Cr	5.19	5.25	5.31
$V_v \times 10^3$	Cu–0.81 at % Cr	6.96	6.97	7.09
Mean free path	Cu–0.58 at % Cr	68.9	70.49	71.8
λ (μm)	Cu–0.81 at % Cr	51.84	53.02	53.00
Particle size	Cu–0.58 at % Cr	0.54	0.55	0.58
(μm)	Cu–0.81 at % Cr	0.55	0.56	0.58

micrographs of particles collected from the Cu–0.81 at % Cr alloy aged for 10,000 minutes are shown in Fig. 5.23. The particles are in the form of sheets and some have lamellar structure, possibly created by the presence of a large number of cracks above the crack that propagated fastest.

These results with two phase metals show that the wear rate is also affected by the coherency of particles to the matrix. In the early stages of precipitation, the particles are coherent, and therefore the stress required to separate the particle from the matrix is large. Therefore, crack nucleation requires large amounts of subsurface deformation in order to develop sufficient interfacial stress between the matrix and the particle. Since the increased

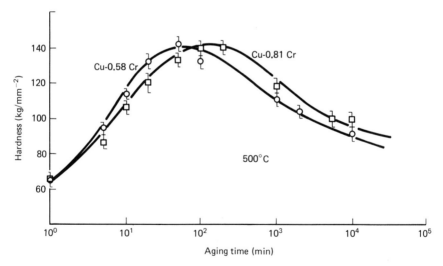

Figure 5.18 Vickers hardness under a 200-g normal load as a function of aging time. The specimens were subjected to an aging treatment at 500°C and were water quenched at the end of the treatment. (From Saka, N.; Pamies-Teixeira, J. J.; and Suh, N. P., "Wear of Two-phase Metals," *Wear*, Vol. 44, 1977, pp. 77–86.)

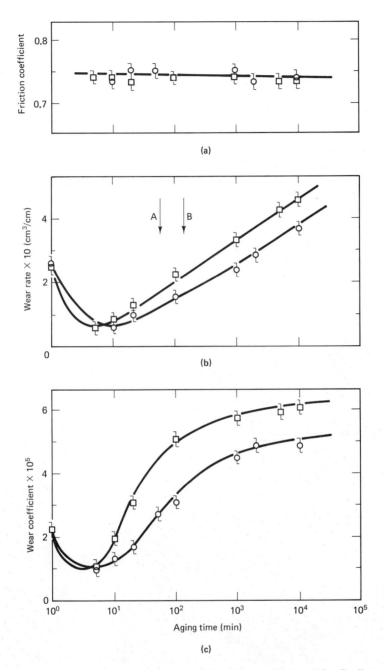

Figure 5.19 Friction and wear properties of precipitation-hardened Cu–Cr alloys as a function of the aging of time: (a) friction coefficient; (b) wear rate; (c) wear coefficient. The normal load was 2 kg and the duration of the tests was 100 minutes at a sliding speed of 200 cm/min. (From Saka, N.; Pamies-Teixeira, J. J.; and Suh, N. P., "Wear of Two-phase Metals," *Wear*, Vol. 44, 1977, pp. 77–86.)

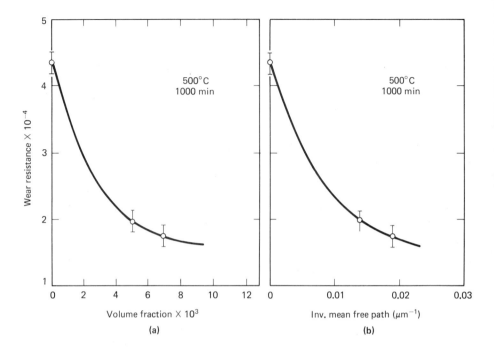

Figure 5.20 Wear resistance (reciprocal of the wear coefficient) versus (a) the volume fraction and (b) the mean free path of Cu–Cr alloys for 1000 minutes of treatment. (From Saka, N.; Pamies-Teixeira, J. J.; and Suh, N. P., "Wear of Two-phase Metals," *Wear*, Vol. 44, 1977, pp. 77–86.)

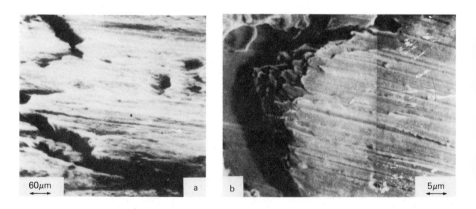

Figure 5.21 Scanning electron micrographs of wear tracks of the precipitation-hardened Cu–0.81 at % Cr alloy for an aging time of (a) 5 minutes and (b) 10,000 minutes. The sliding direction is from left to right. (From Saka, N.; Pamies-Teixeira, J. J.; and Suh, N. P., "Wear of Two-phase Metals," *Wear*, Vol. 44, 1977, pp. 77–86.)

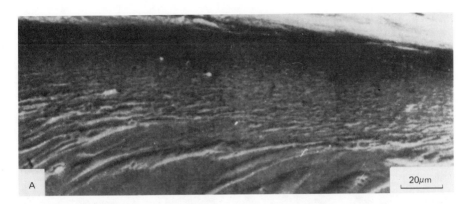

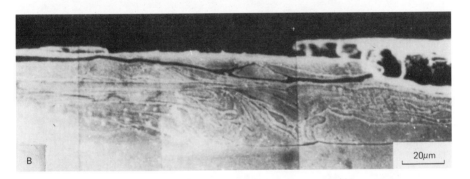

Figure 5.22 Scanning electron micrograph of the subsurface of precipitation-hardened Cu–0.81 at % Cr alloy aged for (a) 5 minutes and (b) 10,000 minutes. (From Saka, N.; Pamies-Teixeira, J. J.; and Suh, N. P., "Wear of Two-phase Metals," *Wear*, Vol. 44, 1977, pp. 77–86.)

hardness decreases the deformation rate, the wear coefficient is decreased. The steep increase in wear coefficient after reaching a minimum value is due to the loss of coherency, requiring less stress for decohesion of a particle from the matrix and is also due to the increased particle size. The deformation rate still decreases, but the deformation required for nucleation decreases even faster. Further, as the interparticle spacing decreases during the early stages of aging, cracks have to propagate smaller distances in order to join with neighboring cracks. This explains why the wear coefficient of Cu–0.81 at % Cr alloy should be greater than that of Cu–0.58 at % Cr because the former has a larger volume fraction of second-phase particles and possibly has a smaller interparticle spacing.

After the maximum hardness is reached, the volume fraction of particles does not increase with aging time any more and coarsening of particles occurs, which increases the interparticle spacing. Crack nucleation at this point tends to be relatively easy and the overall wear rate is controlled by

Figure 5.23 Scanning electron micrographs of wear particles collected from wear tests on the specimen of Cu–0.81 at % Cr aged for 10,000 minutes. (From Saka, N.; Pamies-Teixeira, J. J.; and Suh, N. P., "Wear of Two-phase Metals," *Wear*, Vol. 44, 1977, pp. 77–86.)

the crack propagation rate. In this case the crack growth and the interparticle spacing must be considered. Since the hardness does not change much for the overaged alloys, the number of asperities in contact must be nearly the same and the number of asperities passing by will be about the same. Also, since the matrices of both alloys are exactly the same, the crack growth rate must be the same to a first-order approximation. Assuming that the material is delaminated in successive layers, the crack growth rate is calculated as the ratio of the mean free path to the number of cycles required to remove one layer. Such a calculation leads to a range of values between 4×10^{-4} and 6×10^{-4} μm/cycle[1] for both alloys, and the wear rate depends basically on the mean free path of the interparticle spacing. Thus,

---

[1] These small calculated crack growth rates may be smaller than the actual crack propagation rate for several reasons: (1) rewelding of the crack tip; and (2) the estimated number of asperity contacts being greater than the actual number.

when the overall wear process is controlled by the crack growth rate, the wear coefficient tends to level off toward asymptotic values, since as the aging treatment is continued, the mean free path becomes roughly constant. This explains the leveling off of the wear coefficient when the transition from crack nucleation rate controlled wear to crack propagation controlled wear occurs. The effect of the mean free path is also shown by the difference in the wear rate of Cu–0.81 at % Cr and Cu–0.58 at % Cr. As a higher volume fraction implies a smaller mean free path for the same particle size, the curve for Cu–0.81 at % Cr should be higher than that for Cu–0.58 at % Cr.

5.5 CONCLUDING REMARKS

In this chapter the wear of elastoplastic metals was analyzed and discussed. It was shown that the wear particles are generated by asperity deformation, plowing of the surface by hard particles and asperities, and by delamination wear. Of these three mechanisms, the sliding wear of metals is most severe when it is caused by the delamination mechanisms. Delamination wear is caused by subsurface plastic deformation, crack nucleation, and crack propagation. Delamination wear creates large wear sheets, whereas the wear particles that are generated by asperity deformation and plowing tend to be smaller and thinner than delaminated sheets. However, all wear particles accelerate the delamination process by increasing the frictional force when they are entrapped between the sliding surfaces. In this chapter the wear rate of metals was predicted using the delamination theory, which agrees with experimental results. The effect of microstructure on wear was also discussed. The applications of the theory are discussed in Chapter 9.

The wear processes considered in Chapter 5 are not accompanied by much of a temperature rise. When the interfacial temperature rise is high due to high loads and/or speeds, the wear mechanisms of sliding wear may change from delamination wear to another wear process that is controlled by the chemical behavior of materials. This problem is treated in Chapter 8.

REFERENCES

ABRAHAMSON, E. P., JAHANMIR, S., and SUH, N. P., "The Effect of Surface Finish on the Wear of Sliding Surfaces," *CIRP Annals,* International Institution for Production Engineering Research, Vol. 24, 1975, pp. 513–514.

CLERICO, M., "Sliding Wear Mechanisms of Polymers," in *Fundamentals of Tribology,* N. P. Suh and N. Saka, eds., MIT Press, Cambridge, Mass., 1980.

FLEMING, J. R., and SUH, N. P., "Mechanics of Crack Propagation Delamination Wear," *Wear,* Vol. 44, 1977, pp. 39–56.

GUPTA, P. K., and COOK, N. H., "Statistical Analysis of Mechanical Interaction on Rough Surfaces," *Journal of Lubrication Technology,* Vol. 94, 1972, pp. 19–26.

HAMILTON, G. M., and GOODMAN, L. E., "The Stress Field Created by a Circular Sliding Contact," *Journal of Applied Mechanics,* June 1966, pp. 371–376.

JAHANMIR, S., "A Fundamental Study of the Delamination Theory of Wear," Ph.D. thesis, Department of Mechanical Engineering, MIT, 1977.

JAHANMIR, S., SUH, N. P., and ABRAHAMSON, E. P., II, "Microscopic Observations of the Wear Sheet Formation by Delamination," *Wear,* Vol. 28, 1974, pp. 235–249.

PAMIES-TEIXEIRA, J. J., SAKA, N., and SUH, N. P., "Wear of Copper-Based Solid Solutions," *Wear,* Vol. 44, 1977, pp. 65–75.

RABINOWICZ, E., *Friction and Wear of Materials,* Wiley, New York, 1965.

RABINOWICZ, E., "The Dependence of the Abrasive Wear Coefficient on the Surface Energy of Adhesion," *Wear of Materials—1977,* ASME, New York, 1977, pp. 36–40.

SAKA, N., "Effect of Microstructure on Friction and Wear of Metals," in *Fundamentals of Tribology,* N. P. Suh, and N. Saka, eds., MIT Press, Cambridge, Mass., 1980.

SAKA, N., PAIMES-TEIXEIRA, J. J., and SUH, N. P., "Wear of Two-Phase Metals," *Wear,* Vol. 44, 1977, pp. 77–86.

SPROLES, E. S., JR., GAUL, D. J., and DUQUETTE, D. J., "A New Interpretation of the Mechanism of Fretting and Fretting Corrosion Damage," in *Fundamentals of Tribology,* N. P. Suh, and N. Saka, eds., MIT Press, Cambridge, Mass., 1980.

SUH, N. P., "The Delamination Theory of Wear," *Wear,* Vol. 25, 1973, pp. 111–124.

SUH, N. P., JAHANMIR, S., ABRAHAMSON, E. P., II, and TURNER, A. P. L., "Further Investigation of the Delamination Theory of Wear," *Journal of Lubrication Technology,* Vol. 96, 1974, pp. 631–637.

SUH, N. P., and SIN, H.-C., "On Prediction of Wear Coefficients," *Transactions of the A.S.L.E.* Vol. 26, 1983, pp. 360–366.

SUH, N. P., JAHANMIR, S., FLEMING, J., ABRAHAMSON, E. P., II, SAKA, N., and TEIXEIRA, J. P., *The Delamination Theory of Wear—II,* Progress Report to DARPA (Contract N00014-67-A-0204-0080, NR 229-011), MIT, 1975.

TOHKAI, M., "Microstructural Aspects of Friction," S.M. thesis, Department of Mechanical Engineering, MIT, 1979.

WHITEHOUSE, D. J., "The Effects of Surface Topography on Wear," in *Fundamentals of Tribology,* N. P. Suh and N. Saka, eds., MIT Press, Cambridge, Mass., 1980, pp. 17–52.

6

FRICTION AND WEAR OF POLYMERS AND COMPOSITES

6.1 INTRODUCTION

As stated in Chapter 1, polymers and polymeric composites are important tribological materials because of their self-lubricating qualities, low cost, low noise emission, and excellent wear and corrosion resistance. They are used as journal bearings, ball-bearing cages, sliders, gyroscope gymbals, gears, cams, seals for shafts and other mechanical parts, felt tips for pens, tires, and numerous other applications. These parts are made by injection-molding thermoplastics, compression-molding thermoplastics and thermosetting plastics with fillers and solid lubricants, and by casting and machining reinforced thermoplastics or thermosetting plastics with graphite fibers and fillers.

Many polymers and composites used in tribological applications are viscoelastic–plastic. In these materials the hardness (defined as the load applied divided by the projected indentation area) changes continuously as a function of the indentation time. Therefore, the wear behavior of these materials cannot be characterized in terms of the wear coefficient defined by Eq. (1.12) since it involves hardness. Two alternative means of characterizing the wear characteristics of polymers and composites have been used (Lewis, 1964): wear factor and the Lv limit.[1]

---

[1] In the literature the Lv limit is normally known as the Pv limit. Here, P is replaced by L to use a consistent nomenclature for normal load.

The wear factor of polymeric materials is defined as

$$K' = \frac{V}{LS} = \frac{V}{Lvt}$$

where V is the wear volume, v the sliding speed, L the normal load, and t the sliding time. The wear factor K', unlike the wear coefficient, is not a dimensionless quantity. The Lv *limit* is used to define the onset of catastrophic failure of polymeric bearings due to melting and extrusion. Unfortunately, since the product of the normal load, L, and the velocity, v, is not exactly proportional to the temperature rise, the value of the Lv limit is not a constant, but varies depending on the specific load and the specific velocity. Therefore, the Lv limit has to be specified in terms of a limiting load at a given sliding speed or in terms of a limiting speed at a given load.

Some Lv limits for a number of polymers are given in Table 6.1. Figures 1.2 and 1.3 also present the general tribological behavior of a few polymers. Additional data at low sliding speeds are given in Table 6.2 for high-density polyethylene (HDPE), polyoxymethylene (POM), polymethylmethacrylate (PMMA), and polycarbonate (PC). Table 6.3 gives the wear rate of various polymers at a higher sliding speed (i.e., 180 cm/sec). Additional data on wear factor and Lv limit are also given for various filled polymers in Table 6.4. It should be noted that these data provide only a relative tribological behavior of these materials, since the absolute values depend on the precise test conditions and additives added to these plastics.

The most commonly used thermoplastic bearing materials are:

1. Polytetrafluoroethylene (PTFE) with and without graphite fiber reinforcement
2. Ultrahigh-molecular-weight polyethylene (UHMWPE) with and without fiber reinforcement
3. Polyoxymethylene (POM, sometimes known as acetal)
4. Nylon with and without fillers and fiber reinforcement
5. Fluoroethylene propylene (FEP)

The following thermosetting plastics are also used extensively in tribological applications:

1. Polyimide with and without fiber reinforcement
2. Polyurethane with and without fiber reinforcement and fillers
3. Phenolics with fiber reinforcement and fillers
4. Epoxy with fillers and fiber reinforcement
5. Polyester with fibers and fillers

In some cases, metals such as brass are impregnated with Teflon powder to make bearing materials.

TABLE 6.1 Performance Data for Self-Lubricating Plastic Bearing Materials

Performance Characteristic or Property	Unmodified Polymers					Modified Polymers			
	Nylon	Acetal	Fluorocarbon	Polyimide	Phenolic	Nylon, Graphite Filled	Acetal, TFE Fiber Filled	Fluorocarbon, Wide Range of Fillers	Polyimide, Graphite Filled	Phenolic, TFE Filled
Maximum load/projected area (zero speed) (psi)	4,900	5,200	1,000	10,000	4,000	1,000	1,800	2,000	10,000	4,000
Speed, continuous operation (5-lb load) (max. ft/min)	200–400	500	100	1,000	1,000	200–400	800	1,000	1,000	1,000
Lv for continuous service (0.005-in. wear in 1000 hr)	1,000	1,000	200	300	100	1,000	2,500	2,500 50,000[a]	3,000	5,000
Limiting LV at 100 ft/min	4,000	3,000	1,800	100,000	5,000	4,000	5,500	30,000	100,000	40,000
Coefficient of friction	0.20–0.30	0.15–0.30	0.04–0.13	0.1–0.3	0.90–1.1 250	0.1–0.25	0.05–0.15	0.04–0.25	0.1–0.3	0.05–0.45
Wear factor, $K' \times 10^{-10}$ (in./min/ft-lb-hr)	50	50	2,500	150	2,000	50	20	1–20	15	10
Elastic modulus, bending (10 psi)	0.3	0.4	0.08	0.45	5	0.4	0.4	0.4	0.63	5
Critical temp. at bearing surface (°F)	400	300	500	600	300–400	400	300	500	600	300–400
Resistance to: Humidity	Fair	Good	Excellent	Good	Good	Fair	Good	Excellent	Good	Good
Chemicals	Good	Good	Excellent	Good	Good	Good	Good	Excellent	Good	Good
Density (g/cm³)	1.2	1.43	2.15–2.20	1.42	1.4	1.2	1.54	2.15–2.25	1.49	1.4
Cost index for base material	1.4	1	5	15	—	1.5	6	5	15	—

[a] Exceeds limiting Lv.

Source: Steijn (1967).

TABLE 6.2 Friction and Wear Data of Polymers[a]

Material	Normal Load (g)	Sliding Speed (cm/sec)	Coefficient of Friction	Wear Rate (g/cm)
HDPE	200	3.3	0.31	1.1×10^{-9}
	450	3.3	0.25	2.7×10^{-9}
		16.4	0.25	5.5×10^{-10}
		33.	0.25	6×10^{-10}
POM[b] (Delrin)	200	3.3	0.22	1×10^{-8}
	450	3.3	0.31	1×10^{-8}
PMMA	200	3.3	0.64	4×10^{-8}
	450	3.3	0.68	8×10^{-8}
PC (Lexan 101)	450	3.3	0.6	2.5×10^{-8}
PC (Lexan 121)			0.5	2.4×10^{-8}

[a] Experimental conditions: geometry, pin on disk (AISI 52100 steel pin); test done at room temperature (22°C); atmosphere, air; relative humidity, 65%.
[b] Sometimes known as acetal.

TABLE 6.3 Wear Rates of Pins of Various Polymers at a Load of 400 g[a]

Combination of Materials	Hardness (10^6 g/cm^2)	Wear Rate (10^{-10} cm^3/cm)	Wear Coefficient, K
Teflon (PTFE)	0.5	200	7.5×10^{-5}
Perspex (PMMA)	2.0	14.5	2.1×10^{-5}
Molded Bakelite X5073	2.5	12.0	2.2×10^{-5}
Laminated Bakelite 292/16	3.3	1.8	4.5×10^{-6}
Molded Bakelite 11085/1	3.0	1.0	2.2×10^{-6}
Laminated Bakelite 547/1	2.9	0.4	9×10^{-7}
Polyethylene	0.17	0.3	3.9×10^{-7}
Nylon			4.5×10^{-6}

[a] Experimental conditions: speed, 180 cm/sec; polymer pins were slid against hardened tool steel.
Source: Archard and Hirst (1956).

In this chapter the experimental evidence for friction and wear is examined first. Then a theoretical analysis is presented based on simple models which approximate the wear process of certain thermoplastics and composites. Although much more work needs to be done to understand fully the tribological behavior of polymers and composites, it can be stated conclusively that deformation properties of polymers have profound influence on wear. The sliding wear of glassy polymers (e.g., PMMA below the glass transition temperature of 57 to 68°C) is very different from that of semicrystalline polymers, which are ductile at room temperature. These characteristics are discussed in detail in the following sections. It should be noted that the tribological behavior of polymers is still very poorly understood and a firm scientific base is yet to be established.

TABLE 6.4 Wear Factor and Lv Limit of Polymers

Materials	Lv Limit[a]	K' ($\times 10^{-10}$)[b]	Coefficient of Friction[c]
Acetal filled with PTFE	8,000 (at 10 fpm) 5,500 (at 100 fpm) 2,500 (at 1000 fpm)	20	0.16–0.27
Acetal (POM)	5,000		0.1–0.25
PTFE (no filler)	1,200 (at 10 fpm) 1,800 (at 100 fpm) 2,500 (at 1,000 fpm)	10^3–2×10^4	
PTFE (with glass fiber)	10–17 $\times 10^3$ (at 10–10^3 fpm)	10	0.12–0.26
PTFE (with bronze)	16.5 to 23 $\times 10^3$ (at 10–10^3 fpm)	6	0.14–0.28
PTFE (carbon graphite)	15,000 (at 10–100 fpm) 10,500 (at 1000 fpm)	14	0.11–0.26
PTFE (cadmium oxide)	40,000 (at 10 fpm) 37,500 (at 100 fpm) 23,000 (at 1000 fpm)	20	0.08–0.11
Phenolic–cellulose (filled with PTFE)	40,000 (at 10 fpm) 45,000 (at 100 fpm) 58,000 (at 1000 fpm)	15	0.27–0.41

[a] In "irrational" units of psi—ft/min (fpm).
[b] In "irrational" units of (in.^{3}-min)/(ft-lb-hr); all data obtained at an ambient temperature of 70°F.
[c] Lower values are static coefficient and higher values are obtained at 150 ft/min.
Sources: O'Rourke (1965) and Lewis (1964).

6.2 PHENOMENOLOGICAL OBSERVATIONS ON THE FRICTION AND WEAR BEHAVIOR OF HIGHLY LINEAR SEMICRYSTALLINE POLYMERS: POLYTETRAFLUOROETHYLENE (PTFE), HIGH-DENSITY POLYETHYLENE (HDPE), AND POLYOXYMETHYLENE (POM)

6.2.1 General Remarks

Tribological behavior of polymers is very different depending on the molecular structure, crystallinity, molecular weight, and cross-linking density. Plastics with highly linear molecular configuration, such as PTFE, HDPE, and POM, exhibit unique friction and wear behavior. They are even different from other thermoplastics with nonsymmetric molecular configuration as well as thermosetting plastics and elastomers. The molecular structures of these polymers are shown in Fig. 6.1.

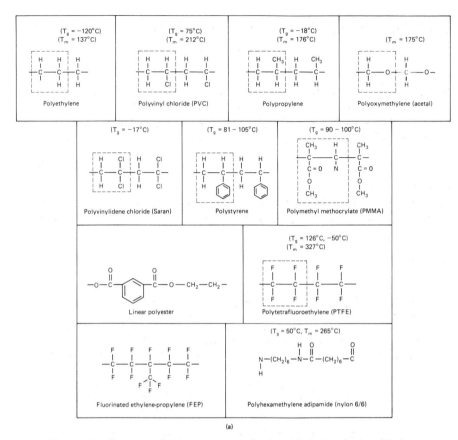

Figure 6.1 Structure of some common polymers: (a) thermoplastics; (b) thermosetting plastics (repeating units blocked off unless only single unit is shown). (From Suh and Turner, 1975.)

Perhaps the most extensively investigated polymeric material for tribological applications is PTFE. In many sliding situations, the coefficient of friction of PTFE is very low, about 0.09. It also has excellent temperature stability, which makes it one of the most preferred materials among all plastics, excluding polyimide. For these reasons PTFE is widely used. What limits its wider application is the relatively poor wear resistance. Because of the extremely high cohesive energy density of the PTFE molecule, the intermolecular strength is not very high. Therefore, PTFE easily shears and yields. Its creep resistance and yield strength, however, can be increased substantially by reinforcing the plastic with glass and graphite fibers.

High-density polyethylene (HDPE) and ultrahigh-molecular-weight polyethylene (UHMWPE) also exhibit low friction coefficients of about 0.25 and 0.1, respectively, in many sliding situations. However, unlike PTFE,

Sec. 6.2 Friction and Wear Behavior of Highly Linear Semicrystalline Polymers

[Structural diagrams: Urea-formaldehyde; Phenol-formaldehyde (bakelite); Cross-linked polyester; Epoxy]

(b)

polyethylene has a low melting point (137°C), which limits its usage to moderate temperatures, low velocity, and low load applications. The dimensional stability of HDPE can be increased significantly by reinforcing it with fibers.

Polyoxymethylene (POM), which is also known as acetal, is a very commonly used bearing material because of its low friction coefficient (about 0.2 to 0.3), reasonably good wear resistance, and a better stability than HDPE (see Table 6.2).

Bearings of PTFE are made by compacting powder and sintering the compact rather than by melt processing, because PTFE cannot easily be melt processed due to its high thermal stability. Similar processes must be used for UHMWPE because of its extremely high viscosity. On the other hand, HDPE and POM can be processed by injection molding and extrusion. A melt-processable fluoropolymer is fluoroethylene polypropylene (FEP), which also exhibits reasonably good tribological properties.

6.2.2 Unique Tribological Properties of Highly Linear Semicrystalline Thermoplastics

Several important experimental observations have been made regarding the tribological behavior of PTFE and HDPE. They may be summarized as follows:

1. Only linear, symmetric semicrystalline polymers without branches and large pendant groups exhibit low coefficients of friction.
2. As soon as sliding commences, thin-film transfer occurs from the polymer to the counterface. The transferred film is highly oriented (Makinson and Tabor, 1964; Kar and Bahadur, 1977; Briscoe et al., 1974; McLaren and Tabor, 1965; Steijn, 1968; Svirodyonok et al., 1973; Jain and Bahadur, 1977; Tanaka and Uchiyama, 1977).
3. When the sliding direction of PTFE reverses, the coefficient of friction increases.
4. The wear sheets of PTFE are very thin films 50 to 200 Å thick and are highly oriented in the sliding direction (Briscoe et al., 1974).

Many different explanations have been advanced to describe the mechanism of the thin-film transfer. Steijn (1968) explained the thin-film-transfer phenomenon as being due to the formation of adhesion junction and drawing of thin films across the sliding surface. Tanaka and Uchiyama (1977) argued that the thin-film transfer of PTFE is due to easy destruction of the special banded structure, but not due to the drawing of molecular chains. Makinson and Tabor (1964) claimed that the easy slipping of crystalline slices causes the thin-film transfer. According to Briscoe et al. (1974), the behavior of the thin-film transfer at a low sliding speed appears to be connected with smooth molecular profile and not with the crystallinity or band structure of PTFE. In the case of the crystalline polymers which transfer thick lumpy films rather than thin films, little orientation was found in the film, although equally strong adhesion junction might have been present.

The process of thin-film formation may be explained in terms of the following hypothesis: When the highly linear semicrystalline polymers (above the glass transition temperature) are deformed by the sliding action, the molecules near the surface orient along the sliding direction due to the large shear strain gradient established near the surface. Then, nearly concurrently, the counterface is covered by these oriented polymers, which are transferred to it due to the chemical or mechanical bonding on its surface. As soon as the transfer occurs, sliding occurs between the same highly oriented polymers. The thin film is formed when these transferred polymeric particles are elongated further during the subsequent shearing process.

A careful examination of worn surfaces reveals very large plastic deformation of the surface and subsurface layer, as discussed in Chapter 4. The deformation and the deformation gradient are largest at the surface and decay rapidly away from the surface. In crystalline materials such as metals, grains deform in such a manner that the slip planes align nearly parallel to the surface. Therefore, it is expected that the surface layer will shear easily and fracture parallel to the surface due to the alignment of the slip planes with the surface. Similarly, in crystalline polymers with no bulky side pendant groups, the crystalline region of the polymer is expected to align parallel to the surface. Two things can happen when part of these molecules adhere to the moving counterface (due to either chemical or mechanical adhesion); the molecules are expected to stretch and orient, or the crystalline platelets are expected to shear off parallel to the surface and elongate when the surface is plowed by the asperities of the opposing surfaces. The peeling process may be facilitated if there are preexisting cracks or if cracks are nucleated during the deformation process.

That the hypothesis for the wear process is reasonable can be demonstrated by devising an experiment that can prevent the process from occurring. To prevent the deformation of the surface layer, the molecular orientation, and the thin-film-transfer process from occurring, the surface of HDPE was cross-linked by subjecting the surface to helium plasma. Such a treatment drastically lowered the wear rate, as discussed in Chapter 9.

6.3 PHENOMENOLOGICAL OBSERVATIONS ON THE TRIBOLOGICAL BEHAVIOR OF OTHER POLYMERS

The tribological behavior of other semicrystalline polymers, such as low-density polyethylene (LDPE), polypropylene (PP), and polyamide, is different from that of highly linear polymers. The wear debris is of the thicker, lump type (Pooley and Tabor, 1972; Bowers et al., 1954; Tanaka and Uchiyama, 1977; Makinson and Tabor, 1964; West and Senior, 1958; James, 1958/59; Warren and Eiss, 1977). This indicates that these molecules do not elongate as much as the highly linear semicrystalline polymers.

The sliding wear of glassy polymers (e.g., PMMA) is caused by crazing and brittle fracture when there is no thermal softening of the surface layer (Puttick et al., 1977; Billinghurst et al., 1966; Matsushige and Baer, 1975; Van Den Boogaart, 1966; Peterson et al., 1974). The wear particles are very large thick lumps.

In glassy polymers cracks can initiate at the surface right behind the slider when the asperity contacts are blunt (see Section 4.2) or at the tip of sharp asperities propagating radially at the surface, axially into the surface from the asperity tip, and laterally at the subsurface. Cracks can also form at the subsurface due to the deformation and fracture resulting from the bulk deformation of the surface due to the contact load. In many glassy polymers the fracture is likely to start from the surface, since flaws are commonly present at the surface, but the subsurface fracture cannot be ignored, especially when sharp asperity contacts slide over the surface. When only a normal load is applied at a contact between a sphere and a semi-infinite solid, the largest shear stress in the body occurs below the surface. The depth depends on the Poisson ratio of the material and is 0.47 times the radius, a, of the contact area when Poisson ratio is 0.3 (see Appendix 4.B). When both normal and tangential loads are present at the asperity contact, the location of the maximum shear stress approaches the surface, as discussed in Chapter 4. In glassy polymers fracture can be initiated from these locations of maximum shear stress or where the largest flaw is present.

6.4 WEAR OF POLYMERIC COMPOSITES

Many composite materials are used in tribological applications, since the properties of bearing and cam materials can be enhanced by placing appropriate components at the right locations. The most frequently used composites are based on polymeric matrices because metal composites are expensive. Many thermosetting and thermoplastic polymers [e.g., epoxy, polyurethane, phenolics, nylon, Teflon (PTFE), acetal, etc.] have been reinforced with fillers and glass and graphite fibers to make them more wear resistant (see Table 6.4). In many commercially available composites the fibers are chopped and randomly oriented, but greater enhancement of the wear resistance can be achieved by orienting long, semicontinuous fibers normal to the surface.

The tribological behavior of thermoplastic-based composites with randomly oriented chopped fibers is very much like those of two-phase metals, which were discussed in the preceding sections. As a result, the delamination theory of wear can explain the observed tribological behavior of these materials. Clerico (1979) tested various bearing materials made of nylon with and without fillers and glass fibers (Table 6.5 shows the materials). These composites were slid against steel and bronze at speeds ranging from 25 to 75 cm/sec. The wear volumes of nylon 6/6 with fillers and with glass fibers are plotted in Fig. 6.2 as functions of sliding distance. The SEM micrographs of worn surfaces are shown in Figs. 6.3 and 6.4. These micrographs show that the composite surfaces became very smooth and exhibit craters and wear sheet formation. Thin-film transfer from the polymer composite to bronze was also observed and many metal wear particles had polymeric films on their surfaces. Subsurface cracks were observed to

Sec. 6.4 Wear of Polymeric Composites

TABLE 6.5 Various Materials Tested by Clerico (1980)[a]

Material	Specific Gravity	Rockwell Hardness	Yield Strength (daN/mm^2)	Tensile Modulus (daN/mm^2)
Nylon 6/6	1.14	82 (R_L)	5.5	170
Nylon 6/6 with fillers	1.11	51 (R_H)	5	165
Nylon 6/6 with 25% glass fibers (random orientation)	1.28	75 (R_H)	10	700
Acetal copolymer	1.41	74 (R_H)	7.3	350
Steel	7.8	66.5 (R_H)	85	21,000
Bronze	8.7	41 (R_H)	13	11,000

[a] Experimental conditions: A cylindrical specimen of polymeric composites was slid against a metal surface by rotating it against a metal cylinder. The metal specimens were made of steel or bronze.

propagate very nearly parallel to the surface until they shear off to the surface, forming thin wear sheets (Figs. 6.5 through 6.7). The depths of subsurface cracks in nylon 6/6 with fillers, nylon 6/6 with fibers, bronze, and steel are shown as a function of the friction coefficient in Fig. 6.8. All these observations are consistent with the postulates of the delamination theory of wear.

According to the delamination theory of wear (Suh, 1973), the subsurface cracks are likely to propagate parallel to the surface in homogeneous materials under combined compressive and shear loading. It follows, then, that if the crack propagation can be eliminated by preventing the shear strain concentration at the crack tip, the wear rate must decrease. To check the validity of this hypothesis, Sung and Suh (1979) investigated the effect of fiber orientation on wear. Three types of fiber-reinforced plastics were studied. One was a unidirectionally oriented composite with graphite fibers in an epoxy matrix fabricated from prepregs. The second composite was composed of unidirectionally oriented Kevlar-49 fibers in the epoxy. The third composite material was a commercial-grade bearing material (RT Duroid 5813), a polytetrafluoroethylene (PTFE)-based material. This composite contained about 16 wt % planar-oriented microglass fibers with two-thirds of the fibers being oriented preferentially in one direction. The ratio of fibers oriented in the three orthogonal directions are approximately 2:1:0. The composition and the properties of these composites are given in Table 6.6.

The experimental results show that the wear rate is least when the fibers are oriented perpendicular to the sliding surface, as shown in Fig. 6.9 for the graphite fiber–epoxy composite. In this case friction was also the least, as shown in the figure. Similar wear results were obtained with the Kevlar–epoxy composite (Fig. 6.10), although the order of frictional

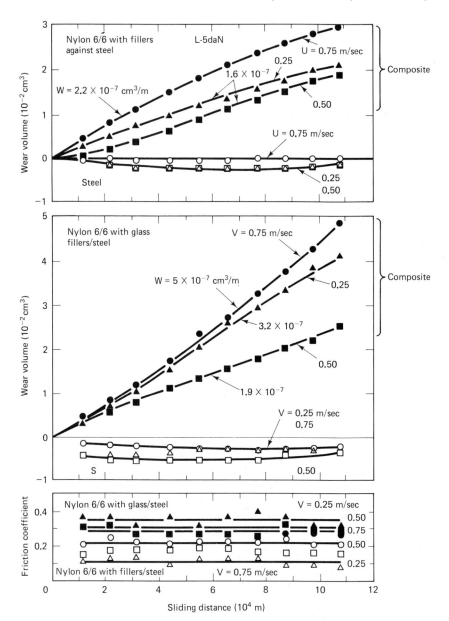

Figure 6.2 Wear volume and friction coefficient of steel-nylon composite pairs in sliding distance. (Reprinted with permission from Clerico, M., "Sliding Wear Mechanisms of Polymers," *Fundamentals of Tribology*, edited by Suh, N. P., and Saka, N., MIT Press, Cambridge, Mass., 1980. Copyright 1980 by MIT Press.)

Sec. 6.4 Wear of Polymeric Composites 235

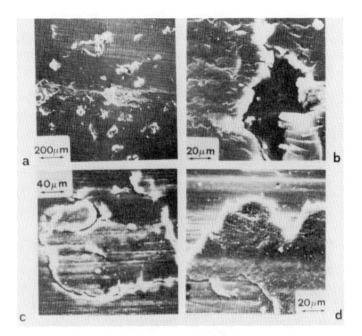

Figure 6.3 Worn surfaces of nylon 6/6 with fillers. Metal pairs (the normal force was daN): (a) worn surface topography of the specimen sliding against bronze at 0.75 m/sec; (b) crater in the specimen sliding against steel at 0.75 m/sec; (c) sheet formation in nylon 6/6 with fillers sliding against steel at 0.25 m/sec; (d) worn surface of steel (the sliding speed was 0.50 m/sec). (Reprinted with permission from Clerico, M., "Sliding Wear Mechanisms of Polymers," *Fundamentals of Tribology*, edited by Suh, N. P., and Saka, N., MIT Press, Cambridge, Mass., 1980. Copyright 1980 by MIT Press.)

force has changed. With the biaxially oriented glass microfiber–MoS_2–PTFE composites the wear rate was again found to be the least when the largest fraction of fibers was oriented normal to the sliding plane (Fig. 6.11). These results indicate that when the fibers are normal to the surface, plastic deformation and crack propagation are both minimized, resulting in low wear rates.

Figure 6.9 shows that the coefficient of friction of graphite fiber–epoxy composites is least when the fibers are normal to the surface, whereas Fig. 6.10 shows that Kevlar–epoxy composite sliding against AISI 52100 steel has the maximum coefficient of friction when the fibers are normal to the surface. The biaxially oriented glass microfiber–MoS_2–PTFE composite sliding on the same steel shows yet another behavior. In this case the frictional force is always constant, probably due to the presence of the solid lubricant (MoS_2) and PTFE at the interface. The frictional behavior must depend on whether or not the fiber can penetrate into the counterface,

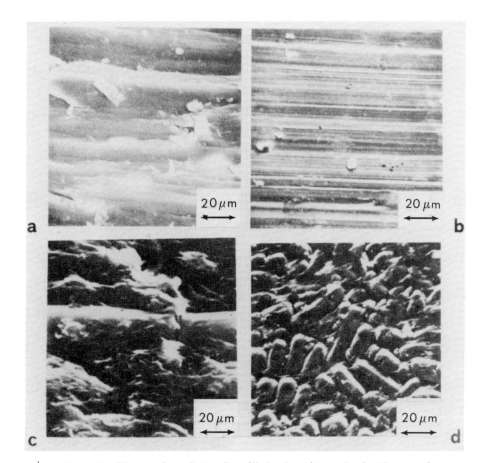

Figure 6.4 Worn surface of glass fiber filled nylon 6/6–metal pairs (the normal force was 15 daN and sliding speed was 0.25 m/sec): (a) specimen surface and (b) bronze counterface; (c) specimen surface and (d) steel counterface. (Reprinted with permission from Clerico, M., "Sliding Wear Mechanisms of Polymers," *Fundamentals of Tribology*, edited by Suh, N. P., and Saka, N., MIT Press, Cambridge, Mass., 1980. Copyright 1980 by MIT Press.)

the number of wear particles that get entrapped at the interface, and the properties of the matrix.

If the matrix material is not brittle but highly viscoelastic and rubbery, a unique frictional behavior is observed. When graphite fiber–polyurethane composites are tested, the wear debris consists of very small and sticky polyurethane particles and fibers. The frictional force increases suddenly when the slider moves over this patch of sticky material. Consequently, the frictional force increases continuously with sliding in these materials, finally reaching a steady state when the surface is completely covered by the wear debris. This is discussed further in Chapter 7.

Sec. 6.5   Basic Mechanisms for Friction in Polymers   237

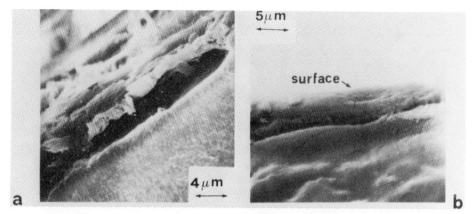

Figure 6.5 Crack propagation in nylon 6/6 with 25% glass fibers: (a) sliding against steel (the normal force was 5 daN, sliding speed was 0.50 m/sec); (b) sliding against bronze the normal force was 25 daN, sliding speed was 0.50 m/sec. (Reprinted with permission from Clerico, M., "Sliding Wear Mechanisms of Polymers," *Fundamentals of Tribology*, edited by Suh, N. P., and Saka, N., MIT Press, Cambridge, Mass., 1980. Copyright 1980 by MIT Press.)

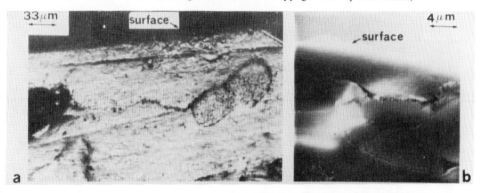

Figure 6.6 Crack nucleation in polymeric composites sliding against bronze: (a) nylon 6/6 with fillers; (b) nylon 6/6 with 25% glass fibers. The normal force was 25 daN, sliding speed was 0.75 m/sec. (Reprinted with permission from Clerico, M., "Sliding Wear Mechanisms of Polymers," *Fundamentals of Tribology*, edited by Suh, N. P., and Saka, N., MIT Press, Cambridge, Mass., 1980. Copyright 1980 by MIT Press.)

6.5 BASIC MECHANISMS FOR FRICTION IN POLYMERS

In Chapter 3 the three basic mechanisms that contribute to the frictional force between metals were discussed. It was shown that the frictional force is the result of plowing the surfaces by particles and asperities, the adhesion between the flat surfaces in contact, and the asperity deformation. Since the relative importance of these three mechanisms in generating the frictional force depends on the test conditions, the topography of the surface, and microstructures of the surface layer, it was shown that the frictional coefficient

Figure 6.7 Sheet formation in (a) nylon 6/6 with fillers, and (b) nylon 6/6 with 25% glass fibers. (Reprinted with permission from Clerico, M., "Sliding Wear Mechanisms of Polymers," *Fundamentals of Tribology*, edited by Suh, N. P., and Saka, N., MIT Press, Cambridge, Mass., 1980. Copyright 1980 by MIT Press.)

can take any value within the friction space. The basic mechanisms of sliding friction of polymers are similar. Plowing adhesion and the deformation of asperities are major contributors to friction between polymeric solids and other materials. However, there are some differences between the frictional behavior of metals and polymers, because the behavior of the former is elastoplastic, whereas that of the latter is viscoelastic–plastic,

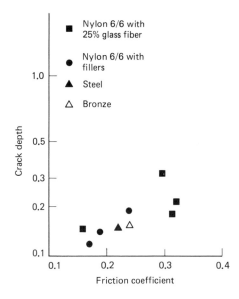

Figure 6.8 Crack depth as a function of friction coefficient. (From Clerico, M., "Sliding Wear of Polymeric Composites," *Wear*, Vol. 53, 1979, pp. 279–301.)

Sec. 6.5   Basic Mechanisms for Friction in Polymers

TABLE 6.6 Properties of Composites

Unidirectional graphite fiber (Thornel 300)–epoxy (SP-228)	
Specific gravity	1.58
Fiber volume	60%
Longitudinal tensile strength	200 klbf-in.$^{-2}$
Longitudinal tensile modulus	18×10^6 lbf-in.$^{-2}$
Compression strength	145 klbf-in.$^{-2}$
Transverse tensile strength	10 klbf-in.$^{-2}$
Transverse tensile modulus	1.2×10^6-lbf-in.$^{-2}$
Interlaminar shear strength	16 klbf-in.$^{-2}$
Unidirectional Kevlar-49 (DuPont)–epoxy	
Specific gravity	1.33
Fiber volume	65%
Longitudinal tensile strength	223 klbf-in.$^{-2}$
Longitudinal tensile modulus	17×10^6 lbf-in.$^{-2}$
Interlaminar shear strength	9 klbf-in.$^{-2}$
Bidirectional glass microfiber–MoS$_2$–PTFE (RT Duroid[a])	
Specific gravity	2.42
Glass fiber	16 wt %
MoS$_2$	15 wt %
Compression modulus[b]	158/183/130 klbf-in.$^{-2}$
Compression strength[b]	3.6/4.4/8.2 klbf-in.$^{-2}$

[a] Registered trademark of Rogers Corporation.
[b] Three values are for the three orthogonal directions.

with low moduli and low melting points. Therefore, the frictional behaviors of polymeric solids are more sensitive to the applied load, temperature, and sliding velocity. Some of these differences will be examined in detail in this section after reviewing typical experimental results.

The coefficients of friction of various polymers sliding against ½-in.-diameter steel balls are shown in Table 6.7. The coefficient of friction is a function of the steel ball size since the amount of entrapment of wear particles at the interface depends on the size of the ball. The static coefficient of friction is higher than the kinetic coefficients of friction, due to the mechanical interlocking of the asperities. It should also be noted that the coefficients of friction of glassy or amorphous polymers are, in general, greater than those of semicrystalline polymers. The effect on the coefficient of friction of changing the normal load from 1000 g to 4000 g is relatively small, although the coefficients decrease with increasing normal load, as discussed in Chapter 1. Because of the low melting point of polymers, the friction coefficient is very sensitive to the interfacial temperature rise, as shown in Fig. 6.12. The friction coefficient rises until the interface temperature reaches the melting point of polymers.

The reason glassy or amorphous polymers give rise to higher frictional forces than do semicrystalline polymers is that these polymers create large chunky wear particles rather than thin films, which contribute a great deal

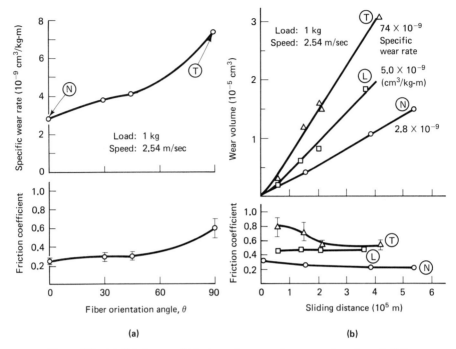

Figure 6.9 (a) Friction coefficients and wear volume as a function of sliding distance in uniaxial graphite fiber–epoxy composite. Sliding against 52100 steel, with fiber orientations normal, longitudinal, and transverse to the sliding direction. (From Sung and Suh, 1979.) (b) friction coefficients and specific wear rate as a function of sliding distance in uniaxial graphite fiber–epoxy composite sliding against 52100 steel, with varying fiber orientations ranging from normal ($\theta = 0$) to transverse ($\theta = 90°$) to the sliding direction. (From Sung, H.-H., and Suh, N. P., "Effect of Fiber Orientation on Friction and Wear of Fiber Reinforced Polymeric Composites," *Wear*, Vol. 53, 1979, pp. 129–141.)

to the frictional force by plowing the sliding interface when they are entrapped. The higher frictional force of glassy polymers is also attributable to their higher flow strength and their inability to align themselves in the easy shear direction, whereas the semicrystalline polymers start shearing very easily once the surface layer deforms plastically, aligning its easy slip planes parallel to the surface.

The fact that the increased normal load tends to lower the frictional coefficient can be partially explained by noting that polymeric materials have low bulk moduli, and therefore the shear strength of these materials is affected by the hydrostatic pressure, which changes the intermolecular spacing. This change in the flow strength of the material with the normal load affects the size of the contact area more sensitively than in the case of metals. Assuming the adhesion mode of frictional force and using the simplest model, the frictional force required to shear the welded junctions formed between two sliding members at temperatures lower than the melting

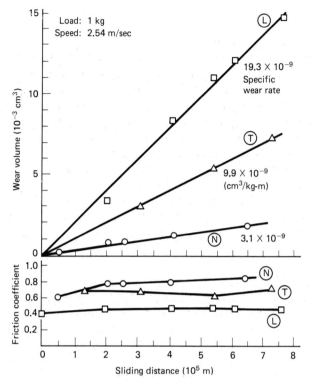

Figure 6.10 Friction coefficients and wear volume as a function of sliding distance in uniaxial Kevlar-49–epoxy composite sliding against 52100 steel with fiber orientations normal (N), longitudinal (L) and transverse (T) to the sliding direction. (From Sung, H.-H., and Suh, N. P., "Effect of Fiber Orientation on Friction and Wear of Fiber Reinforced Polymeric Composites," *Wear*, Vol. 53, 1979, pp. 129–141.)

point may be written as

$$F = \tau A_r \tag{6.1}$$

where τ is the shear strength of the material and A_r is the real area of contact. The flow strength of polymers, τ, depends on the applied hydrostatic pressure, that is,

$$\tau = \alpha + \beta p \tag{6.2}$$

where α and β are material constants and p is the hydrostatic pressure. The constant β denotes the dependence of minimum shear stress for flow on the hydrostatic pressure. Assuming that the hydrostatic pressure in the plastically deformed zone is nearly equal to the hardness, Eq. (6.2) may be written as

$$\tau = \alpha + \beta H \tag{6.3}$$

Substituting Eq. (6.3) into Eq. (6.1) and noting that $A_r = CL/H$, Eq. (6.1)

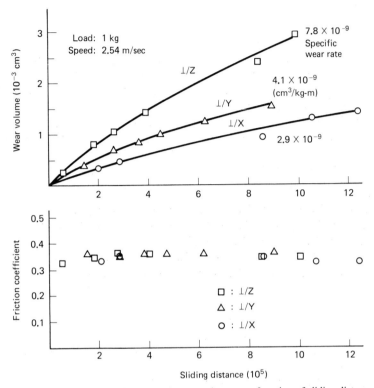

Figure 6.11 Friction coefficients and wear volume as a function of sliding distance in biaxially oriented glass microfiber–MoS_2–PTFE composite, sliding against 52100 steel with sliding planes normal to three orthogonal directions x, y, and z. (From Sung, H.-H., and Suh, N. P., "Effect of Fiber Orientation on Friction and Wear of Fiber Reinforced Polymeric Composites," *Wear*, Vol. 53, 1979, pp. 129–141.)

may be written as

$$F = CL\left(\frac{\alpha}{H} + \beta\right) \quad (6.4)$$

$$\mu = \frac{F}{L} = C\left(\frac{\alpha}{H} + \beta\right)$$

Unlike the case of metals, however, A_r does not increase linearly with the applied normal load L, indicating that the hardness of plastics depends on the magnitude of the applied normal load. The results of indentation experiments show that the diameter of the indented area varies as a function of the applied normal load as

$$d = \left(\frac{L}{k_0}\right)^{1/n} \quad (6.5)$$

Sec. 6.5   Basic Mechanisms for Friction in Polymers

TABLE 6.7   Coefficients of Friction for Steel on Plastics[a]

Plastic	Rockwell Hardness	½-in.-diameter Steel Ball			α[b] (kg/mm^2)	β[c] (kg/mm^2)
		1000 g		4000 g,		
		μ_s	μ_k	μ_k		
Polystyrene	38–40M	0.43	0.37	0.36		
Phenol-formaldehyde	89–90M	0.51	0.44	0.37		
Polyvinyl chloride	52M	0.53	0.38	—		
Polypropylene	105–106M	0.46	0.26	0.24	1.51	0.114
Polycarbonate	41–47M	0.48	0.43	—		
Polymethyl methacrylate	88M	0.64	0.50	0.49	5.13	0.204
Nylon 6/10	105R	0.53	0.38	0.32	4.66	0.258
Nylon 6/6	110R	0.53	0.38	0.36		
Nylon 6	91R	0.54	0.37	0.38		
Polyoxymethylene	118R	0.30	0.17	0.22		
Polyethylene (H.D.)	60R	0.36	0.23	0.21	1.34	0.049
Polyethylene (L.D.)	25R	0.48	0.28	0.26		
Polychlorotrifluoroethylene	112R	0.45	0.27	0.26		
Polytetrafluoroethylene	5R	0.37	0.09	0.10		
Polyimide	118R	0.46	0.34	0.31		
A-B-S resin	105R	0.40	0.27	0.29		
Polyphenylene oxide–styrene blend	118R	0.60	0.46	0.41		

[a] μ_s, static coefficient of friction; μ_k, kinetic coefficient of friction. v (sliding velocity) = 0.001 cm/sec.
[b] Material constant in Eq. (6.2).
[c] Material constant in Eq. (6.2).
Source: Frictional data from Steijn (1967).

where k_0 and n are constants and d is the indentation diameter. Pascoe and Tabor (1956) found n to be 2.7 for PMMA. By definition, d is related to the hardness by

$$d = \frac{2}{\sqrt{\pi}}\left(\frac{L}{H}\right)^{1/2} \quad (6.6)$$

Substituting Eqs. (6.5) and (6.6) into Eq. (6.4), one obtains

$$F = C\frac{\pi\alpha}{4k_0^{2/n}L^{1-2/n}} + \beta L = \mu L$$

$$F = C\frac{\pi}{4}\frac{\alpha d^2}{L} + \beta L = \mu L \quad (6.7)$$

Equation (6.7) states that as L increases, μ decreases. In particular, if $n \gg 2$, μ approaches β as L is made very large. Conversely, μ can never be less than β. Table 6.7 gives the experimentally determined values of α, β, and μ for various plastics sliding against a steel sphere. The foregoing derivation is valid only at low sliding speeds, where the temperature rise

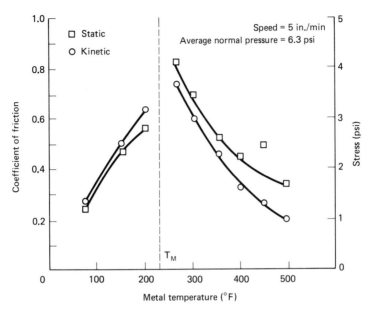

Figure 6.12 Friction coefficient and tangential stress on the sled of low-density polyethylene as a function of temperature sliding against a hard metal. (Reprinted with permission from Chung, C. I., "New Ideas about Solids Conveying in Screw Extruders," *Society of Plastics Engineers Journal*, Vol. 26, No. 5, 1970, pp. 32–44.)

due to the mechanical work is small and does not affect the material parameters significantly.

When the sliding speed and temperature vary, the frictional behavior also changes, due to the dependence of the shear stress of viscoelastic materials on these parameters. Furthermore, an increase in temperature is equivalent to decreasing sliding speeds, and vice versa. This equivalence of the time and temperature effects on the deformation of polymers can be used to interpret the frictional behavior of certain polymeric materials, using the time–temperature superposition principle (see Appendix 4.A). This is a useful concept for a limited number of polymers undergoing small viscoelastic deformation at very low sliding or rolling speeds. In this case the temperature rise due to the heating of asperities during deformation is negligible and the time–temperature superposition may be applied to the frictional behavior under a set of restricted conditions. For example, for rubber sliding at low velocities of less than 1 cm/sec, the superposition principle can be applied to generalize the frictional behavior. At high sliding speeds and for other polymers, the superposition principle cannot be applied quantitatively due to nonlinear effects.

To establish the conditions under which the superposition principle may be applied, the deformation of the asperities may be idealized by

Sec. 6.5 Basic Mechanisms for Friction in Polymers

assuming that a surface layer of thickness Δs undergoes shear deformation due to the traction imposed by the other surface. The coefficient of friction of polymers may then be written as

$$\mu = \frac{F}{L} = \frac{\tau A_r}{L} \tag{6.8}$$

If it is assumed that the effect of hydrostatic pressure is negligible and that A_r/L remains nearly constant, Eq. (6.8) may be written as

$$\mu = \tau(\gamma, t) \tag{6.9}$$

where γ is strain and t is time. The shear strain rate of the deformed surface layer at a reference temperature T_0 may be expressed as

$$\dot{\gamma} = \frac{V_0}{\Delta s} = \frac{\Delta \gamma}{\Delta t_0} \quad \text{at } T = T_0 \tag{6.10}$$

where V_0 is the sliding velocity at $T = T_0$, Δs the characteristic thickness of the deformed layer, and $\Delta \gamma$ the shear strain occurring in an interval Δt_0. Equation (6.10) may be expressed for Δt_0 as

$$\Delta t_0 = \frac{\Delta s}{V_0} \Delta \gamma \quad \text{at } T = T_0 \tag{6.11}$$

If all other variables, except the velocity and temperature, remain constant, the time taken Δt_1 for shear deformation at some other temperature T_1 may be expressed as

$$\Delta t_1 = \frac{\Delta s}{V_1} \Delta \gamma = a(T_1) \Delta t_0 \tag{6.12}$$

where $a(T_1)$ is the shift factor defined in Eq. (6.A3). The substitution of Eq. (6.11) into Eq. (6.12) yields

$$V_0 = a(T_1) V_1 \tag{6.13}$$

According to Eq. (6.13), if the asperity temperature is not affected by sliding and if the product of the thickness of the deformed layer and the strain increment, $\Delta s \Delta \gamma$, is not affected by either the temperature or velocity, the coefficient of friction may be correlated on a master curve as a function of $a(T)V$.

Such a master curve was obtained for various rubbers sliding at speeds lower than 1 cm/sec by Grosch (1963) and Ludema and Tabor (1966). The experimental results of Grosch are shown in Fig. 6.13. Figure 6.13(a) shows the coefficient of friction of the acrylonitrile-butadiene rubber sliding against wavy glass. The master curve and the shift factor are shown in Fig. 6.13(b) and (c), respectively. Grosch also performed frictional tests with rubbers sliding against rough silicon carbide abrasive papers to induce large bulk

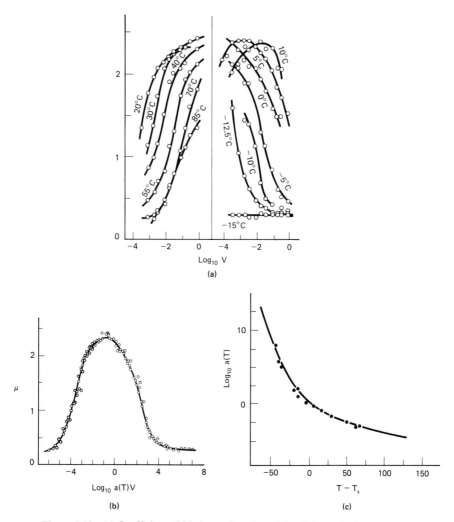

Figure 6.13 (a) Coefficient of friction as function of the sliding velocity at various temperatures of acrylonitrile-butadiene rubber on wavy glass. Curves are shown in two groups for clarity. (b) Master curve of the data shown in (a). Reference temperature is 20°C. (c) Shift factor $a(T)$ versus $(T - T_s)$ for the rubber. The solid line represents the WLF equation given by $\log_{10} a(T) = [-8.86(T - T_s)]/[101.5 + T - T_s]$, where $T_s = 50 + T_g$ (°C) and $T_g = 21.4$°C. (From Grosch, 1963.)

deformation of the rubber as well as the local shear deformation of the surface layer. Based on these experiments, it was concluded that the frictional behavior of rubbers, which is attributable to the energy dissipation during the bulk deformation, is also viscoelastic and that the time–temperature superposition is applicable to this mode of frictional behavior.

Sec. 6.5 Basic Mechanisms for Friction in Polymers

The available experimental evidence on the frictional behavior of polymers other than rubber indicates that, in general, the superposition principle may not be applicable to most polymers, even at very low sliding velocities. Therefore, the technique of superposition must be used judiciously.

The preceding discussion was confined to low sliding speeds. Under normal sliding conditions, however, the velocity and temperature at the contacting asperities are interrelated. Because the work done during sliding raises the interface temperature, it is not possible to relate the frictional behavior to a single parameter such as $a(T)V$. The temperature rise caused by sliding is related to the interface geometry, applied load, and physical properties as well as the sliding velocity.

When a metal sphere rolls on a polymer, a substantial part of the rolling friction is associated with the energy loss due to viscoelastic deformation of polymers. In Fig. 6.14 the rolling friction of a $\frac{3}{16}$-in. steel ball

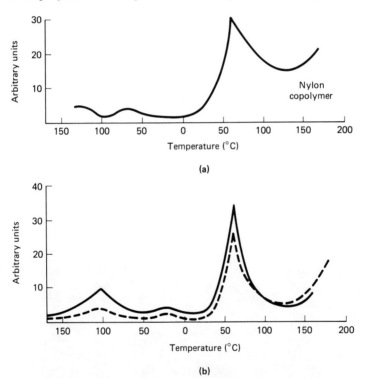

Figure 6.14 (a) Rolling friction μ_r of $\frac{3}{16}$-in. steel ball over the surface of a nylon copolymer, as a function of temperature (load 1050 g). (b) Low-frequency viscoelastic loss data for the same polymer as a function of temperature. Solid line, damping loss or internal friction; dashed line, damping loss corrected for change in modulus ($-150°C < T < +200°C$); damping loss, 200 to 1200 cps). (From Ludema, K. C., and Tabor, D., "The Friction and Viscoelastic Properties of Polymeric Solids," *Wear*, Vol. 9, 1966, pp. 329–348.)

over the surface of a nylon copolymer is compared with the low-frequency (i.e., 1 sec^{-1}) viscoelastic loss data for the same polymer as a function of temperature. The similarity of the curves should be noted. Since the typical coefficient of friction in sliding between steel and plastics is 0.2 to 0.6, which is about an order of magnitude larger than the rolling friction, it can be concluded that the sliding friction is caused primarily by the plastic shearing of the surface rather than by the viscoelastic deformation of the polymer.

6.6 MODEL FOR WEAR OF FIBER-REINFORCED COMPOSITES

In Section 6.4 it was stated that the wear behavior of fiber-reinforced composites depends on the orientation and properties of the fiber, the mechanical behavior of the matrix, and the counterface material. In a brittle matrix such as epoxy, cracks generated in the matrix propagate right through the fiber when the bonding between the fiber and the matrix is strong. In a highly ductile matrix such as polyurethane, the cracks cannot propagate through the matrix and the fiber. Therefore, the fiber bends with the matrix under the asperity contact, and the wear rate of this kind of composite is controlled by the wear rate of the fiber. A model for wear will be presented in this section, assuming that the fiber is graphite and the matrix is polyurethane.

Experimental results show that the ends of worn fibers of a graphite–polyurethane composite are elliptical rather than circular (Fig. 6.15), although

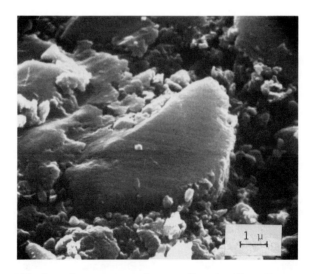

Figure 6.15 Smooth appearance of a worn fiber tip. (From Burgess; 1983.)

Sec. 6.6 Model for Wear of Fiber-Reinforced Composites

the fibers were perpendicular to the surface (Burgess, 1983). Furthermore, the ends of these fibers are extremely well polished. These facts indicate that the fibers were worn gradually in very small increments when the fibers are bent under the slider as schematically illustrated in Fig. 6.16. A wear model developed for this type of composite by Suh and Burgess (1983) will be given in this section, after reviewing the structure of graphite fibers.

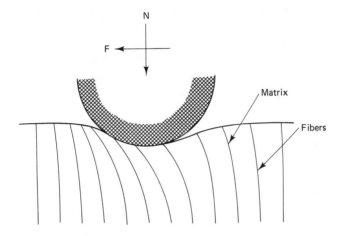

Figure 6.16 Fiber deflection in composite. (From Burgess, 1983.)

Figure 6.17 illustrates schematically the crystallographic structure of a perfect single graphite crystal. The crystal is stacked sheets of carbon atoms. Within the sheet, or basal plane, the carbon atoms are hexagonally ordered and linked by strong covalent bonds. Between sheets the atoms are held together by relatively weak van der Waals forces. The distance between sheets is 3.35 Å.

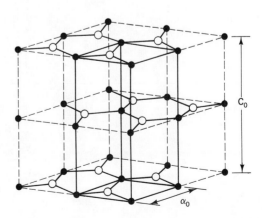

Figure 6.17 Crystallographic structure of graphite.

As a result of this crystallographic structure, the graphite crystal is highly anisotropic. The ratio of intralayer to interlayer bond energies is approximately 29 (Soule and Nezbeda, 1968). Of particular interest is the strength of a single graphite crystal within the basal plane. Williams et al. (1970) calculated the theoretical cohesive strength to be approximately 26 million psi (173 GPa).

Although a detailed and conclusive description of the internal structure of carbon fiber does not exist now, it is clear that carbon fibers consist of fibrils oriented along the fiber axis, and consequently fibers are highly anisotropic. Ruland and coworkers (1967) studied the internal fiber structure by X-ray diffraction and electron microscopy. Their findings led them to construct the fiber model illustrated in Fig. 6.18. The fiber consists of layers of graphite ribbons oriented such that their basal planes are parallel to the fiber axis. The ribbons, or fibrils, are comprised of graphite sheets

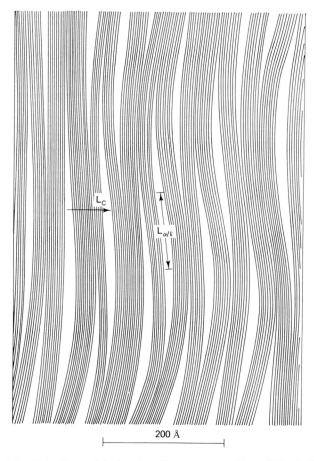

Figure 6.18 Ruland's model of carbon fiber structure. (From Ruland, 1976.)

and are wavy along the fiber length. As the fiber heat treatment temperature is increased, the degree of waviness decreases and the ribbon thickness (20 to 100 Å) increases. Between ribbons, microvoids on the order of 10 Å exist. Dieffendorf and Tokarsky (1975) also studied carbon fiber internal structure by transmission electron microscopy and X-ray diffraction. They also concluded that the fibers exhibit a fibrillar internal structure (Figs. 6.19 and 6.20). For PAN-based carbon fibers of 40 million psi modulus, they reported that the fibrils were typically 13 layer planes thick (approximately 40 Å), 40 Å wide, and "at least microns long." X-ray diffraction measurements indicated that three-fourths of the layer planes were aligned with the fiber axis to within 16°.

Graphite fibers are elastic and brittle. Their tensile strength is flaw-dominated, with voids, inclusions, and microcracks acting as stress concentration and crack nucleation sites. Therefore, failure strengths of graphite fibers are gage-length dependent. However, since in tribological applications

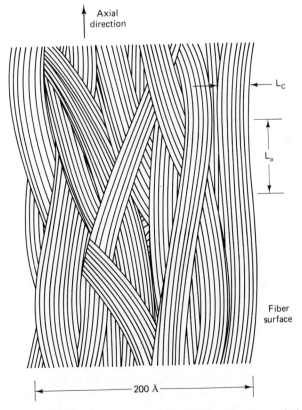

Figure 6.19 Fibrillar structure of carbon fiber. (Reprinted with permission from Dieffendorf, R. J., and Tokarsky, E., "High Performance Carbon Fibers," *Polymer Engineering and Science*, Vol. 15, 1975, pp. 150–159.)

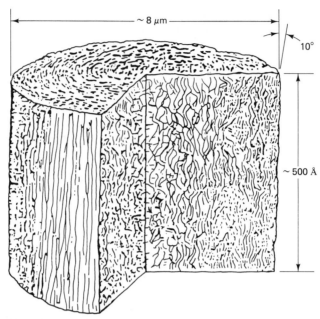

Figure 6.20 Three-dimensional model of carbon fiber. (Reprinted with permission from Dieffendorf, R. J., and Tokarsky, E., "High Performance Carbon Fibers," *Polymer Engineering and Science*, Vol. 15, 1975, pp. 150–159.)

only a very small length of the fiber is highly stressed, a flaw-dependent strength behavior may not be applicable to the fiber wear process; rather, their intrinsic behavior may govern the microfracture process.

Let us now consider the wear process of the graphite-fiber-reinforced elastomer shown in Fig. 6.16. Fibers may undergo substantial elastic bending upon interactions with counterface asperities, because of the low shear modulus of the elastomeric matrix. The fibers are assumed to bend elastically such that the fiber axis is oriented nearly parallel to the sliding direction at the fiber tips (Fig. 6.21). As shown earlier, the fiber is a collection of fibrils oriented predominantly parallel to the fiber axis, as schematically illustrated in Fig. 6.22. For the purpose of the analysis, the fibrils are assumed to have width w_f, thickness t, and infinite length. The top fiber surface is subjected to constant normal and tangential surface tractions. The normal stress component is assumed to be the flow stress, p_m, of the metal counterface. The tangential stress component is μp_m, where μ is the coefficient of friction. The interfibril bonding is assumed to be insignificant in comparison to the strong covalent in-plane bonding of the graphite crystal.

The normal compressive loads are supported by the underlying fibrils. The tangential tractions, however, are borne by the surface fibril. Along the fibril, in the x direction (Fig. 6.23), the tensile load and thus the tensile stress accumulates. Since the tractions are assumed to be constant, the

Sec. 6.6 Model for Wear of Fiber-Reinforced Composites

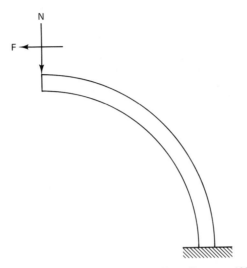

Figure 6.21 Single deflected fiber. (From Burgess, 1983.)

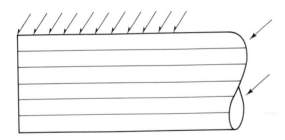

Figure 6.22 Surface traction on fiber tip.

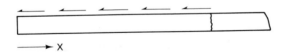

Figure 6.23 Tangential load borne by a fibril. (From Burgess, 1983.)

fibril tensile stress is expressed as a function of x as

$$\sigma = \frac{\mu p_m x}{w_f} \tag{6.14}$$

At some critical length, l_c, the fibril tensile stress exceeds the cohesive strength of the graphite crystal, σ_c, and the fibril fractures, which is found to be

$$l_c = \frac{\sigma_c w_f}{\mu p_m} \tag{6.15}$$

Assuming that one fibril per fiber is removed after the passage of a single asperity, the volume removed, V_{fi}, is

$$V_{fi} = w_f t l_c \tag{6.16}$$

The total number of fibers present in the apparent contact area, N_f, can be approximated as

$$N_f = \frac{A_a}{A_f} \tag{6.17}$$

where A_f is the fiber cross-sectional area and A_a is the apparent area of contact.

The total volume, V_f, of fiber removed after each fiber in the apparent contact area has interacted with one counterface asperity is simply the volume, V_{fi}, of one fibril of critical length times the number of fibers in the apparent contact area:

$$V_f = V_{fi} N_f \tag{6.18}$$

The total number of asperity contacts, N_s, along any line in the apparent contact area after a sliding distance, S, is

$$N_s = \frac{S}{\lambda} \tag{6.19}$$

where λ is the asperity spacing. Thus the total wear volume, V, after a sliding distance, S, is

$$V = V_f N_s$$

The wear rate can be expressed as

$$\frac{V}{S} = \frac{\sigma_c w_f^2 t A_a}{\mu p_m A_f \lambda} \tag{6.20}$$

A numerical prediction of the critical breaking length of the fibril and the magnitude of the wear rate can now be made using Eq. (6.15), Eq. (6.20) and the following parameters: the graphite crystal cohesive strength, σ_c, of approximately 179 GPa (26 million psi); the fibril width, w_f, and thickness, t, of 40 Å; the apparent contact area, A_a, of roughly 1 mm² under a 1-kg load; the friction coefficient of 0.3; for a 52100 steel ball, the flow stress, p_m, of approximately 0.2 GPa (300,000 psi); the fiber diameter of 7 μm; and an asperity spacing of 100 μm.

The critical breaking length, l_c, of the fibril can be obtained by substituting the appropriate values into Eq. (6.15):

$$l_c = \frac{(179 \times 10^9) \text{ N/m}^2 (40 \times 10^{-10}) \text{ m}}{(0.3)(0.2 \times 10^9) \text{ N/m}^2} = 10 \text{ μm} \tag{6.21}$$

The volume wear rate per unit distance slid can also be determined from Eq. (6.20) as

$$\frac{V}{S} = \frac{(179 \times 10^9)\,\text{N/m}^2\,(40 \times 10^{-10})^2\,\text{m}^2\,(40 \times 10^{-10})\,\text{m}\,(1 \times 10^{-3})^2\,\text{m}^2}{(0.3)(0.2 \times 10^9)\,\text{m}\,(\pi/4)(7.0 \times 10^{-6})^2\,\text{m}^2\,(100 \times 10^{-6})\,\text{m}} \quad (6.22)$$
$$= 5.0 \times 10^{-14}\,\text{m}^3/\text{m} = 5.0 \times 10^{-10}\,\text{cm}^3/\text{cm}$$

The experimentally measured wear rate was on the order of $10^{-10}\,\text{cm}^3/\text{cm}$, which is very close to the theoretically predicted values without making any unrealistic assumptions. The agreement supports the hypothesis that the wear rate of graphite-fiber-reinforced polyurethane is controlled by the wear rate of graphite fibers. The agreement between the predicted values and the graphite fiber–epoxy composites shown in Fig. 6.9 is within an order of magnitude, indicating that in this case the graphite fibers *might* also have controlled the wear rates. The details of worn surfaces of graphite–polyurethane composites are given in Chapter 9.

When the fibers are not normal to the surface but are oriented along or across the sliding direction on the surface, the wear rate increases since some of the fibers are pulled out of the matrix by the asperities of the counterface. The model has not taken this kind of occasional fiber pull-out and fracture into account, but still the agreement is reasonably good, indicating that the gradual wear of graphite fibers controls the wear rate in most cases.

6.7 WEAR MECHANISMS IN SINGLE-PHASE POLYMERS

The deformation mechanisms discussed in Chapters 4 and 5 apply to all isotropic elastic or elastoplastic solids when they are subjected to the surface traction that typically exists between sliding surfaces. Single-phase polymers are not exceptions, despite the fact that most polymers also exhibit viscoelastic–plastic properties, since the viscous part of their behavior may be neglected due to the rapid loading of polymers in sliding applications. Therefore, the phenomenological aspects of wear of polymers discussed earlier in this chapter should be explainable in terms of these basic mechanisms. However, there are some important differences between various single-phase polymers and other elastic or elastoplastic materials. In light of these differences, three different mechanisms will be considered in this section that provide plausible explanations for the wear of highly crystalline thermoplastics, glassy polymers, and crystalline polymers with limited ductility.

Highly symmetric and crystalline polymers *without* bulky side-pendant side groups (e.g., PTFE and HDPE) can undergo large plastic deformation without breaking their molecular chains, because these molecules can orient and elongate through "uncoiling" of its molecules. Therefore, the solid made of these molecules can withstand large surface and subsurface deformation under sliding conditions, just as pure ductile metals can, until the molecules can no longer be extended. In these materials, the deformation and the deformation gradient are largest at the surface and decay rapidly away from the surface. As a consequence of the deformation, the molecules align nearly parallel to the surface, and their crystalline platelets form a nearly layered structure. Since these highly stretched molecular layers are held together by weak secondary bond forces, they can easily shear off and separate from the solid surface. These highly elongated "sheets" separated from the surface may further extend due to the sliding motion of the counterface. This peeling process may occur more readily if there are preexisting flaws or if cracks are nucleated during deformation. Then the criteria for wear by thin-film formation may be written, as a first approximation, as (Youn and Suh, 1981)

$$\tau_s \geq k_s$$
$$\sigma_{max} < \sigma_c$$
(6.23)

where τ_s is the tangential component of the surface traction, σ_{max} the maximum principal stress, k_s the shear strength of the surface layer, and σ_c the bulk cohesive strength.

The validity of the simple criterion given by Eq. (6.23) can be checked either by increasing the shear strength of the surface layer, k_s, or by decreasing the magnitude of the applied shear stress, τ_s, without affecting the bulk properties. Experiments done to increase k_s by cross-linking the surface layer with plasma show that the wear rate can be substantially decreased when k_s is greater than τ_s. This technique of reducing the wear of polymers is discussed further in Chapter 9.

Glassy polymers wear by different mechanisms than those discussed in the preceding paragraphs. When glassy polymers (e.g., PMMA and epoxy) are subjected to surface traction, the failure of the surface or subsurface is likely to occur when the maximum tensile stress (the maximum principal stress) exceeds the cohesive strength of the polymer. Cracks are likely to nucleate and propagate right at the asperity contact, as discussed in Chapter 4. A simple criterion that describes the failure of glassy polymers may then be written as

$$\sigma_{max} > \sigma_c$$
(6.24)

When these cracks propagating from various points on the surface intersect, loose, lumpy wear particles are generated.

Finally, the wear mechanisms for polymers with limited ductility, which fracture before completely orienting by extension of molecular chain, are likely to be different from those of either highly crystalline or glassy polymers. These polymers with limited ductility may wear by delamination mechanisms, creating thick wear sheets. In these materials, subsurface cracks may nucleate, even when there are no large second-phase particles, if the shear strain accumulated due to the cyclic load exceeds the critical strain. Once the crack is nucleated, it will propagate due to strain concentration at the crack tips (see Chapter 5).

The foregoing discussions of the wear mechanisms assumed that the interfacial temperature rise is small. When the temperature rise is very large due to either high loads or high sliding speeds, the plastic softens and melts. Then the friction and wear mechanisms will change drastically from those discussed to those involving melting and extrusion of the molten material from the interface.

6.8 CONCLUDING REMARKS

In this chapter it was shown that the molecular structure of unreinforced polymers affects the friction and wear behavior of polymers. Highly linear polymers have low friction coefficients and the wear debris consists of thin films, whereas the glassy polymers exhibit high coefficients of friction and chunky wear particles. These results can be explained in terms of the friction model presented in Chapter 3, where it was shown that whenever plowing occurs due to the presence of chunky wear particles, the frictional force tends to increase. When thin films are transferred to the counterface in the case of highly linear thermoplastics, sliding occurs between highly oriented polymeric surfaces with little plowing.

When polymers are reinforced with short fibers, the wear process is very similar to the delamination wear in two-phase metals. All the micromechanics discussed in Chapter 4 govern the wear process of these composites.

The wear of composites with long fibers is governed primarily by the wear rate of the fibers. A model presented based on this hypothesis predicts the experimentally observed wear rates extremely well.

The frictional behavior of these composites can also be explained in terms of the plowing, adhesion, and asperity deformation. An additional contribution to the frictional force in elastomeric composites may arise if the wear debris is sticky and adheres to the sliding surface, requiring deformation by the asperities of the counterface. When this occurs, the frictional force increases with sliding until a steady state is reached, just as the frictional force increased due to the increase in the entrapped wear debris.

REFERENCES

ARCHARD, J. F., and HIRST, W., "The Wear of Metals under Unlubricated Conditions," *Proceedings of the Royal Society of London*, Series A, Vol. 236, 1956, p. 397.

BILLINGHURST, P. R., BROOKES, C. A., and TABOR, D., "The Sliding Processes as a Fracture-Inducing Mechanism," *Proceedings of the Conference on the Physical Basis of Yield and Fracture,* Oxford, 1966, pp. 253–258.

BOWERS, R. C., CLINTON, W. C., and ZISMAN, W. A., "Frictional Properties of Plastics," *Modern Plastics,* February 1954, pp. 131–225.

BRISCOE, B. J., POOLEY, C. M., and TABOR, D., "Friction and Transfer of Some Polymers in Unlubricated Sliding," in *Advances in Polymer Friction and Wear,* Polymer Science and Technology, Vol. 5A, H. Lee, ed., Plenum Press, New York, 1974, pp. 191–204.

BURGESS, S. M., "Friction and Wear of Composites," S.M. Thesis, MIT, 1983.

CHUNG, C. I., "New Ideas about Solids Conveying in Screw Extruders," *Society of Plastics Engineers Journal,* Vol. 26, No. 5, 1970, pp. 32–44.

CLERICO, M., "Sliding Wear of Polymeric Composites," *Wear,* Vol. 53, 1979, pp. 279–301.

CLERICO, M., "Sliding Wear Mechanisms of Polymers," in *Fundamentals of Tribology,* N. P. Suh and N. Saka, eds., MIT Press, Cambridge, Mass., 1980.

DIEFFENDORF, R. J., and TOKARSKY, E., "High Performance Carbon Fibers," *Polymer Engineering and Science,* Vol. 15, 1975, pp. 150–159.

GROSCH, K. A., "Relation between Friction and Viscoelastic Properties of Rubber," *Proceedings of the Royal Society of London,* Series A, Vol. 274, 1963, pp. 21–39.

JAIN, V. K., and BAHADUR, S., "Material Transfer in Polymer–Polymer Sliding," *Wear of Materials—1977,* ASME, New York, 1977, pp. 487–493.

JAMES, D. I., "Surface Damage Caused by Polyvinyl Chloride Sliding on Steel," *Wear,* Vol. 2, 1958/59, pp. 183–94.

KAR, M. K., and BAHADUR, S., "Macromechanism of Wear at Polymer–Metal Sliding Interface," *Wear of Materials—1977,* ASME, New York, 1977, pp. 501–509.

LEWIS, R. B., "Predicting Wear of Sliding Plastic Surfaces," *Mechanical Engineering,* Vol. 86, October 1964, pp. 32–35.

LUDEMA, K. C., and TABOR, D., "The Friction and Viscoelastic Properties of Polymeric Solids," *Wear,* Vol. 9, 1966, pp. 329–348.

MAKINSON, R. K., and TABOR, D., "The Friction and Transfer of Polytetrafluoroethylene," *Proceedings of the Royal Society of London,* Series A, Vol. 281, 1964, pp. 49–61.

MATSUSHIGE, K., and BAER, E., "The Mechanical Behavior of Polystyrene under Pressure," *Journal of Material Science,* Vol. 10, 1975, pp. 833–845.

MCLAREN, K. G., and TABOR, D., "The Friction and Deformation Properties of Irradiated Polytetrafluoroethylene (PTFE)," *Wear,* Vol. 8, 1965, pp. 3–7.

O'ROURKE, J. T., "Fundamentals of Friction, PV, and Wear of Fluorocarbon Resins," *Modern Plastics,* September 1965, pp. 161–169.

PASCOE, M. W., and TABOR, D., "The Friction and Deformation of Polymers," *Proceedings of the Royal Society of London,* Series A, Vol. 235, 1956, pp. 210–224.

PETERSON, T. L., AST, D. G., and KRAMER, E. J., "Holographic Interferometary of Crazes in Polycarbonate," *Journal of Applied Physics*, Vol. 45, 1974, pp. 4220–4228.

POOLEY, C. M., and TABOR, D., "Friction and Molecular Structure: The Behavior of Some Thermoplastics," *Proceedings of the Royal Society of London*, Series A, Vol. 329, 1972, pp. 251–274.

PUTTICK, K. E., SMITH, L. S. A., and MILLER, L. E., "Stress Field Round Indentation in Polymethyl Methacrylate," *Journal of Physics D: Applied Physics*, Vol. 10, 1977, pp. 617–632.

RULAND, W., "X-ray Studies on Preferred Orientation in Carbon Fibers," *Journal of Applied Physics*, Vol. 38, 1967, pp. 3583–3591.

SOULE, D. E., and NEZBEDA, C. W., "Direct Basal Plane Shear in Single-Crystal Graphite," *Journal of Applied Physics*, Vol. 39, 1968, pp. 5122–5139.

STEIJN, R. P., "Friction and Wear of Plastics," *Metals Engineering Quarterly*, Vol. 7, May 1967, pp. 9–21.

STEIJN, R. P., "The Sliding Surface of Polytetrafluoroethylene, on Investigation with Electron Microscope," *Wear*, Vol. 8, 1968, pp. 193–212.

SUH, N. P., "The Delamination Theory of Wear," *Wear*, Vol. 25, 1973, pp. 111–124.

SUH, N. P., and BURGESS, S. M., "Friction and Wear of Graphite Fiber/Polyurethane Composites," to be published, 1983 (see also Burgess, 1983).

SUNG, N.-H., and SUH, N. P., "Effect of Fiber Orientation on Friction and Wear of Fiber Reinforced Polymeric Composites," *Wear*, Vol. 53, 1979, pp. 129–141.

SVIRIDYONOK, A. I., BELY, V. A., SMURIGOV, V. A., and SAVKIN, V. G., "Study of Transfer in Frictional Interaction of Polymers," *Wear*, Vol. 25, 1973, pp. 301–308.

TANAKA, K., and UCHIYAMA, Y., "Friction, Wear, and Surface Melting of Crystalline Polymers," *Wear of Materials—1977*, ASME, New York, 1977, pp. 499–530.

TANAKA, K., UCHIYAMA, Y., and TOYOOKA, S., "The Mechanism of Wear of Polytetrafluoroethylene," *Wear*, Vol. 23, 1977, pp. 153–172.

VAN DEN BOOGAART, A., "Crazing and Characterization of Brittle Fracture in Polymers," *Proceedings of the Conference on the Physical Basis of Yield and Fracture*, Oxford, 1966, pp. 167–175.

WARREN, J. H., and EISS, N. S., Jr., "Depth of Penetration as a Predictor of the Wear of Polymers on Hard, Rough Surfaces," *Wear of Materials—1977*, ASME, New York, pp. 494–500.

WEST, C. H., and SENIOR, J. M., "Frictional Properties of Polyethylene," *Wear*, Vol. 2, 1958/59, pp. 183–194.

WILLIAMS, W. S., STEFFENS, D. A., and BACON, R., "Bending Behavior and Tensile Strength of Carbon Fibers," *Journal of Applied Physics*, Vol. 41, 1970, pp. 4893–4901.

YOUN, J., and SUH, N. P., "Tribological Characteristics of Surface Treated Polymers," *Proceedings of the Society of Plastics Engineers*, 39th ANTEC, Boston, 1981, pp. 20–23.

APPENDIX 6.A

TIME–TEMPERATURE SUPERPOSITION AND THE SHIFT FACTOR $a(T)$[2]

The concept of time–temperature superposition was originally developed to deal with the creep and stress relaxation behavior of viscoelastic polymers under varying rate of loading and temperature. In these materials the stress relaxation behavior is characterized by the time-dependent relaxation modulus $E_r(t)$, which is defined as

$$E_r(t) = \frac{\sigma(t)}{\varepsilon_0} \qquad (6.A1)$$

where $\sigma(t)$ is the time-varying stress in a tensile specimen strained to ε_0. It should be noted that $E_r(t)$ is a measure of the stress still remaining in the specimen after being held at a constant strain for a time period t, rather than an indication of the stiffness of the material when it is subjected to instantaneous loading at time t. The relaxation modulus of PMMA is shown as a function of time in Fig. 6.A1.

The basic concept of the time–temperature superposition can be illustrated by examining the complete log $E_r(t)$ versus log t curves at various temperatures shown in Fig. 6.A2. Suppose that it is of interest to determine $E_r(t)$ at T_3, but at an accelerated rate. Many plastics behave in such a fashion that the log $E_r(t)$ versus log t curves are similar in shape, regardless of the test temperatures. Then, instead of determining the entire curve (called the *master curve*) at T_3, the test may be accelerated by determining the C portion of the curve at T_3, the B' portion of the curve at T_2, and the A" portion of the curve at T_1. The latter two curves can then be translated horizontally until curves A", B', and C form a smooth continuous curve. The degree of the horizontal shifting, log $a(T)$, depends solely on the relative temperatures and is independent of the time. This is the basis for the time–temperature superposition.

The theoretical basis for the time–temperature superposition may be given by considering how temperature affects the material properties. When materials deform due to the thermally activated motion of molecules under the biasing effect of the applied load, the rate of deformation is a function of temperature. That is, in a thermally activated deformation process, the thermal energy kT (where k is Boltzmann's constant) must overcome the energy barrier ΔE. Since kT at room temperature is of the order of $\frac{1}{40}$ eV, and ΔE is of the order of 1 or 2 eV, the transition from the unfavorable

---

[2] Adapted from N. P. Suh and A. P. L. Turner, *Elements of the Mechanical Behavior of Solids*, McGraw-Hill, New York/Scripta Technica, Washington, D.C., 1975.

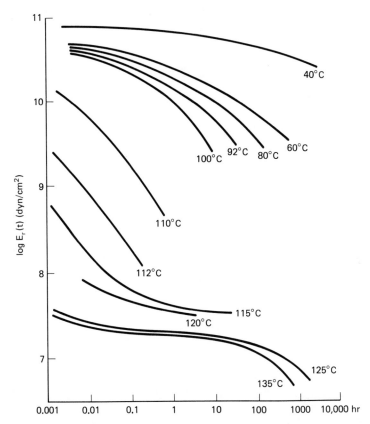

Figure 6.A1 Relaxation modulus for unfractionates polymethyl methacrylate (PMMA). (From Mcloughlin and Tobolsky, 1952.)

sites to the favorable sites cannot occur instantaneously. The probability of such a transition is given by Boltzmann's relation[3]

$$\dot{\varepsilon} = \frac{\Delta \varepsilon}{\Delta t} = A \exp\left(-\frac{\Delta E}{kT}\right) \quad (6.A2)$$

where A is a constant sometimes known as the frequency factor. If the creep rate is governed by the same physical process, ΔE should be the same. The activation energy usually changes when the molecules assume a new state, such as occurs at the glass transition temperature and at the melting point. However, in a small temperature range, ΔE may be assumed to be constant. In this case, Eq. (6.A2) states that the change in temperature affects the creep rate. Since $\dot{\varepsilon} = \Delta \varepsilon / \Delta t$, it can be seen that the effect of

---

[3] Equation (6.A2) is sometimes known as the Arrhenius equation.

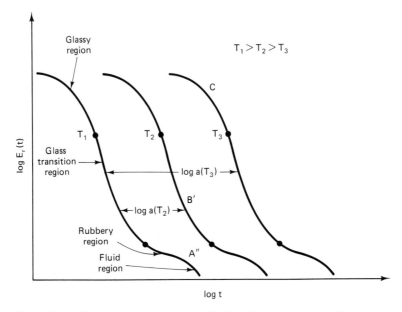

Figure 6.A2 Schematic of master curves for the relaxation modulus of a polymer at different temperatures. Note that the curves are parallel and shifted only along the log (time) axis. (From Suh and Turner, 1975.)

changing the temperature is to alter the time it takes to creep through a given strain $\Delta\varepsilon$. Equation (6.A2) states that a change of 10°C in temperature near room temperature may affect the creep rate by an order of magnitude. That means that if it is known precisely how the temperature change affects the time scale, it is possible to get the long-time creep data by simply raising the temperature.

From Eq. (6.A2) it can be seen that the ratio of the deformation times at two different temperatures for a given creep strain depends only on temperature as

$$\frac{\Delta t}{\Delta t_0} = \frac{\exp(-\Delta E/kT_0)}{\exp(-\Delta E/kT)} = \exp\left[-\frac{\Delta E}{k}\left(\frac{1}{T_0} - \frac{1}{T}\right)\right]$$
$$= a(T) \quad (6.A3)$$

or

$$\Delta t = a(T)\,\Delta t_0$$

The expression $a(T)$ is defined as the shift factor. Equation (6.A3) states that the deformation (or the relaxation) time at temperature T is $a(T)$ times longer than that at temperature T_0. Therefore, the curves shown in Fig. 6.A2 are separated by shifts of log $a(T_2)$ and log $a(T_3)$ from the curve obtained at T_1.

A more rigorous theoretical argument shows that the ratio of a relaxation time at temperature T to the same relaxation time at some other temperature T_0 is given by[4]

$$\frac{\tau(T)}{\tau(T_0)} = \frac{\eta}{\eta_0} \frac{\rho_0 T_0}{\rho T} = a(T) \tag{6A.4}$$

where ρ denotes the density and η the viscosity of the polymer.

Over a limited temperature range, from glass transition temperature T_g to $T_g + 100°C$, the following universal function for the shift factor $a(T)$ for all glassy polymers is found to exist:

$$\log a(T) = \frac{-17.4(T - T_g)}{51.6 + (T - T_g)} \tag{6.A5}$$

where the temperatures are given in degrees Celsius. Equation (6.A5) is known as the WLF (Williams–Landel–Ferry) equation. The WLF equation was originally determined empirically. The form of the WLF equation has been justified by using a theory based on the free-volume concept, which in effect postulates that molecular relaxation depends on the space available for molecular motion. The Arrhenius expression used in deriving Eq. (6.A3) is generally valid at temperatures much higher than T_g.

Time–temperature superposition has limited applicability. Usually, it does not hold rigorously for crystalline polymers over the entire temperature range, since the creep mechanism of such polymers is controlled by both amorphous and crystalline phases at temperatures higher than T_g. Time–temperature superposition is strictly applicable only to linear[5] viscoelastic materials whose stress relaxation times have identical dependence on temperature. This requirement is well approximated by such glassy (amorphous) polymers as polystyrene, polycarbonate, and polymethyl methacrylate. However, one may use time–temperature superposition for all plastics if approximate extrapolation is sufficient for design purposes.

---

[4] J. D. Ferry, "Mechanical Properties of Substances of High Molecular Weight: VI. Dispersion in Concentrated Polymer Solutions and Its Dependence on Temperature and Concentration," *Journal of the American Chemical Society*, Vol. 72, 1950, pp. 3746–3752.

[5] Some investigators claim that time–temperature superposition is also applicable to nonlinear viscoelastic materials.

7

FRICTION AND WEAR DUE TO HARD PARTICLES AND HARD ROUGH SURFACES: ABRASION AND EROSION

7.1 INTRODUCTION

When hard particles are introduced into a sliding interface or when they impinge against solid surfaces at high speeds, severe wear can occur. Under these conditions the hard particles can cut the surface, creating chips, or can cause severe plastic deformation and subsurface damage which eventually creates loose wear particles. When wear is caused by hard particles entrapped between sliding surfaces or by rubbing of the softer surface by the harder rough surface, it is called *abrasive wear*. Wear due to particle impingement is called *erosive wear* due to solid particle impingement. Although both abrasive and erosive wear are caused by hard particles, the erosive wear by solid particle impingement is caused mostly by deformation-induced damage rather than by cutting, whereas the opposite is true in abrasion. Since the deformation-induced damage accumulates slowly, a large number of particle impingements are necessary to create a wear particle in erosive wear. By contrast, abrasive wear can be caused by a few particles. This chapter examines these wear processes in detail.

Abrasive wear poses a difficult and important problem for designers of farm machinery, diesel engines, earth-moving equipment, and armored tanks. When the possibility of abrasive wear is present, it is the most important problem to be solved before dealing with other types of wear, because the abrasive wear coefficient is at least one to two orders of magnitude larger than the other wear coefficients. A common way of eliminating abrasive wear has been the use of air or oil filters of the mechanical, magnetic, and electromagnetic types. These filters are normally capable of removing particles larger than about a micron, but it is difficult to entrap submicron particles. Fortunately, submicron particles do not cause much damage by cutting, although they affect the friction and wear process by plowing and subsurface deformation. Filtering is not sufficient assurance against abrasion, however, since hard particles are also generated by the sliding action when the sliding surface contains hard particles. Many commercially available metals and alloys have hard particles, some of which are deliberately introduced to increase the hardness.

The term "abrasive wear" connotes negative and unbeneficial aspects. However, the abrasive material removal processes, such as grinding, lapping, and abrasive machining with coated abrasive belts, are important industrial technologies for machining and finishing solid parts. In recent years, abrasive machining with coated abrasive belts has become a widely used manufacturing operation in wood and metal processing, replacing more traditional milling, sawing, planing, and grinding. The basic mechanisms involved in abrasive machining and abrasive wear are obviously the same.

Erosive wear has also been a matter of great concern to designers of gas and steam turbines, pipelines, radomes, and polymer processing equipment. Solid particles may be impurity particles or deliberately incorporated fillers and fibers, in gas or liquid streams. The life of many gas turbines and processing machines is critically affected by erosion. During the Vietnam conflict, erosion of helicopter blades and aircraft parts was a severe problem. The problems encountered by Rolls-Royce engineers in developing epoxy–graphite blades for their compressors are well-known examples of how serious erosion problems can be. Just as there were positive aspects of abrasive wear, the removal of materials by erosion has also been utilized as a useful industrial process. Sandblasting, abrasive cutting, and shot peening are some of the examples. Despite the importance of the subject matter and much serious study, the understanding of erosive wear is far from complete, even by the general standard of the tribology field. We cannot yet predict the erosion rate from first principles based on microstructural information.

These two friction and wear processes are essentially governed by mechanical behavior of materials, just as the sliding and fretting friction and wear mechanisms were controlled primarily by mechanical properties, although at high temperatures the thickness of the oxide layer can alter the relative importance of various mechanisms that control abrasion and erosion. In this sense, many basic mechanisms for subsurface damages presented in Chapter 4 are still operative in these situations. The purpose of this chapter is to review the available factual information on abrasive and erosive wear and rationalize it in terms of basic mechanisms and models. Abrasive friction and wear will be taken up first, followed by a detailed exposition of erosive wear.

7.2 FRICTION AND WEAR DUE TO ABRASIVE ACTION

In Chapter 1 a brief review of the early theories for abrasive wear was presented. It was shown that the wear rates predicted by a simple cutting model are much greater than those measured experimentally at least by an order of magnitude. It was demonstrated through a dimensional analysis that the cutting model could not account for 90% of the external work done. These are not the only shortcomings of an oversimplified model. It also cannot account for the effect of particle size, ductility, and the relative hardness in abrasive wear. The goal of this section is to improve the understanding by considering a more realistic model of abrasive friction and wear that can better explain the abrasive phenomena reported in the literature. We begin by reviewing important and relevant experimental results without trying to be exhaustive.

7.2.1 Phenomenological Aspects of Abrasive Wear

The abrasive wear resistance of materials has been determined using a variety of methods. Among the most common methods are rubbing of a specimen surface against a coated abrasive belt (see Fig. 7.1) and the measurement of the comparative wear resistance of flat specimens by lapping them on a lap master plate with abrasive particles of known size (as shown

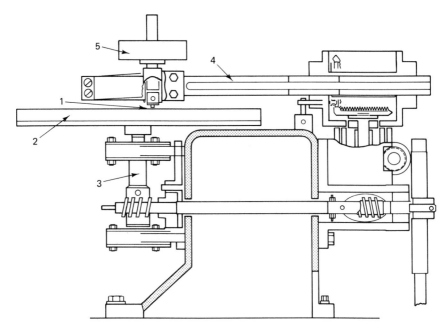

Figure 7.1 Schematic diagram of abrasive wear testing machine. 1, Specimen; 2, disk; 3, shaft; 4, rod; 5, weight. (From Khruschov, 1957.)

in Fig. 7.2). The important parameters that affect the wear resistance in these tests are the hardness, chemical composition, toughness, and the microstructure of the materials; the size of the specimens; the relative hardness of the specimen and the abrasive particles; the geometry of the abrasive (e.g., a coated abrasive belt); the sharpness and the size of abrasive grains; and the environment.

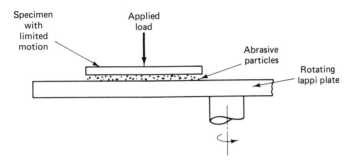

Figure 7.2 Schematic diagram of experimental apparatus for testing relative abrasive wear resistance of materials (three-body abrasive wear).

It does not take advanced education to know that a harder object is required to scratch the surface of a softer material, but it takes more than intuition to understand the exact role of hardness in abrasive wear, when the hardness of the abrasive is much greater than that of the specimen. There is a good correlation between the hardness of single-phase materials and their wear resistance, as shown in Fig. 7.3 (Khruschov, 1957; Richardson, 1967). However, if the hardness of the specimen is increased by cold work or by heat treatment of two phase materials, the wear resistance and the hardness do not correlate as well (see Fig. 7.4). When the hardness of both the specimen and the abrasive particles is comparable, the relative hardness of the specimen and the abrasive particles is very important. No appreciable abrasive action can take place when the hardness of these materials is approximately the same, as shown in Fig. 7.5. Nonetheless, hardness is certainly the most important parameter in determining abrasion resistance. However, it is not the only parameter that affects abrasive wear.

The chemical composition of abrasive particles and specimens is also important. When the abrasive particles and the specimen can react chemically or dissolve in each other, the abrasive particles will rapidly wear and cause minimum abrasive wear. For example, tungsten carbide particles will wear rapidly if they are abraded against steel at high speeds. (This is discussed further in Chapter 8.) In this case, alumina and silicon nitride are much more suitable abrasives, if the goal is to remove the maximum amount of steel.

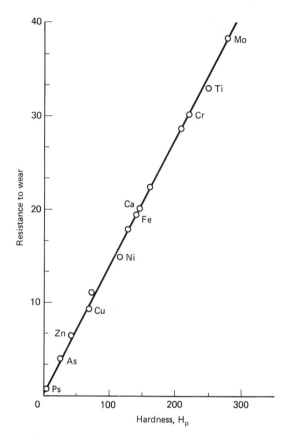

Figure 7.3 Relative wear rate versus hardness of commercially pure metals, (y12 is 1.2% carbon steel and 40 is 0.41% carbon steel). (From Kruschov, 1957.)

Another property that affects the rate of metal removal in abrasive machining is the toughness of grains. For example, alumina has excellent chemical stability for grinding steel, but it does not have sufficient high strength and toughness. These properties can be enhanced by forming a solution of Al_2O_3 and ZrO_2 and by quenching the solution to create a unique microstructure that imparts strength and toughness. Tough, wear-resistant abrasives consisting of Al_2O_3 and MgO composition have also been made using a sintering technique starting from a gel.

That the cutting action ceases when the cutting edges of the abrasive grains become dull is a well-known fact. What is not commonly known, however, is that the size of the specimen has an effect on the abrasion rate. Variations in wear rates up to 30% have been observed when the specimen length is varied (Larsen-Basse, 1972). Some of this difference in wear rates may be attributed to the clogging of the interstices between the abrasive grains by wear debris (Avient et al., 1960).

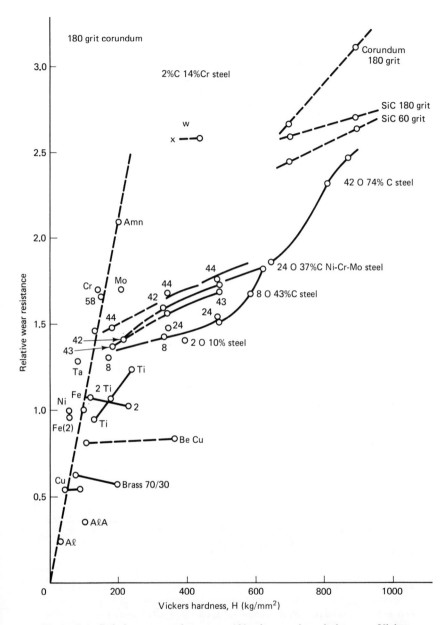

Figure 7.4 Relative wear resistance on 180-grit corundum cloth versus Vickers hardness. Numbers refer to BS 970: 1955, En series steels. (Fe(2)) zone-refined iron; (AMn) austenitic Mn steel; (BeCu) Mallory 73 Be copper; (A1A) A1 alloy BS. 1470: 1955, HS 30. (From Richardson, 1967.)

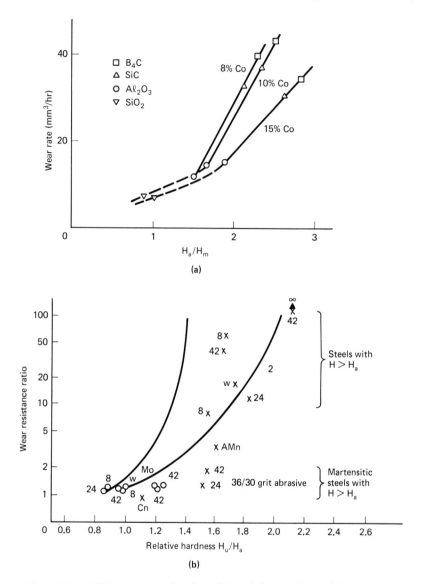

Figure 7.5 (a) Wear rate as a function of the relative hardness of the abrasives to that of WC–Co cermets. (From Larsen-Basse and Tanouye, 1976). (b) Wear resistance ratio and relative hardness H_u/H_a on 36/30 grit abrasive papers. (From Richardson, 1967.)

Sec. 7.2   Friction and Wear Due to Abrasive Action

An interesting characteristic of abrasion is the dependence of wear rate on the particle size, as shown in Fig. 7.6. The volume wear increases rapidly with grit size up to some critical diameter and then increases at a much slower rate (Avient and Wilman, 1962; Nathan and Jones, 1966; Larsen-Basse and Tanouye, 1976) or remains constant (Rabinowicz et al., 1961; Rabinowicz and Mutis, 1965; Mulhearn and Samuels, 1962; Goddard et al., 1959; Avient et al., 1960; Larsen-Basse, 1968). When the grit size is less than 1 μm, wear is no longer caused by the abrasive wear mechanism, but begins to approach the delamination wear. A great deal of work has been done to explain this size effect. This has been attributed to abrasive deterioration (Avient, 1960; Larsen-Basse, 1972; Rabinowicz, 1964; Moore and Douthwaite, 1976; Date and Malkin, 1976) and embedding of abrasives (Johnson, 1970) with the decrease in abrasive size. Others attributed this to the variation in geometry and distribution of abrasive particles (Larsen-Basse, 1968). As will be shown in this chapter, Sin et al. (1979) attributed this to the depth of penetration relative to the tip radius of abrasive particles rather than to the reasons given above.

A special form of abrasive wear of cutting tools can occur in metal cutting when the workplace contains hard particles (Ramalingam and Watson, 1978). Various hard particles can be present in steel, as shown in Table 7.1. As shown in Table 7.2, Al_2O_3 can be abraded by hard carbide particles such as WC at high temperatures (e.g., 800°C) and can even be abraded

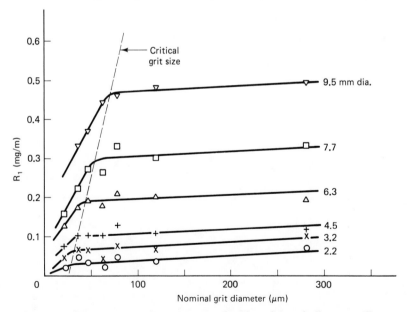

Figure 7.6 Wear rate at unit pressure as a function of the grit diameter. (From Larsen-Basse, J., "Influence of Grit Diameter and Specimen Size on Wear During Sliding Abrasion," *Wear*, Vol. 12, 1968, pp. 35–53.)

TABLE 7.1 Carbides, Nitrides, and Oxygen-Rich Nonmetallic Phases That May Exist in Steels Containing Specified Elements

Element	Dissolved	Carbide	Nitride	Nonmetallic Inclusion
(Fe)		Fe_3C		FeO, FeS
Si	$\underline{Si}$		Si_3N_4	SiO_2, silicates: (Fe, Mn)O–SiO_2, Al_2O_3–SiO_2, (Fe, Mn)O–Al_2O_3–SiO_2, CaO–Al_2O_3–SiO_2, (Mn, Fe)O, MnO–SiO_2, (Mn, Fe)S
Mn	$\underline{Mn}$			
P	$\underline{P}$			
S	$\underline{S}$			Sulfides
Ni	$\underline{Ni}$			
Cr	$\underline{Cr}$	Cr_7C_3, $Cr_{23}C_6$	Cr_2N	Cr_xO_y, (Fe, Mn)O–Cr_2O_3
Mo	$\underline{Mo}$	Mo_2C, M_6C, $M_{23}C_6$	Mo_2N	
W	$\underline{W}$	W_2C, M_6C, $M_{23}C_6$	W_2N	
Ti	$\underline{Ti}$	TiC	(TiN)	Ti_xO_y, (Fe, Mn)O–Ti_xO_y, TiO, TiS, Ti_3S_4, $Ti_4(C, N)_2Si_2$, TiN
Zr	$\underline{Zr}$	ZrC	(ZrN)	$ZrO_2 ZrOS$, ZrS, $Zr_4(C, N)_2S$, ZrN
Hf	$\underline{Hf}$	HfC	(HfN)	HfO_2, HfOS, HfS, HfN
V	$\underline{V}$	VC	VN	FeV_2O_4, V_2O_3, VO
Nb	$\underline{Nb}$	NbC	NbN	$FeNb_2O_6$, NbO_2
Ta	$\underline{Ta}$	TaC	TaN	$FeTa_2O_6$, Ta_2O_6
B	$\underline{B}$	$Fe_3(C, B)$	BN	
Al	$\underline{Al}$		AlN	Al_2O_3, (Fe, Mn)O–Al_2O_3, CaO–Al_2O_3, AlO_xN_y
Cu	$\underline{Cu}$			
Pb	(Pb)			
Y				Oxide sulfide
La				La_2O_3, La(OS), LaS, La_2S_3
Ce				Ce_2O_3, Ce(OS), CeS, Ce_2S_3
U				UO_2, US

Source: Data from Ramalingam and Watson (1978).

TABLE 7.2 Hardness at Specified Temperatures of the Phases That May Be Present in a Steel

Phase	Hardness H_V (kgf/mm^2)		
	400°C	600°C	800°C
Iron	45	27	10
Iron and interstitials	90	27	10
TiO	1300	1000	650
FeO	350	210	50
MgO	320	220	130
NiO	200	140	100
MnO	120	60	45
Al$_2$O$_3$	1300	1000	650
SiO$_2$	700	2100	300
ZrO$_2$	650	400	350
TiO$_2$	380	250	160
MgAl$_2$O$_4$	1250	1200	1050
ZrSiO$_4$	400	290	140

Source: Data from Ramalingam and Watson (1978) and Westbrook (1966).

by SiO$_2$ at 600°C. Carbide grains, especially WC, are less likely to be abraded by other hard particles, but an entire carbide grain from a cemented carbide tool may be pulled out after the binder phase around the particle is removed. High-speed-steel (HSS) tools can be abraded by hard particles in steel even though similar hard phases are also present in HSS tools.

Abrasive wear involves the generation of chips when the material abraded is elastoplastic or viscoelastic–plastic. Glassy materials are also abraded, but the wear particles are not generated by cutting mechanism. In this case, crack propagation and fracture processes create dustlike debris. However, even normally brittle crystalline solids (e.g., Al$_2$O$_3$) may be cut by abrasive particles when the abrasion is confined to a very small region that does not have any flaws.

As discussed in Chapter 1, abrasive wear involves a great deal of subsurface plastic deformation, in addition to cutting of the surface and the generation of chips. The amount of plastic deformation increases as the abrasive particles wear and develop flat surfaces. This will be discussed later.

The ductility of the material affects the abrasive wear rate as shown by Fig. 7.7. The wear coefficients of ductile materials such as OFHC copper and pure nickel are less than a half of the wear coefficients of such relatively brittle materials as polymethyl methacrylate (PMMA) and AISI 1095 steel. This indicates that some of the chips do not separate from the surface during the engagement by an abrasive grain, and that the subsurface cracks developed due to plastic deformation may contribute to the material removal.

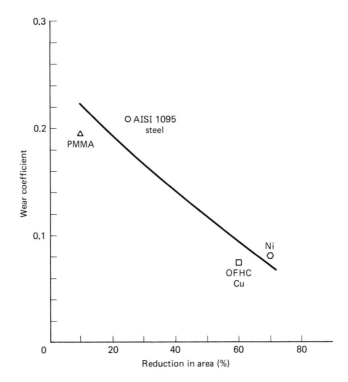

Figure 7.7 Wear coefficient as a function of reduction in area of materials tested for 60 grit. Applied load was 39.2 N.

The wear coefficients range from 0.1 to 0.2, which is one to three orders of magnitude greater than those of sliding wear.

7.2.2 Critical Experimental Observations

To understand the friction and wear mechanisms under abrasive sliding conditions, several critical experiments were performed recently (Sin, 1978; Sin et al., 1979), which will be described before presenting the analytical models.

Polymethyl methacrylate (PMMA), commercially pure nickel, AISI 1095 steel, and OFHC copper were chosen for study. The choice of the materials was based on several considerations. PMMA was chosen for its transparency, which enables direct optical observation. Commercially pure nickel was chosen because it work-hardens rapidly. Spherodized AISI 1095 steel was chosen to investigate the role of spherical hard second-phase particles, whereas annealed OFHC copper was chosen because of its low shear strength.

The specimens were prepared as follows. Before heat treatment of metal samples, ends of each sample were ground on abrasive papers to

Sec. 7.2   Friction and Wear Due to Abrasive Action   275

TABLE 7.3   Experimental Materials

Material	Treatment	Vickers Hardness (kg/mm^2)	Density (g/cm^3)
PMMA	—	17.5	1.18
OFHC Cu	Recrystallized 500°C, 1 hr	44.0	8.90
Ni	Annealed 800°C, 1 hr	88.5	8.90
AISI 1095 steel	Spherodized 900°C, 30 min; oil-quenched; 400°C, 1 hr	472.0	7.85

ensure uniform contact. Commercially pure nickel samples were encapsulated in Vycor tubes and then annealed at 800°C for an hour. To get small grain size the cold-worked OFHC copper specimens were encapsulated and recrystallized at 500°C for an hour. AISI 1095 steel was heat-treated at 900°C for half an hour, then oil quenched and tempered at 400°C for an hour to yield spherodized structure. The hardness and density of the materials are given in Table 7.3 and the microstructures are shown in Fig. 7.8.

Abrasion experiments were conducted on a pin-on-disk setup similar to that shown in Fig. 7.1. Specimens were abraded on commercial silicon carbide abrasive papers. The specimen held in a loading arm followed a

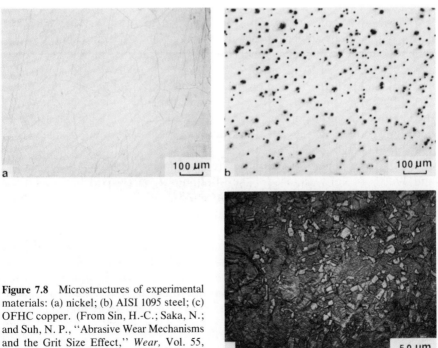

Figure 7.8 Microstructures of experimental materials: (a) nickel; (b) AISI 1095 steel; (c) OFHC copper. (From Sin, H.-C.; Saka, N.; and Suh, N. P., "Abrasive Wear Mechanisms and the Grit Size Effect," *Wear*, Vol. 55, 1979, pp. 163–190.)

spiral track on a 20-cm-diameter abrasive disk and therefore always passed over the fresh abrasive. The total distance slid was 4 m. The normal load was applied by a dead-weight and the tangential force was measured continuously.

Grits of 600, 320, 180, 120, and 60 were used and the normal load was varied between 4.9 N (0.5 kg) and 39.2 N (4 kg). The angular velocity of the disk was held constant so that the sliding velocity was varied from 2 to 6 cm/sec along the track. The specimens were weighed to an accuracy of 0.01 mg before testing. After testing, the specimens were brushed lightly and the weight was again measured. The weight loss and the tangential force were determined as an average of at least four measurements for each exprimental condition. In the case of OFHC copper and AISI 1095 steel, 4/0 and 3/0 emery papers were also used to investigate the wear behavior extensively.

After abrasion tests the surface of worn specimens were observed microscopically. Selected samples were electroplated with nickel to preserve the edges and cut by a diamond saw parallel to the sliding direction and perpendicular to the worn surface. Then the specimens were mounted, polished, and etched with standard etchants and observed in optical and scanning electron microscopes to determine the subsurface deformation. Wear particles were collected during the test and examined in an optical microscope.

PMMA samples were 6 mm long and their ends were mechanically polished on various abrasive papers and finally with 0.05-μm α-Al_2O_3 powder. The specimen was loaded under the objective of the microscope, while abrasive paper was fixed on the stage. The stage was moved slowly and steadily, and grooves formed on the surface of the specimen were observed through the specimen directly. The displacement of the stage, therefore the length of the grooves, was carefully controlled. To prevent interference between grooves, the stage was moved a distance of the order of the average spacing between contacting particles.

Abrasive papers of five grit sizes and loads between 4.9 N (0.5 kg) and 39.2 N (4 kg) were used. The total number of grooves formed on the surface was counted and their widths were measured. When the total groove number was large, at least 100 grooves were selected randomly and their widths were determined.

To simulate the abrasive action, experiments were also conducted with single-point conical tools. A pin-on-cylinder type of apparatus was used. A conical diamond tool was held in a dynamometer mounted on the carriage of a lathe. Normal and tangential forces were measured and monitored by a two-channel recorder. The tests were carried out under a normal load of 4.4 N (0.45 kg) and at the cutting speed of about 1.6 cm/sec. Tool angles were varied between 70 and 140°. Weight loss was determined, and cutting particles were collected and examined in an optical microscope.

Figure 7.9 shows the friction coefficient as a function of the abrasive grit diameter D_g for the materials tested, sliding under 4.9 N (0.5 kg) to 39.2 N (4 kg) load on different grades of dry silicon-carbide abrasive papers. It can be seen that the friction coefficient initially increases with grit diameter slightly, and later levels out and becomes substantially constant. It can also be seen that the friction coefficient does not vary much with materials,

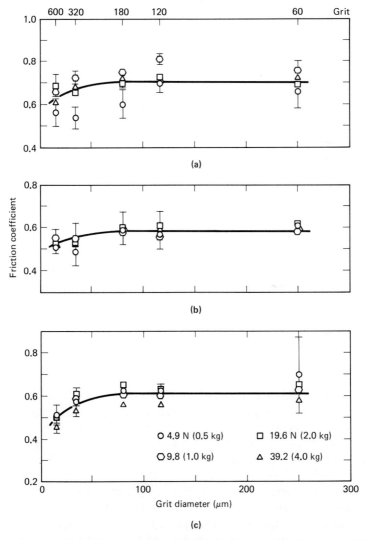

Figure 7.9 Friction coefficient as a function of the abrasive grit diameter for different normal loads: (a) PMMA; (b) nickel; (c) AISI 1095 steel. (From Sin, H.-C.; Saka, N.; and Suh, N. P., "Abrasive Wear Mechanisms and the Grit Size Effect," *Wear*, Vol. 55, 1979, pp. 163–190.)

and for metals it does not depend much on the applied load. For PMMA the friction coefficient exhibits large scatter compared to metals.

The wear rates and the wear coefficients as a function of the grit diameter are shown in Figs. 7.10 through 7.13. These figures clearly show the influence of abrasive grit size on wear. As the abrasive grit diameter is increased, the wear rate increases rapidly until a critical grit size is reached and later becomes independent of grit diameter or increases only slowly. It can be seen that the slope of the latter stage depends on the material and the normal load. For PMMA with the applied load of 39.2 N (4 kg), the slope is quite large. The critical grit diameter for all the materials tested is about 80 μm.

Wear particles were collected during the test to investigate their size and shape. Figure 7.14 shows the collected wear particles. It can be seen that the abrasive particles as well as wear particles were removed during abrasion. The particles are essentially in the form of microchips and their sizes vary with grit diameter. For finer grits wear particles generated during abrasion were attached to the abrasive surface; these are shown in Fig. 7.15.

Metallographic examination of the surface and the subsurface of worn specimens was conducted to investigate the mechanism. Figure 7.16 shows that the surfaces of worn specimens are completely covered with many grooves. It can be seen from the figure that large plastic deformation of groove materials by plowing took place as a result of abrasion. The figure also shows the variation of the number of grooves and their widths with grit diameter. These micrographs indicate that plowing is dominant during abrasion, even for the (4/0) abrasive.

Optical and scanning electron micrographs of the subsurfaces, shown in Figs. 7.17 and 7.18, clearly indicate that large subsurface deformation took place as a result of abrasion. For the coarser grits the deformation zone is clearly observable, but for the finer it is hardly observable probably because it is too shallow.

Figure 7.19 shows the scanning electron micrographs of the surface of the abrasive. It can be seen that the shapes and orientations of the abrasive particles are irregular. They are azimuthally and randomly oriented pyramids rather than cones with a wide range of included angles. From the figure the total number of abrasive particles was counted. For similar packing condition, it would be expected that if the grains are similar in shape, the total number of particles would vary inversely as the square of the mean grain diameter. Results for the variation of the number of abrasive particles per unit area with grit diameter are shown in Fig. 7.20.

Figure 7.21 shows the grooves formed on the surface of the PMMA sample. It can be seen that for each grit diameter a wide range of groove sizes is obtained. In Fig. 7.22 the total number of contacts is plotted for each grit size as a function of the applied load. The number of contacting

Sec. 7.2 Friction and Wear Due to Abrasive Action

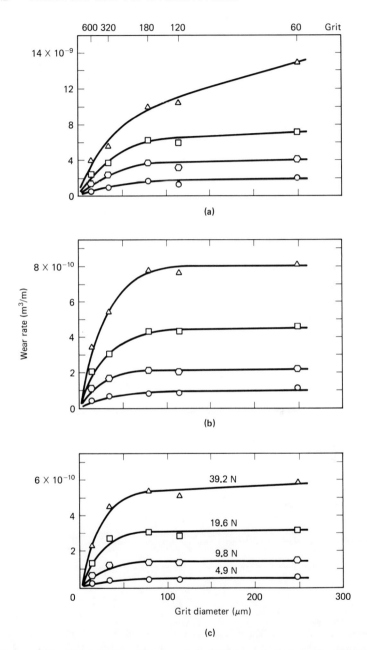

Figure 7.10 Wear rate as a function of the abrasive grit diameter for different normal loads: (a) PMMA; (b) nickel; (c) AISI 1095 steel. (From Sin, H.-C.; Saka, N.; and Suh, N. P., "Abrasive Wear Mechanisms and the Grit Size Effect," *Wear,* Vol. 55, 1979, pp. 163–190.)

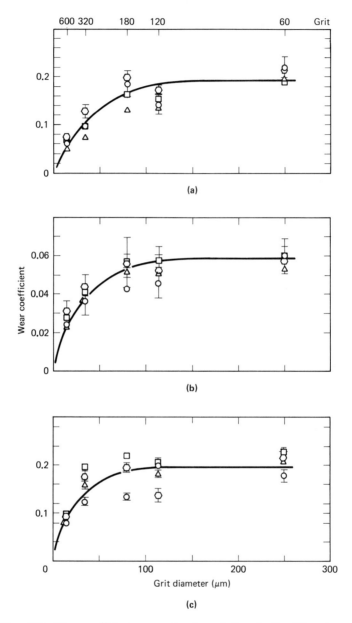

Figure 7.11 Wear coefficient versus abrasive grit diameter for different normal loads: (a) PMMA; (b) nickel; (c) AISI 1095 steel. (From Sin, H.-C.; Saka, N.; and Suh, N. P., "Abrasive Wear Mechanisms and the Grit Size Effect," *Wear*, Vol. 55, 1979, pp. 163–190.)

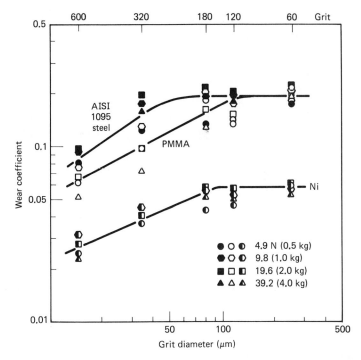

Figure 7.12 Logarithm of wear coefficient as a function of the logarithm of abrasive grit diameter. (From Sin, 1978.)

abrasive particles increases approximately linearly with the applied load, which is responsible for the linear dependence of the wear rate on the applied load. The variation of the total number of contacting particles with grit diameter for each applied load is shown in Fig. 7.23, where it can be seen that the number of contacting particles varies inversely as the square of the grit diameter.

Figure 7.24 shows the variation of the average groove width with the applied load for each grit size. A wide range of groove widths is observable for each grit size, especially for the coarse grit. For the range of loads investigated, the applied load has little influence on the groove width. The average groove width is shown as a function of the grit diameter for each applied load in Fig. 7.25. The average groove width increases linearly with the abrasive grit diameter.

Using the data of the average groove widths and the number of contacting particles, the real contact area for all grit sizes was calculated. The real contact area can be related to Nw^2, where N is the total number of contacts and w is the average groove width. Results for the variation of Nw^2 with applied load are shown in Fig. 7.26. The theoretical value was calculated from the hardness–real contact area relationship. Figure 7.27 shows the

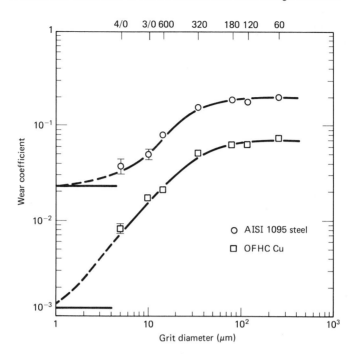

Figure 7.13 Wear coefficient as a function of the abrasive grit diameter up to 4/0 paper. The normal load was 39.2 N. Horizontal lines on the left represent the sliding wear coefficient. (From Sin, H.-C.; Saka, N.; and Suh, N. P., "Abrasive Wear Mechanisms and the Grit Size Effect," *Wear*, Vol. 55, 1979, pp. 163–190.)

ratio of the contact area calculated theoretically to Nw^2 as a function of the grit diameter. Results agree with the simple theory within 50%.

Figure 7.28 shows the variation of the removal coefficient as a function of tool angle (i.e., twice the rake angle). The removal coefficient was calculated in the same way as the wear coefficient. The coefficient decreases with increase of tool angle, and it is extremely small at the tool angle of 140° (i.e., negative rake angle of 70°). In Fig. 7.29 the ratio of the removal volume to the theoretically calculated groove volume is plotted as a function of the "attack" angle, which is the angle of the slope of the cone surface (i.e., $\pi/2$ minus the negative rake angle γ). When the attack angle is greater than 45°, the ratio does not increase with the attack angle.

In all cases except for the 140° tool, chips were observed. Chips are shown in Fig. 7.30. Figure 7.31 shows the grooves formed on the specimens.

7.3 ABRASIVE FRICTION AND WEAR MECHANISMS

In Chapters 1 and 3, the friction and wear coefficients due to plowing of the surface by abrasive particles were analyzed using upper-bound solutions.

Sec. 7.3 Abrasive Friction and Wear Mechanisms

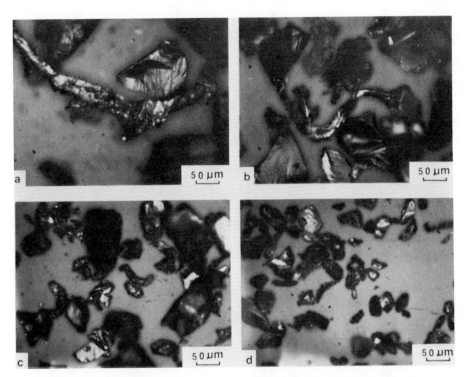

Figure 7.14 Abrasive wear particles of OFHC copper for different grit sizes: (a) 60; (b) 120; (c) 180; (d) 320. Experimental conditions are same as in Fig. 7.13. (From Sin, H.-C.; Saka, N.; and Suh, N. P., "Abrasive Wear Mechanisms and the Grit Size Effect," *Wear*, Vol. 55, 1979, pp. 163–190.)

The models assumed that the abrasive particle shear the surface leaving a wear groove which has exactly the same shape as the projected cross section of the abrasive particle (see Appendix 3.C). The solutions have shown that the friction due to plowing by conical abrasives with spherical tips is a function of the penetration of the surface by abrasive particles. The coefficient of friction varied from 0 to 1, depending on the ratio $w/2r$, where w is the penetration depth and r is the radius of curvature of the abrasive grain. It was also shown in Chapter 1, by dimensional analysis, that if the simple cutting mechanism were true, the wear coefficient for abrasive wear should be equal to approximately 1, indicating that a large fraction of energy required for plowing is dissipated by subsurface plastic deformation during the chip generation process. In this section, analyses based on the simple upper-bound model presented in Chapter 3 and a slip-line field will be presented.

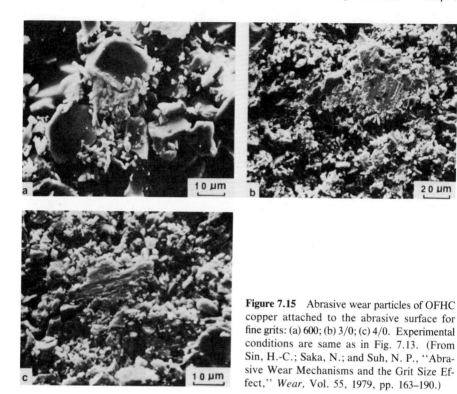

Figure 7.15 Abrasive wear particles of OFHC copper attached to the abrasive surface for fine grits: (a) 600; (b) 3/0; (c) 4/0. Experimental conditions are same as in Fig. 7.13. (From Sin, H.-C.; Saka, N.; and Suh, N. P., "Abrasive Wear Mechanisms and the Grit Size Effect," *Wear*, Vol. 55, 1979, pp. 163–190.)

7.3.1 Upper-Bound Solutions for Friction Due to Plowing by Abrasives

As shown in Appendix 3.C, the friction coefficient due to plowing can be modeled assuming that the abrasive particle is in the shape of a cone with a hemispherical tip (Fig. 7.32). The plowing component of friction coefficient, μ_p, may then be derived for the case when the penetration is confined to the spherical tip (i.e. when $w \leq 2r \sin \theta$) as

$$\mu_p = \frac{2}{\pi} \left\{ \left(\frac{w}{2r}\right)^{-2} \sin^{-1} \frac{w}{2r} - \left[\left(\frac{w}{2r}\right)^{-2} - 1 \right]^{1/2} \right\} \quad (7.1a)$$

and for the case when the penetration is so deep that the depth is greater than the diameter of the spherical tip (i.e., $w \geq 2r \sin \theta$) as

$$\mu_p = \frac{2 \tan \theta}{\pi} \left[1 - \frac{4(\tan \theta - \theta)}{\tan \theta} \left(\frac{w}{r}\right)^{-2} \right] \quad (7.1b)$$

Figure 7.32 shows the friction coefficient of this combined model as a function of w/r, the ratio of the groove width to the tip radius.

Sec. 7.3 Abrasive Friction and Wear Mechanisms   285

Figure 7.16 Surfaces of worn OFHC copper specimen for different grits: (a) 60; (b) 180; (c) 600; (d) 4/0. Experimental conditions are same as in Fig. 7.13. (From Sin, H.-C.; Saka, N.; and Suh, N. P., "Abrasive Wear Mechanisms and the Grit Size Effect," *Wear*, Vol. 55, 1979, pp. 163–190.)

As the figure shows, this simple model can explain the effect of the depth of penetration on friction coefficient, which is due to the relative bluntness of the abrasive particles. This effect of penetration cannot be explained by a model that treats the abrasive particle as a sphere or a cone. On the other hand, this upper-bound model cannot explain the subsurface plastic deformation and the energy dissipation that occurs during abrasive wear.

7.3.2 Slip-Line Field Solution

The slip-line field used for the analysis of the chip formation and subsurface deformation process is shown in Fig. 7.33. It shows an abrasive particle moving relative to the softer surface with velocity U. The slip-line field is developed assuming that the plane strain condition approximates the abrasive wear situation reasonably well. This slip line was originally developed by Abebe and Appl (1981) to investigate the cutting of metals with negative-rake-angle tools. It should be noted that as the negative rake angle, γ, becomes larger, the chip formation process ceases and only the subsurface deformation takes place. To allow for the chip formation process,

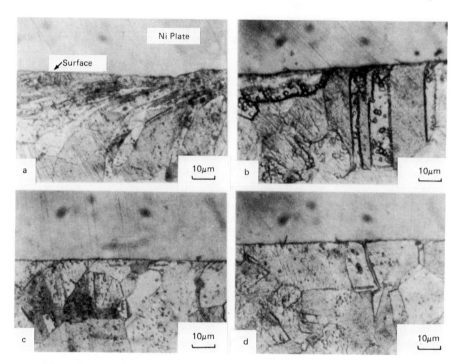

Figure 7.17 Optical micrographs of the OFHC copper subsurface for different grits: (a) 60; (b) 180; (c) 320; (d) 600. Experimental conditions are same as in Fig. 7.13. (From Sin, H.-C.; Saka, N.; and Suh, N. P., "Abrasive Wear Mechanisms and the Grit Size Effect," *Wear*, Vol. 55, 1979, pp. 163–190.)

a dead metal zone, BJC, is assumed to form which sticks to and acts as part of the abrasive particle. Along the boundaries, AB, BJ, and JC, the material being plowed flows over the abrasive particle. The interfacial stress along these boundaries is controlled by the local shear stress existing along each of the boundary segments (i.e., the angles η_1, η_2, and η_3 are dependent on the interfacial friction at these surfaces). The normal stress along the boundary AI is zero (i.e., AI is the stress-free boundary in the chip). Beyond AI the chip does not undergo any further plastic deformation. In the centered fan fields IBK, IGF, and JED, the shear angle and the hydrostatic pressure change from a radial line to radial line, whereas the stresses in the fields AIB, HIG, $IJEF$, and JDC are uniform in each region. The three angles η_1, η_2, and η_3 are directly related to the interfacial friction conditions through the relations

$$\eta_i = \frac{1}{2} \cos^{-1} \frac{\tau_i}{k} \qquad i = 1, 2, 3 \qquad (7.2)$$

where τ_i is the interfacial shear stress.

Sec. 7.3 Abrasive Friction and Wear Mechanisms

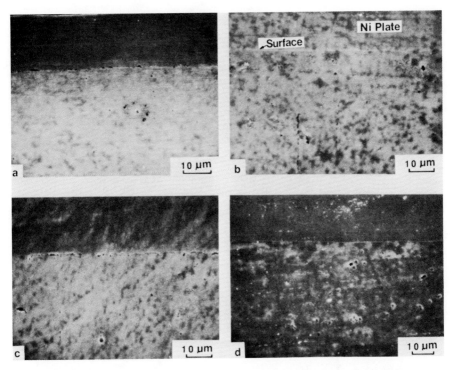

Figure 7.18 Scanning electron micrographs of the AISI 1095 steel subsurface for different grits: (a) 120; (b) 600; (c) 3/0; (d) 4/0. Experimental conditions are same as in Fig. 7.13. (From Sin 1978.)

The horizontal (i.e., tangential) component of the plowing force, F, is given as (see Appendix 7.A for details)

$$\frac{F}{k(JC)} = \frac{\sin \beta_1}{\sin \beta_2} \left[\frac{\sqrt{2} \cos (\eta_2 + \theta/2) \sin f_1}{\sin \theta} + S_2 \cos (\beta_2 - \gamma) \right.$$

$$\left. + \theta \frac{\sin (\beta - 2\gamma)}{\sin (\theta/2)} + C_2 \sin (\beta_2 - \gamma) \right] \quad (7.3)$$

$$- S_3 \cos \beta_3 + \sin \beta_4$$

where $\beta = \eta_1 - \eta_2 + \theta + \gamma$

$\Delta + \phi = \dfrac{\pi}{2} + \eta_2 + \eta_3$

$\beta_1 = \beta + \Delta - \gamma$
$\beta_2 = \beta - \gamma - \theta/2$

$f_1 = \dfrac{\pi}{4} + \eta_1 - \gamma$

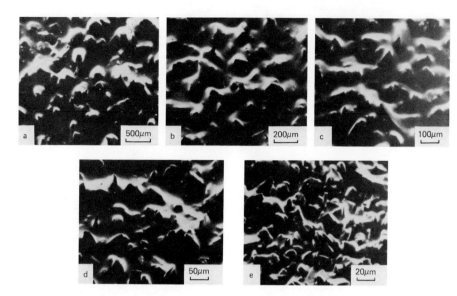

Figure 7.19 Scanning electron micrographs of the abrasive surface: (a) 60; (b) 120; (c) 180; (d) 320; (e) 600. (From Sin, H.-C.; Saka, N.; and Suh, N. P., "Abrasive Wear Mechanisms and the Grit Size Effect," *Wear*, Vol. 55, 1979, pp. 163–190.)

$$\beta_3 = \beta + \Delta - 2\gamma$$
$$\beta_4 = \beta + \Delta - 2\gamma - 2\eta_3$$
$$S_2 = 1 + \sin 2\eta_2$$
$$C_2 = \cos 2\eta_2 - 2$$
$$S_3 = 1 + 2\theta + 2\phi$$

Similarly, the vertical component of the plowing force, L, is given by

$$\frac{L}{k(JC)} = \frac{\sin \beta_1}{\sin \beta_2}\left[\frac{\sqrt{2}\cos(\eta_2 + \theta/2)\cos f_1}{\sin \theta} - S_2 \sin(\beta_2 - \gamma)\right.$$
$$\left. + \theta\frac{\cos(\beta - 2\gamma)}{\sin(\theta/2)} + C_2 \cos(\beta_2 - \gamma)\right] \quad (7.4)$$
$$+ S_3 \sin \beta_3 + \cos \beta_4$$

The coefficient of friction, $\mu = F/L$, due to plowing can be expressed as

$$\mu = \frac{\tilde{S}_1\left[\tilde{C}_1 \sin f_1 + S_2 \cos(\beta_2 - \gamma) + \theta\frac{\sin(\beta - 2\gamma)}{\sin \theta/2} + C_2 \sin(\beta_2 - \gamma)\right] - S_3 \cos \beta_3 + \sin \beta_4}{\tilde{S}_1\left[\tilde{C}_1 \cos f_1 - S_2 \sin(\beta_2 - \gamma) + \frac{\theta \cos(\beta - 2\gamma)}{\sin \theta/2} + C_2 \cos(\beta_2 - \gamma)\right] + S_3 \sin \beta_3 + \cos \beta_4}$$

$$(7.5)$$

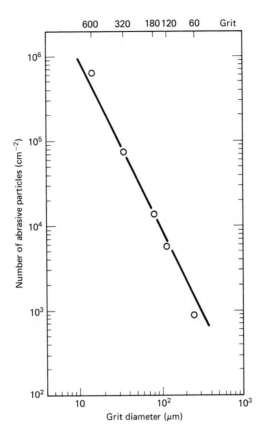

Figure 7.20 Number of abrasive particles per unit area of the abrasive surface as a function of the grit diameter. (From Sin, H.-C.; Saka, N.; and Suh, N. P., "Abrasive Wear Mechanisms and the Grit Size Effect," *Wear,* Vol. 55, 1979, pp. 163–190.)

where

$$\tilde{S}_1 = \frac{\sin(\eta_1 - \eta_2 + \Delta + \theta)}{\sin(\eta_1 - \eta_2 + \theta/2)}$$

$$\tilde{C}_1 = \frac{\sqrt{2}\cos(\eta_2 + \theta/2)}{\sin\theta}$$

The geometric condition at point J requires that

$$\Delta = \frac{\pi}{2} + \eta_2 + \eta_3 - \phi \tag{7.6}$$

In order to form HI, the angle θ must satisfy

$$\gamma - \eta_1 < \theta < \frac{1}{2}\left(\frac{\pi}{4} + \gamma - \eta_1\right) \tag{7.7}$$

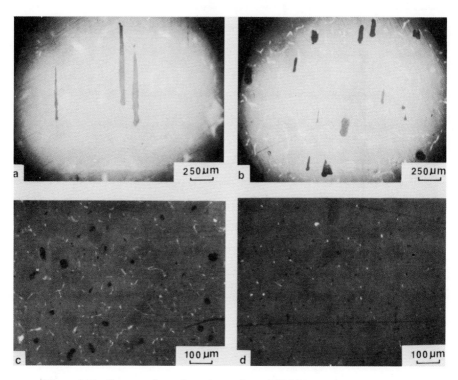

Figure 7.21 Grooves formed on the surface of PMMA specimen: (a) 60 grit, 4.9 N; (b) 120 grit, 19.6 n; (c) 320 grit, 39.2 N; (d) 600 grit, 19.6 N. (From Sin, 1978.)

Equation (7.7) indicates that there can be an infinite number of solutions within the limits set by the equation. However, the friction coefficient can be bound between a maximum and a minimum value. At a given value of θ, for the velocity to be the same along the β line HGFEDC, θ must satisfy (see the hodograph given in Appendix 7.A)

$$\phi = \eta_3 - \gamma + \eta_1 + \theta - \sin^{-1}\left[\sqrt{2}\sin\left(\frac{\pi}{4} - \eta_1 - 2\theta + \gamma\right)\sin\eta_3\right] \quad (7.8)$$

For the "dead zone" BJC to form, θ must also satisfy

$$\phi \geq \eta_1 + \eta_3 + \theta - \frac{\pi}{2} \quad (7.9)$$

These geometric relationships, Eqs. (7.6) through (7.9), must be satisfied in addition to the equilibrium conditions, Eqs. (7.3) and (7.4), for the slip-line field shown in Fig. 7.33 to be valid.

Sec. 7.3   Abrasive Friction and Wear Mechanisms

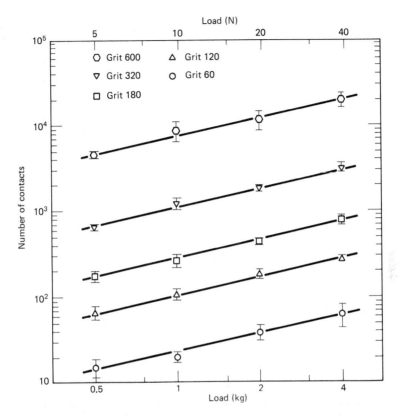

Figure 7.22 Number of contacting abrasive particles versus applied normal loads for different grits. PMMA specimen of 6.35 mm diameter was used. (From Sin, H.-C.; Saka, N.; and Suh, N. P., "Abrasive Wear Mechanisms and the Grit Size Effect," *Wear*, Vol. 55, 1979, pp. 163–190.)

Equations (7.3) through (7.5) indicate that the frictional force depends on the negative rake angle, γ, and the frictional conditions at the abrasive grain–chip interface, η_1, η_2, and η_3. Because of the geometric similarity, the coefficient of friction associated with plowing by the sharp abrasive with straight surface should be independent of the depth of penetration. The maximum normal load and the resultant friction force that hard abrasive particle may support are determined by the yield strength of the softer surface being plowed as well as by the depth of penetration. As the normal load on the abrasives at sliding interface increases, the load will be distributed over a larger number of asperities as each asperity penetrates deeper into the surface (see Fig. 7.22).

The maximum and minimum friction coefficient versus the asperity angle is plotted in Figs. 7.34a and 7.34b. The solutions are for the case of the same friction condition prevailing at the abrasive–chip interface (i.e., $\eta_1 = \eta_2 = \eta_3$). In general, the coefficient of friction decreases with increase

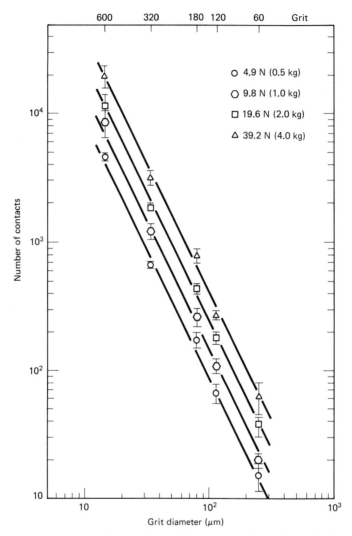

Figure 7.23 Number of contacting particles as a function of the grit diameter for different normal loads. PMMA specimen of 6.35 mm diameter was used. (From Sin, H.-C.; Saka, N.; and Suh, N. P., "Abrasive Wear Mechanisms and the Grit Size Effect," *Wear,* Vol. 55, 1979, pp. 163–190.)

in the asperity angle. The figures show the friction coefficients for several different interfacial conditions, τ_i/k. When the asperity angles are less than 70°, the friction coefficient is always greater than 0.3 and reaches a peak value between 0.83 and 1. As the asperity angle approaches 65 to 70°, all the friction coefficient curves predict the same coefficient of friction of about 0.25 to 0.3. When the asperity angle is greater than 70°, the friction coefficient is less than 0.25. The coefficient of friction reaches very low

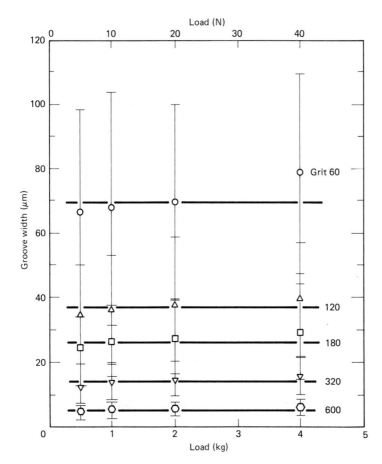

Figure 7.24 Groove width formed on PMMA specimen surface as a function of normal load for different grits. (From Sin, H.-C.; Saka, N.; and Suh, N. P., "Abrasive Wear Mechanisms and the Grit Size Effect," *Wear,* Vol. 55, 1979, pp. 163–190.)

values when the asperity angle is very close to 90°. When the asperity angle is very large, no chips are formed and the wear mechanism is no longer by the chip generation process. The transition from the abrasive wear to delamination wear occurs when subsurface deformation, crack nucleation, and crack propagation processes can generate wear particles faster than the rate of chip generation.

In Chapter 3 the coefficient of friction due to plowing by spherical particles was plotted as a function of $w/2r$ (see Fig. 3.12). Similarly, Fig. 7.32 shows that the minimum friction coefficient depends on the depth of penetration when the abrasive particle is shaped as a cone with a spherical tip. Similar results are observed for the maximum value of friction. When the asperity tip has a finite radius of curvature, the friction coefficient is a

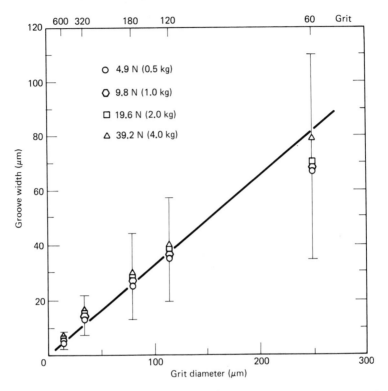

Figure 7.25 Groove width formed on PMMA specimen versus abrasive grit diameter. (From Sin, H.-C.; Saka, N.; and Suh, N. P., "Abrasive Wear Mechanisms and the Grit Size Effect," *Wear,* Vol. 55, 1979, pp. 163–190.)

function of penetration, whereas when the asperity is sharp and has a straight slope, the friction is independent of the penetration depth, as shown by the analysis presented in this section. However, the solution presented in this section can be compared with the solution presented in Chapter 3 by approximating the shape of the sharp asperity with a spherical shape as shown in Fig. 7.35. The asperity is approximated with a sphere of radius r, which is a reasonable approximation when the asperity angle γ is large. The results shown in Fig. 7.34 are replotted in Fig. 7.36 using this approximation. The magnitude of the coefficient of friction increases from very low values of 0.001 to 0.83 as $w/2r$ increases from 0 to 1. These results are similar to the results obtained in Chapter 3 with a more approximate model.

The solution obtained in using the slip-line field shown in Fig. 7.33 is correct only when the assumed slip-line field is valid [i.e., as long as the geometric relationships given by Eqs. (7.6) through (7.9) are satisfied]. The dashed curve in Fig. 7.34 indicates the limit of the validity of the assumed slip-line field. Outside the dashed line, different slip-line fields must be

Sec. 7.3 Abrasive Friction and Wear Mechanisms

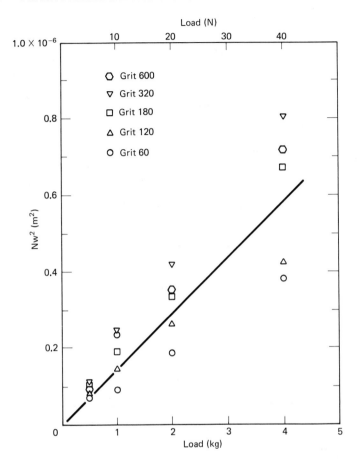

Figure 7.26 Number of contacting particles times the square of the average groove width as a function of the applied normal load (Nw^2 is proportional to the real contact area). (From Sin, H.-C.; Saka, N.; and Suh, N. P., "Abrasive Wear Mechanisms and the Grit Size Effect," *Wear*, Vol. 55, 1979, pp. 163–190.)

generated. However, the solution presented in Fig. 7.34 covers many typical sliding conditions.

7.3.3 Wear Due to Abrasive Actions

A large fraction of the external work done is dissipated to deform the subsurface plastically, as indicated by the slip-line field shown in Fig. 7.33. This increases as the asperity angle increases. The effective asperity angle increases with the increase in the grit diameter because the relative as well as the absolute depth of the penetration decreases. This is the reason the abrasive wear rate is a function of the grit diameter, which was shown in Fig. 7.10. That is, the increase in the relative bluntness of the indenting

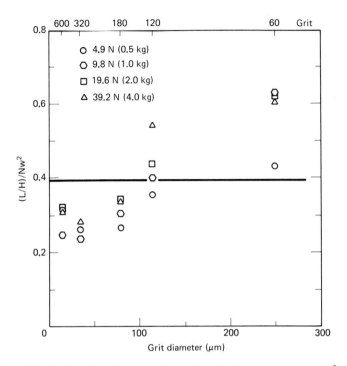

Figure 7.27 Ratio of the theoretically calculated real contact area to Nw^2 as a function of the abrasive grit diameter. The horizontal line shows the theoretical value. (From Sin 1978.)

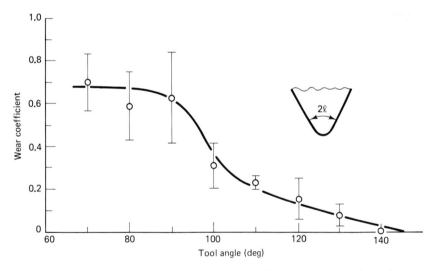

Figure 7.28 Removal coefficient as a function of conical tool angle in cutting tests of AISI 1095 steel. The applied load was 4.4 N. (From Sin, H.-C.; Saka, N.; and Suh, N. P., "Abrasive Wear Mechanisms and the Grit Size Effect," *Wear,* Vol. 55, 1979, pp. 163–190.)

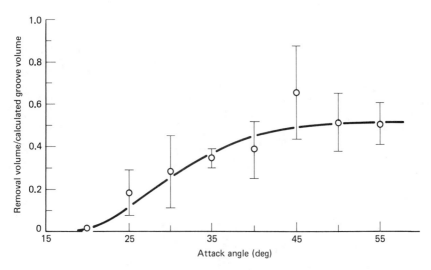

Figure 7.29 Ratio of the volume removed to the calculated groove volume as a function of attack angle. Experimental conditions are same as in Fig. 7.28. (From Sin, H.-C.; Saka, N.; and Suh, N. P., "Abrasive Wear Mechanisms and the Grit Size Effect," *Wear,* Vol. 55, 1979, pp. 163–190.)

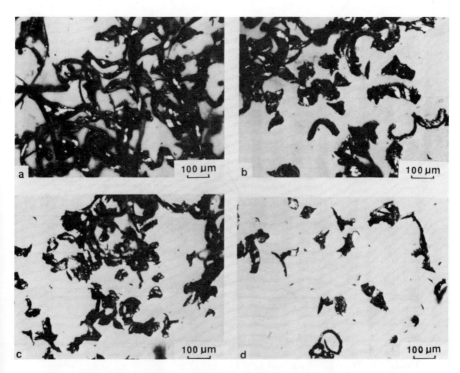

Figure 7.30 Micrographs of AISI 1095 steel chips: (a) 70°, (b) 90°, (c) 110°, (d) 130° conical tool. (From Sin, H.-C.; Saka, N.; and Suh, N. P., "Abrasive Wear Mechanisms and the Grit Size Effect," *Wear,* Vol. 55, 1979, pp. 163–190.)

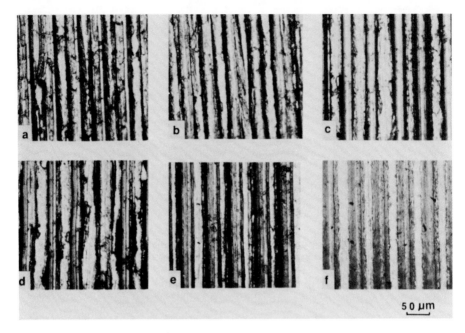

Figure 7.31 Grooves formed on AISI 1095 steel specimens: (a) 70°, (b) 90°, (c) 100°, (d) 110°, (e) 120°, (f) 140° conical tool. (From Sin, 1978.)

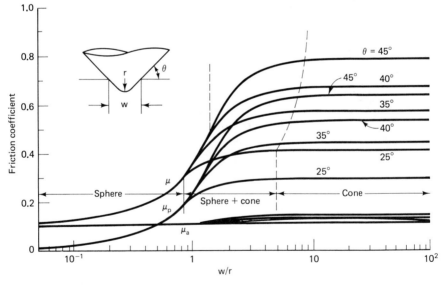

Figure 7.32 Friction coefficient as a function of the ratio of the groove width to the tip radius of conical particle for different cone angles. (From Sin, H.-C.; Saka, N.; and Suh, N. P., "Abrasive Wear Mechanisms and the Grit Size Effect," *Wear,* Vol. 55, 1979, pp. 163–190.)

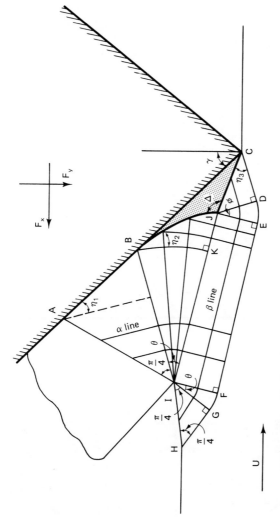

Figure 7.33 Slip-line field for plowing by an asperity with a negative rake angle γ. Note that γ is equal to $(\pi/2 - \theta)$ of Fig. 7.32. (From Abebe and Appl, 1981, and Komvopoulos et al., 1985.)

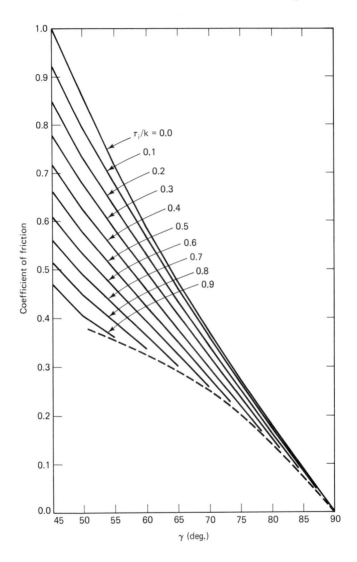

Figure 7.34a Maximum values of the coefficient of friction versus asperity angle. (From Komvopoulos et al., 1985.)

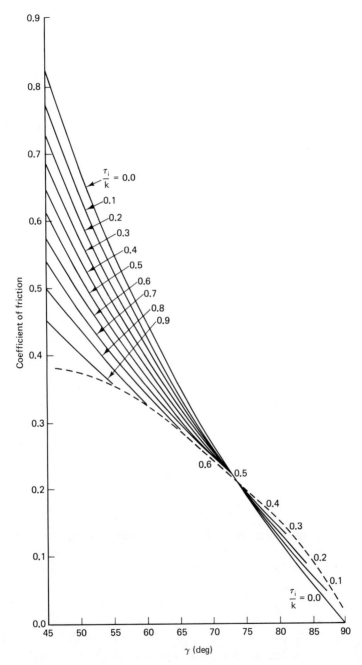

Figure 7.34b Minimum values of the coefficients of friction versus asperity angle. (From Komvopoulos et al., 1985.)

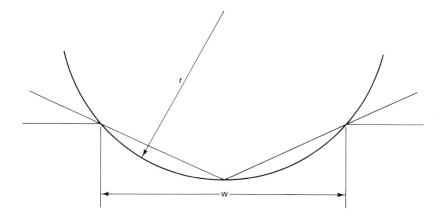

Figure 7.35 Approximation of a sharp asperity by a spherical particle.

part of the abrasive particle with decreased diameter is the major cause for the dependence of friction and wear coefficients on the grit size.

The wear coefficient was defined in Chapter 1 as

$$K = \frac{3VH}{LS} \tag{1.13}$$

where H is the hardness, V the volume worn, L the normal load, and S the distance slid. For the wear geometry shown in Fig. 7.33, the ratio V/LS may be written as

$$\frac{V}{LS} = \frac{U_{AB}(AI)\sin[(\pi/4) + \eta_1]}{UL} \tag{7.10}$$

where L is given by Eq. (7.4). From the hodograph the relationship between U_{AB} and U can be found from trigonometric relations as

$$U_{AB} = \frac{\sin(\eta_1 + \theta - \gamma)\cos(\eta_2 - \theta/2)}{\cos \eta_1 \cos(\eta_2 + \theta/2)} U \tag{7.11}$$

The angle θ is obtained from the frictional force analysis given in the preceding section and corresponds to the case of the minimum tangential force. Substituting Eq. (7.10) into Eq. (1.13), the wear coefficient may be expressed as

$$K = \frac{3HU_{AB}(AI)\sin[(\pi/4) + \eta_1]}{UL} \tag{7.12}$$

Figures 7.37a and 7.37b are plots of the theoretical maximum and minimum wear coefficients for various interfacial shear strengths and the rake angle of the abrasive particles. The rake angle covers fairly sharp to dull abrasives. When the included angle of the abrasive is greater than 150°, the wear coefficient is relatively small at all values of τ_i/k and rapidly decreases with the increase in the rake angle. When the included angle is

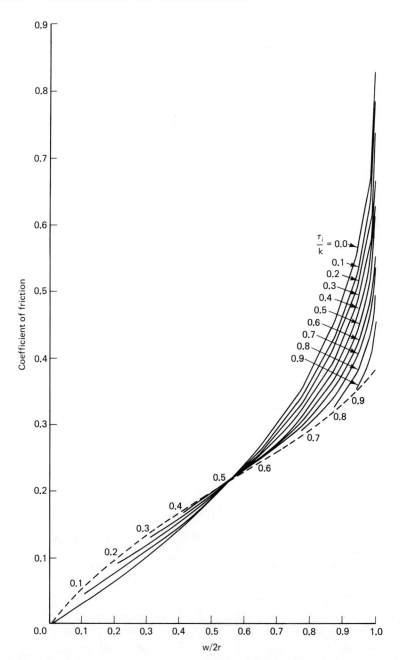

Figure 7.36 Minimum values of the coefficient of friction versus depth of penetration to wear particle diameter ratio. (From Komvopoulos et al., 1985.)

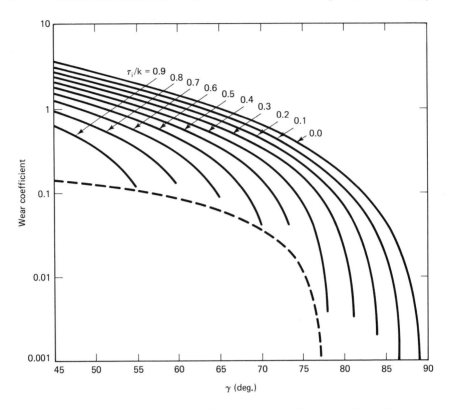

Figure 7.37a Maximum wear coefficient versus asperity angle. (From Komvopoulos et al., 1986.)

less than 150°, the wear coefficient can vary between about 0.1 to a value greater than unity, depending on the angle and the shear strength τ_i/k. At 90°, the wear coefficient ranges from 0.1 to 3.7. At this angle little subsurface plastic deformation occurs, whereas at large included angles subsurface plastic deformation occurs. It is probable that at large included angles the abrasive action will result in a wear rate that is nearly the same as the wear rate due to the delamination wear process.

The results obtained using the slip-line field indicate that the simple cutting model for abrasive wear discussed in Chapter 1 may be a reasonable model for sharp abrasives, whereas the model is inadequate for abrasives with large included angles. The results also illustrate the importance of the interfacial stress, which may provide an explanation as to why the experimental data vary from test to test. They also provide an important insight to abrasive machining: Use sharp abrasives and lubricants to maximize the material removal rate.

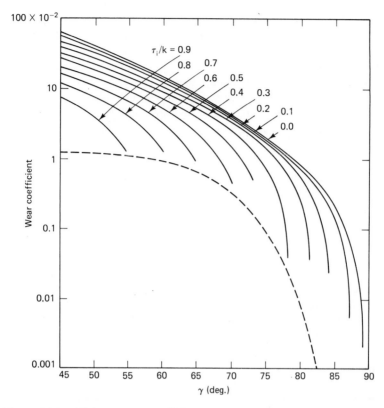

Figure 7.37b Minimum wear coefficient versus asperity angle. (From Komvopoulos et al., 1986.)

7.4 EROSIVE WEAR BY SOLID PARTICLE IMPINGEMENT

When a solid surface is gradually worn away by the action of gases and particles, it is called *erosion*. Erosion of materials can occur under four different conditions: (1) impingement of solid particles against a solid surface, (2) impingement of liquid droplets against a solid surface, (3) flow of hot gases over a solid surface, and (4) cavitation at a solid surface in liquid media. Except the erosion due to hot gases, all other erosion mechanisms depend a great deal on the mechanical behavior of solids. The erosion that is of greatest interest to tribologists is that due to the solid particle impingement, the subject covered in this section.

Erosion due to solid particle impingement occurs in many applications. The erosion of gas and steam turbine blades, infrared windows, pipelines (especially around elbows), polymer processing machinery that handles rapidly flowing resin solid particle slurries, and synthetic fuels processing equipment are but a few of the undesirable consequences of erosion by solid particle impingement. Erosion mechanisms are also usefully employed in such applications as abrasive machining and grit blasting. It is well established that in these applications the erosive behavior is a sensitive function of the ductility of the surface, the liquid media that transport the solid particles, and the impingement angles. These phenomenological aspects of erosive wear have been known for many years through extensive study (see the monograph by Wahl and Hartstein, 1946). Since the late 1950s, a good number of papers have been published dealing with the mechanistic aspect of erosion due to solid particle impingement in addition to the phenomenological aspects.

The paper most often cited as being the pioneering work on the mechanistic aspect of erosion is due to Finnie (1958). He proposed a cutting model to describe the erosion of ductile materials which has formed the basis for much of the recent thinking on the topic. More recent studies have shown, however, that the cutting model is applicable only when the impingement angle is *very* shallow (i.e., nearly tangent to the solid surface) and that in most cases the erosive wear is caused by deformation-induced subsurface damage. The basic mechanisms of erosion are very similar to those responsible for delamination wear (Suh, 1977; Jahanmir, 1980). However, the erosion rate cannot yet be predicted from the first principles, although several models provide a scheme for correlating experimental results. Some of these models will be presented following a review of the experimental data.

7.4.1 Phenomenological Aspects of Erosive Wear: Ductile Materials

The experimental data reported (Finnie et al., 1967; Smeltzer, Gulden and Compton, 1970; Sheldon, 1977; Tilly and Sage, 1970; Sheldon, 1970; Sheldon and Finnie, 1966a) will be summarized here. Typical erosion data for ductile and brittle solids are shown in Fig. 7.38, which have been obtained using a variety of different experimental arrangements shown in Fig. 7.39. The ductile solids have very low wear rates when the impingement angle is very shallow, which reaches a peak value at an impingement angle of about 20°. It then decreases to an intermediate value as the impingement angle increases. The erosion behavior of brittle solids is different from that of ductile solids. The erosion rate is small at low impingement and increases to the maximum rate when the impingement angle is perpendicular to the surface. As the particle size decreases, even nominally, brittle solids have a maximum erosion rate at angles smaller than 90°.

Erosion has been found to be proportional to a power of the impingement velocity. The reported numerical values of the exponent range from about 1.7 to 2.8 for ductile metals (see Fig. 7.40) and 1.4 to 5.1 for brittle solids.

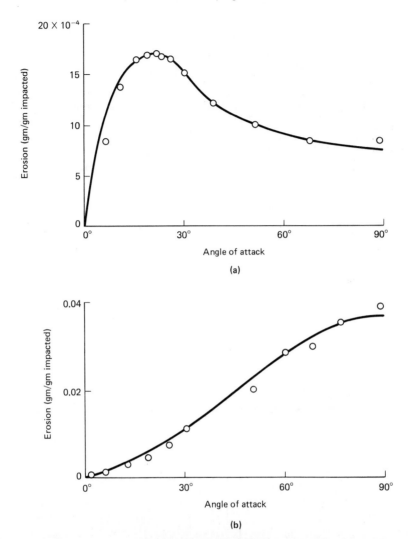

Figure 7.38 Erosion versus angle of attack: (a) aluminum eroded by 210-μm alumina particles at 494 ft/sec; (b) glass plates eroded by 210-μm alumina particles at 354 ft/sec. (From Neilson, J. H., and Gilchrist, A., "Erosion by Stream of Solid Particles," *Wear*, Vol. 11, 1968, pp. 111–122.)

Rebound velocities of 5 to 20% have been observed, indicating that the transfer of kinetic energy is 96 to 99.8%.

A general increase in erosive wear has been observed with increasing particle size. It has been reported that erosion is a linear function of particle diameter and that it also is proportional to a power of particle diameter. There is a general agreement, however, that erosion increases with increasing particle size up to a critical value but remains insensitive to further increase

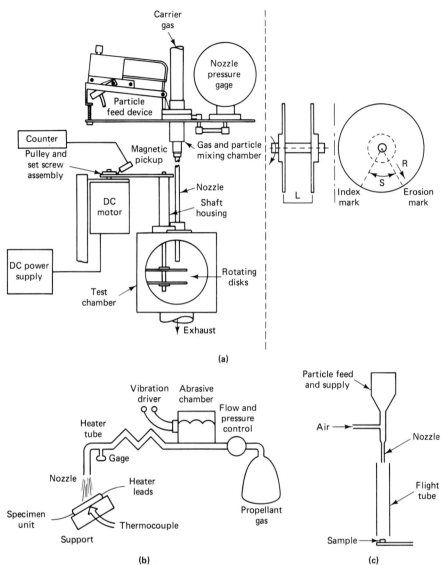

Figure 7.39 Variety of experimental arrangements of erosion test devices. Schematic diagram of erosion tester with rotary particle-velocity-testing device attached. L, Disk separation; S, arc length between erosion marks; R, radius-to-arc distance; v, angular velocity (rev^{-1}); $v_p = 2\pi RvL/S$, particle velocity. [(a), From Levy, A. V., "The Solid Particle Erosion Behavior of Steel as a Function of Microstructure," *Wear,* Vol. 68, 1981, pp. 269–287. (b) From Young, J. P., and Ruff, A. W., "Particle Erosion Measurements on Metals," *Journal of Engineering Materials and Technology,* Vol. 99 H, 1977, pp. 121–125. Reprinted by permission of the American Society of Mechanical Engineers. (c) From Ives, L. K., and Ruff, A. W., "Electron Microscopy Study of Erosion Damage in Copper," *Erosion: Prevention and Useful Applications,* ASTM STP 664 (W. F. Adler, Ed.), American Society for Testing and Materials, 1979, pp. 5–35. Reprinted, with permission, from ASTM, 1916 Race Street, Philadelphia, PA 19103. Copyright 1979, ASTM.

Sec. 7.4 Erosive Wear by Solid Particle Impingement 309

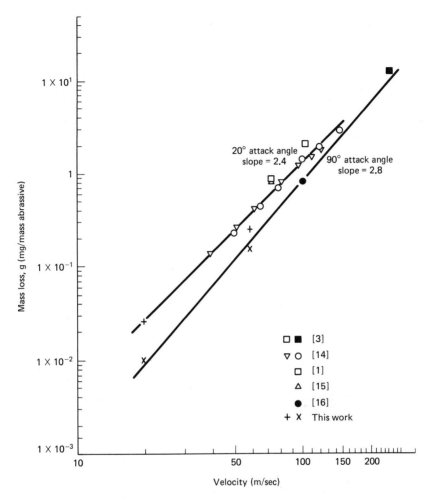

Figure 7.40 Collected erosion results for copper at attack angles of 20 and 90°. Straight lines represent a least-squares fit to the data. (From Ives, L. K., and Ruff, A. W., "Electron Microscopy Study of Erosion Damage in Copper," *Erosion: Prevention and Useful Applications,* ASTM STP 664 (W. F. Adler, Ed.), American Society for Testing and Materials, 1979, pp. 5–35. Reprinted, with permission, from ASTM, 1916 Race Street, Philadelphia, PA 19103. Copyright 1979, ASTM.)

in particle size. With sufficiently small particles, it has been found that normally brittle materials will erode in a typically ductile fashion. It should be noted that this dependence on the particle size is the same as that for abrasive wear. In general, hard particles are more erosive than soft particles.

Although there are irregularities in the early stages of erosion, it increases with increasing total mass impacted. The relationship in most cases is linear. However, the efficiency of the erosion process is quite low. One part of material eroded requires 250 to 1000 parts by weight of abrasive particles. Erosion rate per unit mass *increases* with decrease in particle flux.

Erosion is strongly temperature dependent. A reduction in erosion on the order of 50% in aluminum–titanium alloy and stainless steel has been obtained when the temperature of the solid was raised from room temperature to 700°F. One exception to the general temperature effect is that the erosion of very hard materials such as nickel tends to increase with increasing temperature, especially as the nominal impingement angle decreases.

Erosion tends to decrease with increasing toughness and hardness. It also increases with the melting temperature of the workpiece in single-phase metals. However, in two-phase metals it is insensitive to both heat treatment and cold working, and therefore the hardness of two-phase metals does not correlate with their erosion resistance.

7.4.2 Microstructural Observations of Erosive Wear: Ductile Metals

When particles initially impact a stress-free surface, the surface undergoes large plastic deformation creating indentation craters, plowing grooves, or cutting craters, as shown in Fig. 7.41 (Hutchings, 1979; Bellman and Levy, 1981; Suh, 1977). These craters are observed at all impingement angles. At very high impact velocities, the metal flows like molten liquid. Figure 7.42 shows the surfaces of 2024-T4 aluminum impacted by silicon carbide particles at 32, 156, and 310 m/sec. The impingement angle was 45° and the average particle size was 254 μm. These micrographs indicate

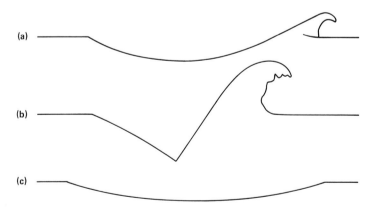

Figure 7.41 Sections through impact craters showing typical shapes (impact direction left to right): (a) indentation crater formed by a sphere; (b) crater formed by plowing with extrusion; (c) cutting crater. (From Hutchings, I. M., "Mechanisms of the Erosions of Metals by Solid Particles," *Erosion: Prevention and Useful Applications,* ASTM STP 664 (W. F. Adler, Ed.), American Society for Testing and Materials, 1979, pp. 59–76. Reprinted, with permission, from ASTM, 1916 Race Street, Philadelphia, PA 19103. Copyright 1979, ASTM.)

Sec. 7.4   Erosive Wear by Solid Particle Impingement   311

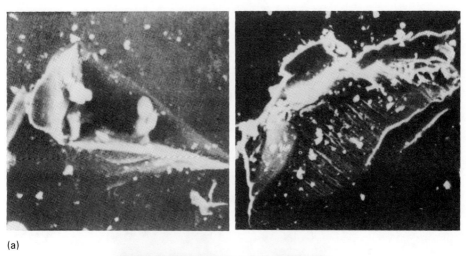

(a)

(c)

Figure 7.42 Surface of 2024-T4 aluminum impacted by silicon carbide particles at an impingement angle of 45°. Impact velocity: (a) 32 m/sec; (b) 156 m/sec; (c) 310 m/sec.

that very large indentations have occurred, creating craters, and in some cases extrusions. In the process of forming craters, platelets of metal that are locally attached to the crater rim are extruded (see Fig. 7.43). Figure 7.44 shows aluminum specimens impacted at 280 m/sec at room temperature and at 150 m/sec at 200°F. The specimens impacted at the higher speed shows a sign of melting. The specimen impacted at 200°F shows typical mud cracks, indicating that the surface has undergone oxidation at a high temperature or has melted.

Figures 7.45 and 7.46 show cross-sectional views of the subsurface cracks in copper and steel. It should be noted that the cracks tend to be

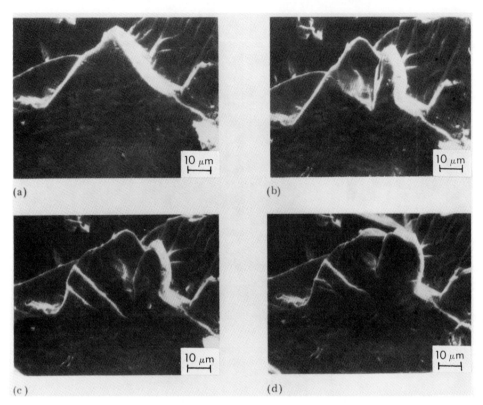

Figure 7.43 "Extruded" chip of the eroded material. Material 1100-0 aluminum. (a) 0.1 g, (b) 0.2 g, (c) 0.3 g, (d) 0.4 g of impacting particles. (From Bellman, R., Jr., and Levy, A., "Erosion Mechanism in Ductile Metals," *Wear*, Vol. 70, 1981, pp. 1–27.)

perpendicular to the direction of impact and parallel to the direction of material flow, indicating that the cracks are shear cracks similar to the cracks observed in delamination wear. It should also be noted that subsurface cracks exist at a finite distance away from the surface. The subsurface cracks extend deeper into the specimen when it is impacted by particles traveling perpendicular to the surface. The mechanics of the surface deformation is such that we have a maximum deformation at the surface. Since crack nucleation is not favored whenever there is a high hydrostatic pressure, cracks will not nucleate right underneath the impacted surface. Crack propagation is a function of crack tip sliding displacement when the surface is under the combined loading of compression and shear.

Just as in delamination wear the distribution and the concentration of hard particles in a soft matrix affect the erosion rate of two-phase metals. It can be appreciated that if there are a large number of second-phase particles in the matrix, there will be many nucleation sites for subsurface cracks and therefore the wear rate will increase, since the distance the

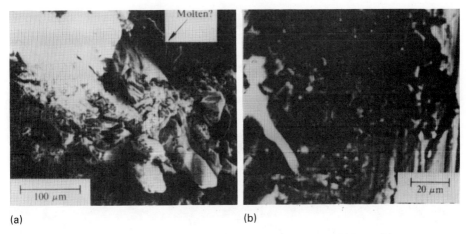

Figure 7.44 Surfaces of 2024 aluminum impacted by silicon carbide particles at an impingement angle of 90°: (a) temp. 17°F, velocity 280 m/sec; (b) temp. 200°F, velocity 150 m/sec.

cracks must travel to link up with neighboring cracks will be smaller with the decrease in the interparticle spacing. However, when the number of particles is so small that the ductile phase is large, very large plastic deformation of the matrix phase can occur, which increases the crack propagation rate per impact and thus the erosion rate. Therefore, the optimum concentration of hard particles corresponds to that which minimizes the overall crack propagation rate (i.e., the product of the number of crack nucleation centers and the crack propagation rate of each crack).

7.4.3 Phenomenological and Microscopic Aspects of Erosive Wear by Solid Particle Impingement: Brittle Elastic Solids

As stated earlier, the maximum erosion of brittle solids is observed at an impingement angle of 90°, but it is particle-size dependent. When Corning 7740 Pyrex glass was impacted with 240-grit Al_2O_3 powder (mean particle size $\simeq$ 30 μm), the maximum erosion occurred at 90°, whereas with 400-grit powder the corresponding impingement angle was 80° (Sargent et al., 1979). The erosion rate decreased with the time of erosion, indicating that the impingement angle changed from 90 to smaller angles with the wear of original glass surface. The erosion rate increased with the particle velocity to the nth power, where n varied from 2.2 to 2.7.

Gulden (1979) investigated the erosion behavior of hot-pressed silicon nitride (HP Si_3N_4), reaction-bonded silicon nitride (RB Si_3N_4), glass-bonded alumina (GB Al_2O_3), and hot-pressed magnesium fluoride (HP MgF_2). These specimens were eroded using natural quartz of 10 and 385 μm diameter at five different velocities. HP Si_3N_4 was also impacted with silicon carbide

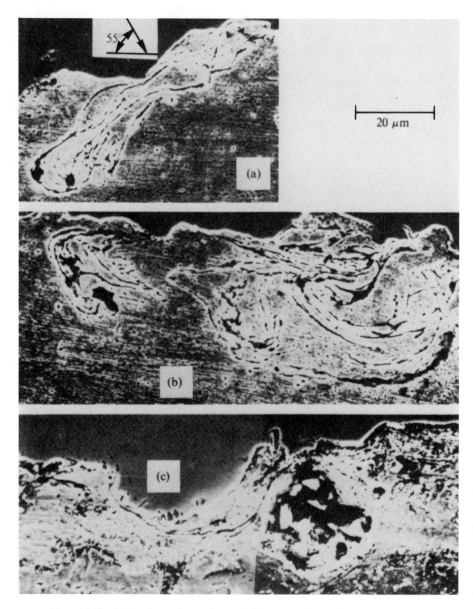

Figure 7.45 Subsurface mirocracks. (a) Copper (cold-rolled and annealed at 375°C for 12 hours); abrasive 254-μm SiC; velocity 105 m/sec. (b) The same as (a) except the direction of the impact was 90° to the surface. (c) Steel (cold-rolled AISI 1020 steel annealed at 70°C for 53½ hours); otherwise, the same as (b). (Reprinted with permission from Suh, N. P., "Microstructural Effects in Wear of Metals," in R. I. Jaffe and B. A. Wilcox, Eds., *Fundamental Aspects of Structural Alloy Design,* pp. 565–593. Copyright 1977 by Plenum Publishing Corporation, New York.)

Sec. 7.4   Erosive Wear by Solid Particle Impingement   315

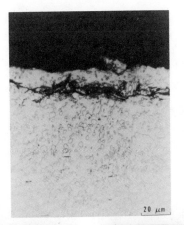

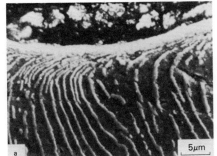

Figure 7.46 (top) Scanning electron micrograph of cross section of spheroidized 1075 steel after multiple-particle erosion with SiC particles 240 μm in diameter (V_p = 61 m/sec (200 ft/sec); α = 30°); (bottom) scanning electron micrographs of coarse pearlite 1075 steel after multiple erosion with SiC particles 240 μm in diameter (V_p = 61 m/sec; α = 30°). (From Levy, A. V., "The Solid Particle Erosion Behavior of Steel as a Function of Microstructure," *Wear*, Vol. 68, 1981, pp. 269–287.)

(SiC) under the same particle size–velocity conditions. Both single-impact and multiple-impact tests were performed.

Impact damages done on HP Si_3N_4 specimen by a single particle of SiC and on MgF_2 by quartz revealed a highly deformed surface crater and radial and lateral cracks emanating from the contact area, which has undergone plastic deformation (see Fig. 7.47). The diameter of the material removed was as much as three times the measured contact diameter. The diameter of material removed from RB Si_3N_4 was essentially the same as the estimated contact diameter and there were extensive subsurface radial cracks. Erosion of HP Si_3N_4 impacted with quartz occurs by minor chipping, which was an order of magnitude less than the contact radius without any secondary cracking (see Fig. 7.47). Erosion of glass-bonded Al_2O_3 is a two-step process involving plastic deformation of the glass and chipping of Al_2O_3 grains. It

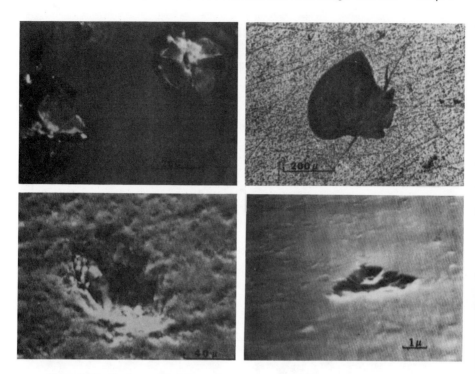

Figure 7.47 Single particle impacts for systems which exhibit consistent particle size–velocity functions with erosion. (a) Hot-pressed MgF_2 impacted with 273-μm quartz: velocity 188 m/sec; contact radius -27 μm; $E = R^4V^4$ (three impacts). (b) Hot-pressed Si_3N_4 impacted with 305-μm SiC; velocity 176 m/sec; contact radius -22 μm; $E = R^4V^4$. (c) Reaction-bonded Si_3N_4 impacted with 385-μm quartz: velocity 174 m/sec; contact radius -35 μm; $E = R^4V^4$ (higher velocities). (d) Hot-pressed Si_3N_4 impacted with 273-μm quartz: velocity 188 m/sec; contact radius -25 μm; $E = R^3V$. (From Gulden, M. E., "Solid Particle Erosion of High-Technology Ceramics (Si_3N_4, Glass Bonded Al_2O_3, and MgF_2)," *Erosion: Prevention and Useful Applications,* ASTM STP 664 (W. F. Adler, Ed.), American Society for Testing and Materials, 1979, pp. 101–122. Reprinted, with permission, from ASTM, 1916 Race Street, Philadelphia, PA 19103. Copyright 1979, ASTM.)

is interesting to note that ring cracks were not observed in any of these tests.

Gulden's multiple-erosion tests showed that the erosion rates were within a factor of 2 of the single-impingement test results. The erosion rates were functions of the particle size and velocity, as shown in Fig. 7.48. The relationships may be expressed as

$W \propto R^4V^4$ (for MgF_2 eroded by quartz, RB Si_3N_4 by quartz, and HP Si_3N_4 by SiC)

$W \propto R^3V$ (for HP Si_3N_4 eroded by quartz)

$W \propto R^3V^{1 \text{ to } 3}$ (for glass-bonded Al_2O_3 eroded by quartz)

Sec. 7.4 Erosive Wear by Solid Particle Impingement   317

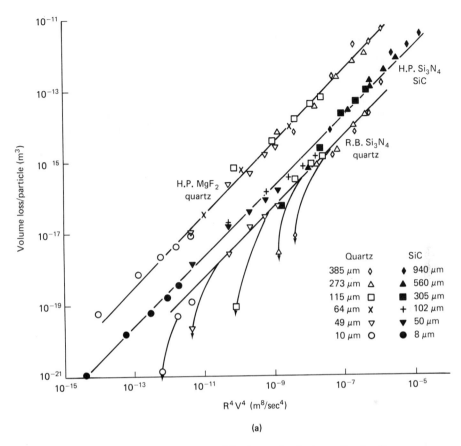

Figure 7.48 (a) Erosion versus particle size and velocity to the fourth power (velocity varied between 40 and 285 m/sec);

(continued)

It should be noted that MgF_2 and Si_3N_4 are single-phase materials, whereas GB Al_2O_3 were multiphase materials. It is also interesting to note that the erosion with quartz did not reduce the strengths of HP Si_3N_4 and GB Al_2O_3, whereas there was a major decrease in the strength of RB Si_3N_4. These results may indicate that the cracks are more easily arrested in HP Si_3N_4 and GB Al_2O_3, probably at the grain boundary.

Extensive erosion data at 90° impingement were obtained by Hansen (1979). Over 200 materials were tested using 27-μm alumina particles at an impingement velocity of 170 m/sec (558 ft/sec). The particle flow rate was 5 g/min and the test lasted 3 minutes. The test was done in high-purity nitrogen atmosphere. The test results are given in Figs. 7.49 through 7.52 in terms of the relative erosion factor (REF), which is defined as the material eroded in volume divided by that of Stellite 6B. Figure 7.49 shows the erosion rate of various metals. Nearly all metals and metal alloys except

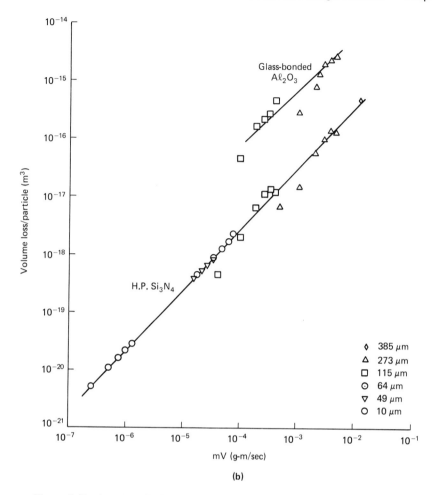

Figure 7.48 (*continued*) (b) erosion by quartz versus a measure of particle momentum (velocity varied between 40 and 285 m/sec). (From Hansen, J. S., "Relative Erosion Resistance of Several Materials," *Erosion: Prevention and Useful Applications* (W. F. Adler, Ed.), American Society for Testing and Materials, 1979, pp. 148–162. Reprinted, with permission, from ASTM, 1916 Race Street, Philadelphia, PA 19103. Copyright 1979, ASTM.)

tungsten and molybdenum have similar erosion rates at room temperature. Their erosion rates were higher at 700°C in most cases. The wear rates of these metals were typically much greater than the pore-free ceramics (see Fig. 7.50). Figure 7.51 shows the erosion rate of WC–Co cermets. It is clear that the erosion rate increases with the metal binder content, as shown in Fig. 7.52. Tables 7.4 through 7.7 list the relative erosion factors of many solid materials and coated materials. Abbreviations used are found in Table 7.8.

Sec. 7.4 Erosive Wear by Solid Particle Impingement

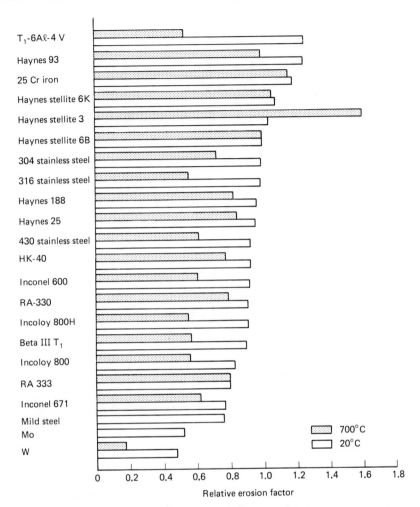

Figure 7.49 Commercially available metals (90° impingement). (From Hansen, J. S., "Relative Erosion Resistance of Several Materials," *Erosion: Prevention and Useful Applications* (W. F. Adler, Ed.), American Society for Testing and Materials, 1979, pp. 148–162. Reprinted, with permission, from ASTM, 1916 Race Street, Philadelphia, PA 19103. Copyright 1979, ASTM.)

7.4.4 Fundamental Mechanisms of the Erosive Wear by Solid Particle Impingement: Ductile Materials

The erosion mechanisms for ductile, elastoplastic solids are basically different from those responsible for the erosion of brittle elastic solids. Plastic deformation-induced damage causes erosive wear of the ductile solid, whereas the brittle solid wears due to crack propagation under tensile stress. The erosion mechanisms of these two classes of solids will be examined here separately, starting with ductile solids.

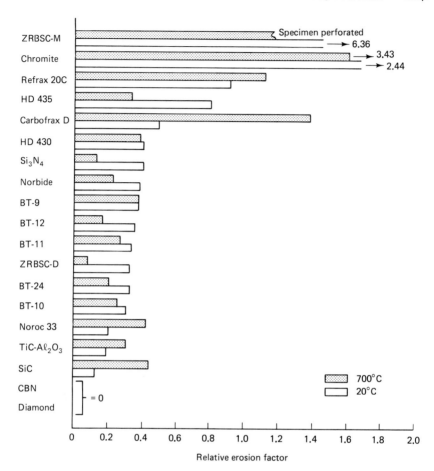

Figure 7.50 Ceramics (90° impingement). (From Hansen, J. S., "Relative Erosion Resistance of Several Materials," *Erosion: Prevention and Useful Applications* (W. F. Adler, Ed.), American Society for Testing and Materials, 1979, pp. 148–162. Reprinted, with permission, from ASTM, 1916 Race Street, Philadelphia, PA 19103. Copyright 1979, ASTM.)

Consider a solid particle impinging on a surface of an elastoplastic solid at high velocity, as shown in Fig. 7.53. Several different things can happen when the particle impacts the surface, depending on the material properties, velocity, impingement angle, the shape and the size of the particle, and the environment. Among the several possibilities are:

1. The solid particle may fragment upon impact when the acoustic impedance and the hardness of the surface are very large.
2. The solid particle may cut the surface, creating a chip, when the impingement angle is very small or if the particle is very sharp. The

Sec. 7.4 Erosive Wear by Solid Particle Impingement

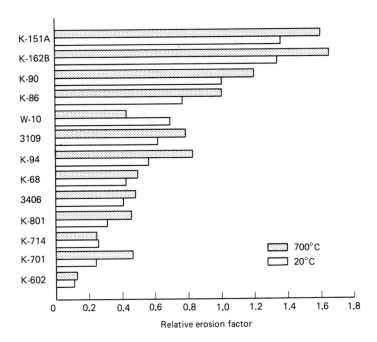

Figure 7.51 Kennametal cemented carbides (90° impingement). (From Hansen, J. S., "Relative Erosion Resistance of Several Materials," *Erosion: Prevention and Useful Applications* (W. F. Adler, Ed.), American Society for Testing and Materials, 1979, pp. 148–162. Reprinted, with permission, from ASTM, 1916 Race Street, Philadelphia, PA 19103. Copyright 1979, ASTM.)

chip formation mechanisms are very similar to the abrasive wear mechanisms considered in this chapter.

3. The solid particle may simply deform the surface and penetrate the material, creating and propagating subsurface cracks by mechanisms very similar to the delamination wear mechanisms. This is likely to occur at large impingement angles.

4. At an intermediate impingement angle, severe plastic deformation of the surface [discussed under (3)] and partial extrusion of the solid by the chip formation process [discussed under (2)] may simultaneously occur.

The fragmentation of the solid particle will not be considered here, since when it happens there may be only limited erosion damage. Erosion due to cutting will be considered first, followed by discussion of deformation-induced erosion.

The case of small impingement angle and sharp abrasive particle. Finnie (1958) was the first to analyze the erosion due to cutting, which may be valid when the erosion is due to sharp particles impinging

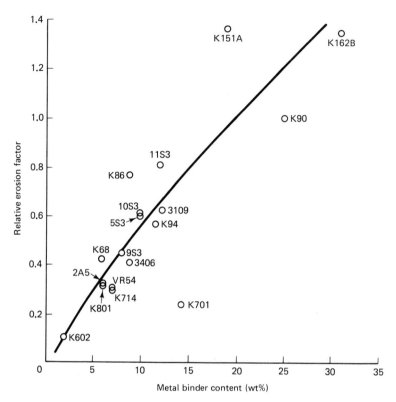

Figure 7.52 Relative erosion resistance of cemented carbides as a function of metal binder content (room temperature, 90° impingement). (From Hansen, J. S., "Relative Erosion Resistance of Several Materials," *Erosion: Prevention and Useful Applications* (W. F. Adler, Ed.), American Society for Testing and Materials, 1979, pp. 148–162. Reprinted, with permission, from ASTM, 1916 Race Street, Philadelphia, PA 19103. Copyright 1979, ASTM.)

on the surface at very small angles (i.e., small negative or positive rake angle). Finnie's idealized particle and its coordinate system are shown in Fig. 7.53. The particle is impinging at the surface with velocity U at an impingement angle of α. To simplify the analysis, he assumed that the ratio of vertical to horizontal force components, J, is constant, which is valid only when the particle shape is pyramidal, as shown in the figure, and when the particle does not rotate (see the analysis in Section 7.C). The length l over which the abrasive grain contacts the surface may be considerably greater than the depth of cut y_t. The ratio l/y_t will be denoted by ψ. If b is the width of the cut, the equation of motion for the particle can be written as

$$m\ddot{x} + p\psi by = 0$$
$$m\ddot{y} + pJ\psi by = 0 \qquad (7.13)$$
$$I\ddot{\phi} + p\psi b\phi = 0$$

Sec. 7.4 Erosive Wear by Solid Particle Impingement

TABLE 7.4 Room-Temperature Erosion Test Results[a,b]

Material	Manufacturing Method	Composition	REF[c]
99P	ps	99Al$_2$O$_3$ (Krohn)	12.49
ZRBSC-M	hp	ZrB$_2$-SiC-graphite (N)	6.36
Chromite	ps	(UCAR)	2.44
K151A	ps	19Ni binder (K)	1.37
K162B	ps	25Ni + 6Mo binder (K)	1.35
98D	ps	98 Al$_2$O$_3$ (Krohn)	1.29
Ti-6A1-4V	w	—	1.26
Haynes 93	c	17Cr-16Mo-6.3Co-3C-bal Fe (Stellite)	1.25
Graphite-air	w	1.4C-1.9Mn-1.25Si-1.9Ni-1.5Mo-bal Fe (TRB)	1.19
25Cr iron	c	25Cr-2Ni-2Mn-0.5Si-3.5C-bal Fe (OGC)	1.19
Stellite 6K	w	30Cr-4.5W-1.5Mo-1.7C-bal Co (Stellite)	1.08
Stellite 3	c	31Cr-12.5W-2.4C-bal Co	1.04
K90	ps	25 binder (K)	1.01
Stellite 6B	w	30Cr-4.5W-1.5Mo-1.2C-bal Co (Stellite)	1.00
304 SS	w	17Cr-9Ni-2Mn-1Si-bal Fe	1.00
316 SS	w	17Cr-12Ni-2Mn-1Si-2.5Mo-bal Fe	0.99
Haynes 188	w	22Cr-14.5W-22Ni-0.15C-balCo (Stellite)	0.97
Haynes 25	w	20Cr-15W-10Ni-1.5Mn-0.15C-bal Co (Stellite)	0.96
430 SS	w	17Cr-1Mn-1Si-0.1C-bal Fe	0.93
HK-40	c	26Cr-20Ni-0.4C-bal Fe	0.93
Inconel 600	w	76Ni-15.5Cr-8Fe (HA)	0.92
RA 330	w	19Cr-35Ni-1.5Mn-1.3Si-bal Fe (RA)	0.91
Refrax 20C	ps	SiC-Si$_3$N$_4$ bond (Carbor)	0.91
Incoloy 800H	w	32.5Ni-21Cr-0.07C-46Fe (HA)	0.91
Beta III Ti	w	11.5Mo-6Zr-4.5Sn-bal Ti	0.90
Incoloy 800	w	32.5Ni-46Fe-21Cr (HA)	0.83
HD 435	—	Recrystallized SiC(N)	0.80
RA 333	w	25Cr-1.5Mn-1.3Si-3Co-3Mo-3W-18Fe-bal Ni (RA)	0.80
K86	ps	8.8Co binder (K)	0.78
Inconel 671	w	50Ni-48Cr-0.4Ti (HA)	0.77
Lucalox	—	Densified Al$_2$O$_3$ (GE)	0.76
Mild steel	w	0.15C-bal Fe	0.76
W10	ps	90W-10(Ni, Cu, Fe) (K)	0.70
3109	ps	12.2 binder (K)	0.62
K94	ps	11.5 binder (K)	0.57

(continued)

TABLE 7.4 (continued)

Material	Manufacturing Method	Composition	REF[c]
Mo	w	—	0.52
Carbofrax D	ps	SiC ceramic bond (Carbor)	0.49
W	w	(GE)	0.48
K68	ps	5.8 binder (K)	0.43
3406	ps	7.8 binder (K)	0.42
HD 430	—	Recrystallized SiC (N)	0.40
Si_3N_4	hp	(N)	0.40
Norbide	hp	B_4C (N)	0.38
BT-9	ps	$2MgO-25TiB_2-3.5WC-bal\ Al_2O_3$ (OGC)	0.37
BT-12	ps	$1.5MgO-49TiB_2-3.5WC-bal\ Al_2O_3$ (OGC)	0.35
BT-11	ps	$1.7MgO-38TiB_2-3.5WC-bal\ Al_2O_3$ (OGC)	0.33
ZRBSC-D	hp	ZrB_2-SiC (N)	0.32
VR-54	ps	WC-7Co binder (F)	0.32
BT-24	ps	$2MgO-30TiB_2-3.5WC-bal\ Al_2O_3$ (OGC)	0.32
K801	ps	6Ni binder (K)	0.32
BT-10	ps	$2MgO-30TiB_2-3.5WC-bal\ Al_2O_3$ (OGC)	0.30
K714	ps	6Co + 1Cr binder (K)	0.26
K701	ps	10.2Co + 4Cr binder (K)	0.25
CA 306	ps	WC-6Co binder (Carmet)	0.23
Noroc 33	hp	Si_3N_4-SiC(N)	0.20
$TiC-Al_2O_3$	ps	(B and W)	0.19
895	ps	WC-6Co binder (Carb)	0.19
SiC	hp	(N)	0.12
K602	ps	<1.5 binder (K)	0.11
SiC	—	(GE)	0.05
CBN	—	(GE)	0
GE diamond	—	(GE)	0

[a] Experimental conditions: 90° impingement, 27-μm Al_2O_3 particles, 5-g/min particle flow, 170-m/sec particle velocity, 3-min test duration, N_2 atmosphere.

[b] Abbreviations are listed in Table 7.8.

[c] REF (relative erosion factor) = $\dfrac{\text{volume loss material}}{\text{volume loss Stellite 6B}}$.

Source: Hansen (1979).

m, I, and r are, respectively, the mass, mass moment of inertia, and the distance from the center to the edge of the particle, and p is the constant horizontal component of contact stress. Several assumptions were made in writing the equation. The moment applied by the vertical component of the contact stress is neglected and the penetration y_t is assumed to be equal to the vertical motion of the particle y. The origin of the x and y coordinates is taken to be at the center of gravity of the particle. The

TABLE 7.5 700°C Erosion Test Results[a,b]

Material	Manufacturing Method	Composition	REF[c]
ZRBSC–M	hp	ZrB_2–SiC graphite(N)	>5.00
99P	ps	$99Al_2O_3$ (Krohn)	>4.00
Chromite	ps	(UCAR)	3.43
K162B	ps	25Ni + 6Mo binder (K)	1.67
K151A	ps	19Ni binder (K)	1.62
Stellite 3	c	31Cr–12.5W–2.4C–bal Co (Stellite)	1.61
Carbofrax D	ps	SiC–ceramic bond (Carbor)	1.38
895	ps	WC–6Co binder (Carb)	1.32
K90	ps	25 binder (K)	1.21
25 Cr iron	c	25Cr–2Ni–2Mn–0.5Si–3.5C–bal Fe (OGC)	1.16
Refrax 20C	ps	$SiC–Si_3N_4$ bond (Carbor)	1.15
98D	ps	$98Al_2O_3$ (Krohn)	1.12
Stellite 6K	w	30Cr–4.5W–1.5Mo–1.7C–bal Co (Stellite)	1.06
K86	ps	8.8Co binder (K)	1.03
Stellite 6B	w	30Cr–4.5W–1.5Mo–1.2C–bal Co (Stellite)	1.00
Haynes 93	c	17Cr–16Mo–6.3Co–3C–bal Fe (Stellite)	1.00
Haynes 25	w	20Cr–15W–10Ni–1.5Mn–0.15C–bal Co (Stellite)	0.85
K94	ps	11.5 binder (K)	0.84
Haynes 188	w	22Cr–14.5W–22Ni–0.15C–bal Co (Stellite)	0.83
RA 333	w	25Cr–1.5Mn–1.3Si–3Co–3Mo–3W–18Fe bal Ni (RA)	0.80
3109	ps	12.2 binder (K)	0.80
RA 330	w	19Cr–35Ni–1.5Mn–1.3Si–bal Fe (RA)	0.79
HK–40	c	26Cr–20Ni–0.4C–bal Fe	0.78
304 SS	w	17Cr–9Ni–2Mn–1Si–bal Fe	0.73
Inconel 671	w	50Ni–48Cr–0.4Ti (HA)	0.62
430 SS	w	17Cr–1Mn–1Si–0.1C–bal Fe	0.62
Inconel 600	w	76Ni–15.5Cr–8Fe (HA)	0.61
Lucalox	—	Densified Al_2O_3 (GE)	0.57
Beta III Ti	w	11.5Mo–6Zr–4.5Sn–bal Ti	0.57
Incoloy 800	w	32.5Ni–21Cr–46Fe (HA)	0.57
316 SS	w	17Cr–12Ni–2Mn–1Si–2.5Mo–bal Fe	0.56
Ti–6A1–4V	w	—	0.54
Incoloy 800H	w	32.5Ni–21Cr–0.07C–46Fe (HA)	0.54
K68	ps	5.8 binder (K)	0.50
VR–54	ps	WC–7Co binder (F)	0.50
3406	ps	7.8 binder (K)	0.49
K701	ps	10.2Co + 4Cr binder (K)	0.47

(continued)

TABLE 7.5 *(continued)*

Material	Manufacturing Method	Composition	REF[c]
K801	ps	6Ni binder (K)	0.46
SiC	hp	(N)	0.44
W–10	ps	90W–10(Ni, Cu, Fe) (K)	0.44
Noroc 33	hp	Si_3N_4–SiC (N)	0.42
HD 430	—	Recrystallized	0.38
CA 306	ps	WC–6Co binder (Carmet)	0.36
BT–9	ps	2MgO–25TiB_2–3.5WC–bal Al_2O_3 (OGC)	0.36
HD 435	—	Recrystallized SiC (N)	0.32
TiC–Al_2O_3	ps	(B and W)	0.30
BT–11	ps	1.7MgO–30TiB_2–3.5WC–bal Al_2O_3 (OGC)	0.26
K714	ps	6Co + 1Cr binder (K)	0.25
BT–10	ps	2MgO–30TiB_2–3.5WC–bal Al_2O_3 (OGC)	0.25
Norbide	hp	B_4C (N)	0.21
BT–24	ps	2MgO–30TiB_2–3.5WC–bal Al_2O_3 (OGC)	0.20
W	w	(GE)	0.17
BT–12	ps	1.5MgO–49TiB_2–3.5WC–bal Al_2O_3 (OGC)	0.16
K602	ps	<1.5 binder (K)	0.13
Si_3N_4	hp	(N)	0.12
ZRBSC–D	hp	ZrB_2–SiC (N)	0.07
SiC	—	(GE)	0.02
Diamond	—	(GE)	0
CBN	—	(GE)	0

[a] Experimental conditions: 90° impingement, 27 μm Al_2O_3 particles, 5-g/min particle flow, 170-m/sec particle velocity 3-min test duration, N_2 atmosphere.

[b] Abbreviations are listed in Table 7.8.

[c] REF (relative erosion factor) = $\dfrac{\text{volume loss material}}{\text{volume loss Stellite 6B}}$.

Source: Hansen (1979).

initial conditions are $x = y = 0$ at $t = 0$ and $\dot{x} = U \cos \alpha$, $\dot{y} = U \sin \alpha$, and $\phi = \alpha$ at $t = 0$.

Equation (7.13) can be solved for the assumed initial conditions as

$$x = \frac{U \sin \alpha}{\beta J} \sin \beta t + (U \cos \alpha)t - \frac{U \sin \alpha}{J} t$$

$$y = \frac{U \sin \alpha}{\beta} \sin \beta t \qquad (7.14)$$

$$\phi = \frac{mr \, U \sin \alpha}{\beta J I} (\sin \beta t - \beta t) + \dot{\phi}_0 t$$

Sec. 7.4 Erosive Wear by Solid Particle Impingement

TABLE 7.6 Room-Temperature Erosion Test Results on Coated Materials[a,b]

Material	Composition and Coating Method	REF[c]
Borofuse Stellite 31	25Cr–10.5Co–2Fe–7.5W–0.5C–bal Co w/diffused B (MDC)	1.40
Ni–Cr–B	Plasma 0.5C–4Si–16Cr–4B–4Fe–2.4Cu–2.4Mo–2.4W–bal Ni (CWS)	1.32
Borofuse Stellite 6	29Cr–4.5W–1C–bal Co w/diffused B (MDC)	1.29
Cr_2O_3	Plasma Cr_2O_3–5SiO_2–3TiO_2 (CWS)	1.23
WC	Plasma 35(WC + 8Ni)–11Cr–2.5B–2.5Fe–2.5Si–0.5C–bal Ni (CWS)	1.11
Borofuse Stellite 3	31Cr–12.5W–2.4C–bal Co w/diffused B (MDC)	0.92
W	Pure CVD coating (RMRC)	0.53
Borofuse MT–104	0.5Ti–0.08Zr–0.03C–bal Mo w/diffused B (MDC)	0.30
Borofuse PM moly	Mo w/diffused B (MDC)	0.25
SiC	CVD SiC on C converted to SiC	0.06
SiC	Pure CVD coating	0.05
Borofuse WC	WC w/diffused B (MDC)	0.02
TiB_2	Electrodeposited over Ni (CPMRC)	0
18B–11	TiB_2 electrodeposited over 310 SS (UT)	0
19A–13	TiB_2 electrodeposited over 310 SS (UT)	0

[a] Experimental conditions: 90° impingement, 27-μm Al_2O_3 particles, 5-g/min particle flow, 170-m/sec particle velocity, 3-min test duration, N_2 atmosphere.
[b] Abbreviations are listed in Table 7.8.
[c] REF (relative erosion factor) = $\dfrac{\text{volume loss material}}{\text{volume loss Stellite 6B}}$.

Source: Hansen (1979).

where

$$\beta = \left(\frac{p\psi bJ}{m}\right)^{1/2}$$

Neglecting the initial angular velocity ϕ_0, the maximum rotation (ϕ at $\beta t = \pi$) in terms of the maximum depth of cut (y at $\beta t = \pi/2$) can be obtained as

$$\phi_{max} = \frac{\pi mr}{JI} y_{max} \qquad (7.15)$$

Taking $I = mr^2/2$, $J = 2$, and $y_{max}/r = 0.1$, ϕ_{max} was shown to be less than 18°. Then the shape of the cut may be expressed as

$$\begin{aligned} x_t &= x + r\phi \\ y_t &= y \end{aligned} \qquad (7.16)$$

TABLE 7.7 700°C Erosion Test Results on Coated Materials[a,b]

Material	Composition and Coating Method	REF[c]
Ni–Cr–B	Plasma 0.5C–4Si–16Cr–4B–4Fe–2.4Cu–2.4Mo–2.4W–bal Ni (CWS)	2.79
WC	Plasma 35(WC + 8Ni)–11Cr–2.5B–2.5Fe–2.5Si–0.5C–bal Ni (CWS)	2.06
Borofuse Stellite 6	29Cr–4.5W–1C–bal Co w/diffused B (MDC)	1.40
Borofuse Stellite 31	25Cr–10.5Co–2Fe–7.5W–0.5C–bal Co w/diffused B (MDC)	1.37
Borofuse Stellite 3	31Cr–12.5W–2.4C–bal Co w/diffused B (MDC)	0.83
Borofuse WC	WC w/diffused B (MDC)	0.72
Borofuse PM moly	Mo w/diffused B (MDC)	0.28
W	Pure CVD coating (RMRC)	0.25
Borofuse MT 104	0.5Ti–0.08Zr–0.03Cr–bal Mo w/diffused B (MDC)	0.19
SiC	Pure CVD coating	0
SiC	CVD on SiC on C converted to SiC	0
TiB$_2$	Electrodeposited on Ni (CPMRC)	0
18B–11	TiB$_2$ electrodeposited on 310 SS (UT)	0
19A–13	TiB$_2$ electrodeposited on 310 SS (UT)	0

[a] Experimental conditions: 90° impingement, 27-μm Al$_2$O$_3$ particles, 5-g/min particle flow, 170-m/sec particle velocity, 3-min test duration, N$_2$ atmosphere.
[b] Abbreviations are listed in Table 7.8.
[c] REF (relative erosion factor) = $\dfrac{\text{volume loss material}}{\text{volume loss Stellite 6B}}$.

Source: Hansen (1979).

The volume worn per unit width is the area swept by the tip of the abrasive grain. The cutting action stops when the horizontal motion of the particle tip ceases ($\dot{x}_t = 0$) or when the particle leaves the surface ($y_t = y = 0$). From Eq. (7.14), this occurs when $\beta t_c = \pi$ and $\cos \alpha t_c = 1 - J/3 \tan \alpha$. The former applies to lower values of α than the latter. Two conditions are satisfied simultaneously at $\alpha = \tan^{-1}(J/6)$. The weight removed from the surface of density ρ by a single particle may be expressed as

$$W = \rho b \int_0^{t_c} y_t \, dx_t \simeq \rho b \int_0^{t_c} y \, d(x + r\phi) \qquad (7.17)$$

Substituting Eq. (7.14) into Eq. (7.17), W may be expressed as

$$W = \frac{3\rho b U^2 \sin^2 \alpha}{\beta^2 J} \left[\frac{J \cot \alpha}{3} - \frac{3}{4} - \frac{\cos 2\beta t_c}{4} - \left(\frac{J \cot \alpha}{3} - 1 \right) \cos \beta t_c \right]$$

$$+ \frac{r\dot{\phi}_0 \rho b U \sin \alpha}{\beta^2} (1 - \cos \beta t_c) \qquad (7.18)$$

TABLE 7.8 Abbreviations Used in Tables 7.4 through 7.7

B and W	Babcock and Wilcox
c	cast
Carb	Carbology Systems Dept., General Electric Corp.
Carbor	Carborundum Co.
Carmet	Carmet Co., Allegheny Ludlum Steel Corp.
CPMRC	College Park Metallurgy Research Center
CWS	CWS corp.
F	Fansteel, Inc.
GE	General Electric
HA	Huntington Alloy Products Div., International Nickel Co.
hp	hot-pressed
K	Kennametal, Inc.
Krohn	Krohn Ceramics Corp.
MDC	Materials Development Corp.
N	Norton
OGC	Oregon Graduate Center
ps	pressed and sintered
RA	Rolled Alloys Corp.
RMRC	Rolla Metallurgy Research Center
Stellite	Stellite Div., Cabot Corp.
TRB	Timken Roller Bearing Co.
UCAR	Union Carbide Corp.
UT	United Technologies Corp.
w	wrought

Source: Hansen (1979).

The erosion W due to the total mass M of abrasive particles can be obtained by multiplying Eq. (7.18) by M/m. Then W may be expressed as

$$W = \frac{\rho}{p\psi} \frac{MU^2}{J} \left(\sin 2\alpha - \frac{6}{J} \sin^2 \alpha \right) \tag{7.19}$$

when $\beta t_c = \pi$ (or $\tan \alpha \leq K/6$), or

$$W = \left(\frac{\rho}{p\psi} \frac{MU^2}{J} \right) \frac{J \cos^2 \alpha}{6} \tag{7.20}$$

when $\cos \beta t_c = 1 - J/3 \tan \alpha$ (or $\tan \alpha \geq J/6$). According to this cutting model, maximum erosion occurs at $\tan 2\alpha = J/3$.

This model predicts that the erosion rate is proportional to the velocity squared, whereas the experimental data shown in Fig. 7.40 indicate that this is not the case. A typical exponent for velocity is greater than 2.3, indicating that the model may not be an accurate description of the actual physical phenomenon.

In the foregoing analysis it was assumed that J is a constant. However, the previous analysis for abrasion showed that the ratio J is a function of the rake angle of the particle. Finnie (1958) justified this assumption by noting that the angle for maximum erosion is not sensitively dependent on

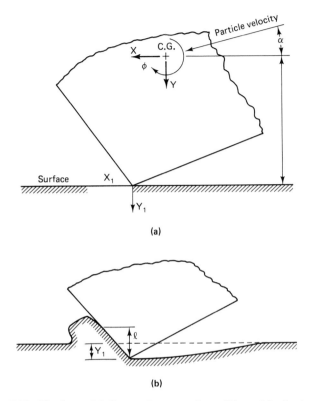

Figure 7.53 Finnie model for erosive wear by solid particle impingement: (a) coordinate system; (b) idealized picture of an abrasive grain removing material. (From Finnie, I., "The Mechanism of Erosion of Ductile Metals," *Proceedings of the Third National Conference of Applied Mechanics*, 1958, pp. 527–532. Reprinted by permission of the American Society)

J; that is, when J is equal to 1.5, 2.0, and 2.5, the corresponding angles are 13°15′, 16°51′, and 19°44′. Hutchings et al. (1976) experimental results show that this assumption may not be valid due to the rotation of the solid particles during impingement.

The general shape of the predicted erosion curve according to Eqs. (7.19) and (7.20) is comparable to the experimental results shown in Fig. 7.54. However, the absolute magnitude of the erosion rate cannot be predicted on the basis of these equations, since the predicted wear rates are generally much lower than the experimentally determined wear rates, unless the factor ψ is assumed to be so small as to be unreasonable. This discrepancy is due to the fact that not all impinging particles create chips. In fact, for geometric reasons, only a limited number of the particles create chips in a typical erosion situation.

Even when cutting predominates, the surface traction applied by the particle is expected to induce plastic deformation, crack nucleation, and

Sec. 7.4 Erosive Wear by Solid Particle Impingement   331

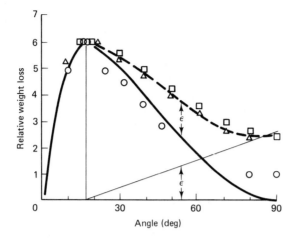

Figure 7.54 Predicted and measured relative weight loss. Maximum weight loss taken as the same in all cases. Experimental points: △, copper, ○, aluminum, □, SAE 1020 steel.—predicted by Eqs. (7.14) and (7.20);—predicted curve increased by amount ε. (From Finnie, I., "The Mechanism of Erosion of Ductile Metals," *Proceedings of the Third National Conference of Applied Mechanics*, 1958, pp. 527–532. Reprinted by permission of the American Society of Mechanical Engineers.)

crack propagation. When the material removed by cutting is much more rapid than the rate of accumulation of subsurface damage, the material removal rate will be determined by cutting rate. However, as the frequency of chip generation decreases (either because the impingement angle is not small enough or because the effective negative rake angle of the impingement particle is too large), the wear rate will be predominantly controlled by the subsurface crack propagation process. At an intermediate range the cutting process may also be aided by the existence of subsurface cracks.

Erosion at moderate and large impingement angles. Erosion at typical impingement angles and by common erosive particles is not expected to occur by the cutting mechanism, as assumed by Finnie. Rather, the wear is expected to be caused by subsurface crack nucleation and crack propagation induced by the large plastic deformation around the particle as it penetrates the surface. Crack nucleation is expected to occur more readily in two-phase metals than in single-phase metals due to the displacement incompatibility between the matrix and the particle when plastic deformation of the matrix occurs. The number of crack nucleation sites is expected to be proportional to the number of second-phase particles larger than a critical size. Jahanmir (1980) considered the crack nucleation rate under the condition of solid particle impingement and showed that it takes only a few impacts to nucleate a crack around a hard particle. Therefore, it is expected that, in two-phase metals, the crack propagation rate will be the rate-determining factor for the erosion wear rather than the crack nucleation rate. This is similiar to the case of sliding wear. In these materials the wear rate is expected to be dependent on the average distance cracks have to propagate to link up with neighboring cracks (which depends on the number of crack nucleation sites) and the rate of crack extension per impact. It was shown in Chapters 4 and 5 that the crack extension per cycle of asperity loading during delamination wear is very small, being controlled by the crack tip sliding displacement under the combined loading of compression and shear. To predict the erosive wear rate when the impingement angle is large and when the rake angle of the particle is very negative, the crack propagation per impact by the abrasive particle must be determined from first principles. Such an analysis has not yet been done.

Empirical erosion equations. In view of the discrepancy between Finnie's model and the experimental results, a number of empirical relationships have been advanced (Bitter, 1963; Nielson and Gilchrist, 1968). These empirical relationships are useful in predicting the erosion rates at various impingement angles once a small number of data points are available. However, they do not provide any physical insight into the basic mechanisms involved and the means of preventing erosive wear. One such relationship, due to Nielson and Gilchrist (1968), is as follows:

$$W = \frac{M(U^2 \cos^2\alpha - V_r^2)}{2\phi} + \frac{M(U \sin\alpha - K)^2}{2\varepsilon} \quad \text{for } \alpha < \alpha_0$$

$$W = \frac{MU^2 \cos^2\alpha}{2\phi} + \frac{M(U \sin\alpha - K)^2}{2\varepsilon} \quad \text{for } \alpha > \alpha_0 \quad (7.21)$$

where ϕ, ε, V_r, and K are empirically determined constants. ϕ may be interpreted as the kinetic energy absorbed by the surface to release one unit of eroded mass, ε is the corresponding parameter for "deformation wear," V_r is the residual parallel component of particle velocity at small angles of attack, and K is the velocity component normal to the surface below which no erosion takes place in hard materials.

These equations have been used by first determining the empirical constants using at least four sets of experimental data obtained at different impingement angles and velocities. Once the constants are determined, they can be used to "predict" erosion rates under different conditions.

7.4.5 Fundamental Mechanisms of the Erosive Wear by Solid Particle Impingement: Brittle Materials

In Chapter 4 it was shown that brittle solids fracture when the normal tensile stress exceeds the cohesive strength of the solid. It was shown that some brittle solids fracture in a brittle manner without involving plastic deformation, normally from preexisting flaws. This typically occurs when the indenter is larger than a critical size. For example, when the indent is blunt and large, ring cracks form outside the contact area without involving any plastic deformation. On the other hand, when the indenter is smaller than the critical size, many nominally brittle solids first undergo local plastic deformation at and near the tip of the sharp abrasive particles and then cracks emanate from the plastically deformed zone. These cracks propagate from the center of the contact area radially along the surface, laterally in the subsurface region, and also circumferentially. Brittle fracture often is accompanied by a wide statistical variation even under identical erosive conditions, due to the random distribution and size of preexisting flaws.

The erosion of brittle solids caused by ring cracks around large contact areas was treated as a statistical failure problem by Oh et al. (1972) and also by Sheldon and Finnie (1966a,b). The erosion caused by sharp particles in materials that can undergo limited plastic deformation was analyzed semiempirically by Evans and Wilshaw (1976) and Evans et al. (1976). These approaches will be reviewed in this section, after a discussion on the transition from ring cracking to elastoplastic cracking of brittle solids.

Transition from ring cracking to plastic indentation cracking. It was stated earlier that there is a critical radius above which ring cracks form before the contact area can undergo plastic deformation, whereas when the radius of the contact area is smaller than the critical value, local microplastic indentation occurs from which cracks emanate. The critical radius will be determined here following the work of Evans and Wilshaw (1976).

The radial stress at the surface around a large indenter decreases as a function of distance from the center of the contact area. Therefore, the Hertzian fracture load, L_f, may be related to the indenter radius, R, as

$$L_f \simeq B_1 R^n \tag{7.22}$$

where n and B_1 are constants. n is nearly equal to 1 and B_1 depends on the fracture toughness and the preexisting flaws of the material.

In Chapter 4 it was shown that the peak compressive stress below the indenter, σ_0, is related to the applied load, L, as

$$\sigma_0 = \frac{3L}{2\pi a^2} \tag{7.23}$$

where a is the radius of the contact area. a is related to R as

$$a = \left(\frac{4kLR}{3E}\right)^{1/3} \tag{7.24}$$

where k is a constant and E is Young's modulus. Eliminating a from Eqs. (7.23) and (7.24) and equating σ_0 to the hardness of the material, H, yields the plastic load, L_y, as

$$L_y = B_2 \left(\frac{k}{E}\right)^2 H^3 R^2 \tag{7.25}$$

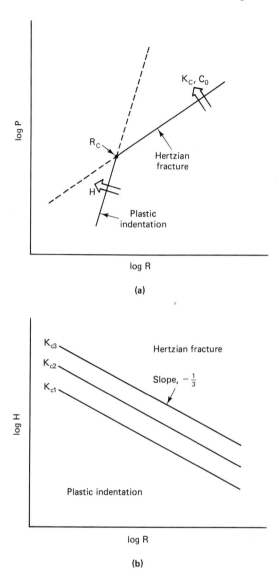

Figure 7.55 (a) Effects of sphere diameter on the indentation load for Hertzian fracture, and for plastic indentation; (b) regimes of Hertzian fracture and plastic indentation expressed in terms of hardness and sphere diameter, with the fracture toughness as a boundary locator. (From Evans and Wilshaw, 1976.)

where B_2 is a constant. The critical radius, R_c, is obtained by equating L_y to L_f as

$$R_c = B\left(\frac{E}{k}\right)^2 \frac{1}{H^3} \quad (7.26)$$

where B is B_1/B_2. When $R < R_c$, plastic deformation occurs at the point

Sec. 7.4 Erosive Wear by Solid Particle Impingement

of loading, whereas when $R > R_c$, ring cracks form around the area of contact. Figure 7.55 schematically illustrates the transition of the failure mode as a function of the load and the radius of the contact area.

Loading due to impact of spheres. Since erosion occurs due to the impact of spheres with brittle solid, the force generated at the impact point must be known.

When R is greater than R_c, the impact will be purely elastic and the pressure acting between an elastic sphere and an elastic half-space can readily be computed (Timoshenko and Goodier, 1970). Consider the impact of two elastic spheres. As soon as impact occurs, a normal load L will develop between the spheres. As the spheres compress each other, the load L will increase. The equations of motion of each sphere may be written as

$$m_1 \frac{dv_1}{dt} = -L$$
$$m_2 \frac{dv_2}{dt} = -L \quad (7.27)$$

where m and v are mass and the velocity of the mass center, respectively. The subscripts denote sphere 1 and sphere 2. Denoting the distance the spheres approach each other by α, the equations of motion may be written as

$$\ddot{\alpha} = -L \frac{m_1 + m_2}{m_1 m_2} \quad (7.28)$$

where $\dot{\alpha} = v_1 + v_2$

Since the spheres remain in contact for a very long time compared to the transit time of elastic waves through the sphere, the contact load L can be related to the quasi-static deflection α as

$$L = \frac{4}{3\pi} \frac{1}{k_1 + k_2} \sqrt{\frac{R_1 R_2}{R_1 + R_2}} \alpha^{3/2} = n\alpha^{3/2} \quad (7.29)$$

where

$$k_1 = \frac{1 - v_1^2}{E_1}$$

$$k_2 = \frac{1 - v_2^2}{E_2}$$

Substituting Eq. (7.29) into Eq. (7.28) and integrating the magnitude of approach at the instant of maximum compression (i.e., $\dot{\alpha} = 0$), α_1 may be written as

$$\alpha_1 = \left(\frac{5}{4} \frac{\dot{\alpha}_m^2}{n} \frac{m_1 m_2}{m_1 + m_2} \right)^{2/5} \quad (7.30)$$

where α_m is the initial velocity of approach of two spheres at the beginning of the impact. The duration of the impact is given by

$$t = 2.94 \frac{\alpha_1}{\dot{\alpha}_m} \tag{7.31}$$

When one of the spheres has an infinite radius, the maximum contact load L_m may be expressed as

$$L_m = n\alpha_1^{3/2} = n^{2/5} \frac{5}{4} U^2 m^{3/5} \tag{7.32}$$

where R = radius of the sphere
$n = (4/3\pi)[1/(k_1 + k_2)]\sqrt{R}$
$m = \frac{3}{4}\pi R^3 \rho$
ρ = density of the sphere
U = initial velocity of the sphere

The solution given by Eq. (7.32) is valid up to appreciable velocities of the sphere (e.g., for contact times at least as small as 10^{-5} sec).

The impact load developed when the contact is plastic can be determined similarly. The peak load, L_m, is approximately given by Evans and Wilshaw (1976) as follows:

For $L_m > 10L^*$:

$$\exp\left[2.7 \frac{G}{\sigma_y}\left(\frac{L_m}{R^2 G}\right)^{1/3}\right]\left[21 \frac{L_m}{R^2 G}\left(\frac{G}{\sigma_y}\right)^3 - 23 \left(\frac{L_m}{R^2 G}\right)^{2/3}\left(\frac{G}{\sigma_y}\right)^2\right.$$
$$\left. + 16\left(\frac{L_m}{R^2 G}\right)^{1/3}\frac{G}{\sigma_y} - 123\right] = 3 \times 10^3 \left(\frac{U}{C_s}\right)^2 \frac{G_s}{G}\left(\frac{G}{\sigma_y}\right)^5 - 7$$

For $L_m < 10L^*$:

$$\frac{L_m}{R^2 G} = 4.6\left[\left(\frac{U}{C_s}\right)^2 \frac{G_s}{G}\right]^{3/5} \tag{7.33}$$

where $L^* = 1.8 \times 10^{-2}(\sigma_y^3 R^2 / G^2)$
G = shear modulus of the half-space
G_s = shear modulus of the sphere
C_s = wave speed in the sphere
σ_y = yield stress of the half-space

Equation (7.32) may be used to determine the contact load when the radius of the contact area is larger than the critical radius, and Eq. (7.33) when the radius is smaller than the critical radius, if the material can undergo plastic deformation on a micro scale. Figure 7.56 shows a plot of $(L_m/R^2 G)$ versus $(v_s/C_s)^2(G_s/G)$ for various values of σ_y/G. The figure indicates that the quasi-static elastic results provide reasonably accurate solutions at low velocities, but at higher speeds the elastic assumption yields a higher force

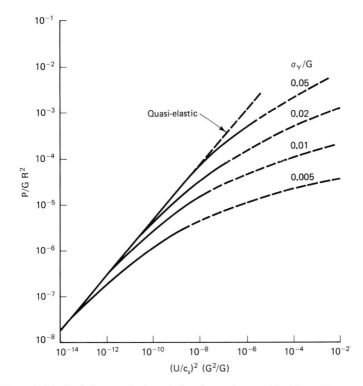

Figure 7.56 Peak force–velocity relation for various σ_y/G. (From Evans and Wilshaw, 1976.)

than the plastic indentation model, the difference increasing with the decrease in the yield strength.

Ring crack formation and erosion when $R > R_c$. Brittle fracture in the form of ring cracks without microindentation can occur when the contact area is larger than the critical radius given by Eq. (7.26). Figure 7.57 shows the shape of ring cracks forming around an indenter. If the outermost ring crack radius is a^*, the volume damaged by a single eroding particle V_p may be expressed as (Oh et al., 1972)

$$V_p \propto (a^*)^2 h F(\sigma_a) \tag{7.34}$$

where h is the depth of the primary ring crack, $F(\sigma_a)$ is the probability of the particle generating ring cracks at the surface, and σ_a is the maximum tensile stress generated by the impact. Assuming that the depth d of the primary ring crack is proportional to the depth of indentation by the sphere, from the Hertz equations it can be shown that

$$h \propto R\sigma_a^2 \propto L^{2/3} R^{-1/3} \propto R U^{4/5}$$
$$\sigma_a \propto U^{2/5} \tag{7.35}$$

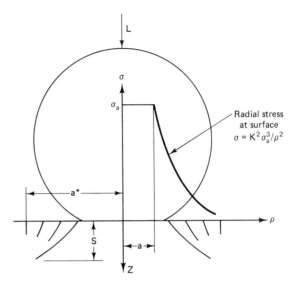

Figure 7.57 Coordinate system used in study of ring cracking. (From Oh et al., 1972.)

Then Eq. (7.34) may be written as

$$V_p \propto R U^{4/5} (a^*)^2 F(\sigma_a) \tag{7.36}$$

The probability function used by Oh et al. (1972) is the Weibull distribution given by

$$F(\sigma) = \begin{cases} 0 & \text{when } \sigma \leq \sigma_u \\ 1 - \exp\left(-V \dfrac{\sigma - \sigma_u^m}{\sigma_0}\right) & \text{when } \sigma > \sigma_u \end{cases} \tag{7.37}$$

where V is the volume of the fractured material, and σ_u, σ_0, and m are assumed to be material constants.

Figure 7.58 shows the experimental results obtained by Sheldon and Finnie (1966a,b) replotted by Oh et al. (1972). The experimental results were obtained by impacting graphite, magnesia and glass with steel balls. The agreement between the theoretical and experimental results were good for graphite and magnesia but poor for Pyrex glass (see Fig. 7.56). When the small angular erosive particles are used to impact many of these brittle materials, Eq. (7.36) does not predict the experimentally observed erosion rates, probably because of plastic microindentation (Sargent et al., 1979).

Erosion at $R < R_c$. When sharp erosive particles impact the surface, the radius of contact may be small and plastic microindentation can occur from which cracks can propagate. If we assume that the material removal occurs by lateral fracture when the adjacent lateral cracks interact, the

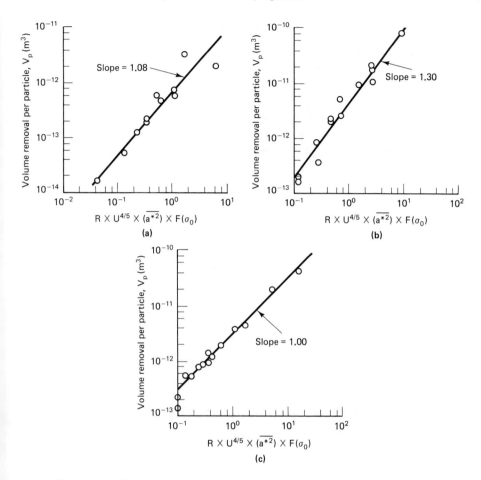

Figure 7.58 Experimentally observed values of volume removed per particle (for a wide range of velocity and particle size) as a function of the predicted damaged volume: (a) magnesia; (b) glass; (c) graphite. (From Oh et al., 1972.)

volume of the material removed by each particle may be expressed as (Evans and Wilshaw, 1976)

$$V_p = \pi(C_l)^2 h \quad (7.38)$$

where C_l and h are the length and the depth of the lateral crack (see Fig. 7.59). When the spacing of the indenters r_c is slightly greater than $2C_l$, loose wear particles will be generated.

From the fracture mechanics it can be shown that the stress intensity factor K (see Chapter 4) can be expressed as

$$\frac{K}{\sqrt{a}} = f_1(c/a) \int_{-c/a}^{c/a} f_2[(c/a), (r/a)] \sigma \, d(r/a) \quad (7.39)$$

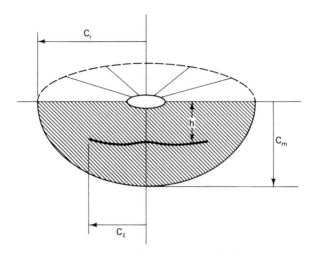

Figure 7.59 Radial, lateral, and median cracks emanating from plastically deformed microindentation due to loading by a sharp abrasive. (From Evans and Wilshaw, 1976.)

where c is the crack half-length, a is the contact radius, r is the distance from the loading point, and f_1 and f_2 are dimensionless functions. The stress field, σ, decays from a maximum value at the point of loading as a function of distance. Since the maximum stress must correspond to the yield strength, σ_y, σ may be expressed as

$$\sigma = \sigma_y[f_3(R_y/a)f_4(r/a)f_5(\nu)f_6(\mu)] \tag{7.40}$$

where ν and μ are the Poisson's ratio and the coefficient of friction, respectively. Substituting Eq. (7.40) into Eq. (7.39) and integrating, we obtain

$$\frac{K}{\sigma_y\sqrt{a}} = F_1(c/a)F_2(R_y/a)F_3(\nu)F_4(\mu) \tag{7.41}$$

The uniaxial yield stress is related to the hardness of material as

$$H = \phi\sigma_y \tag{7.42}$$

where ϕ is a proportionality constant. Therefore, Eq. (7.41) may be written in terms of hardness. Expressing Eq. (7.41) for $F_1(c/a)$, we obtain

$$F_1(c/a) = \frac{K_c\phi}{H\sqrt{a}} F_2^{-1}(R_y/a) F_3^{-1}(\nu) F_4^{-1}(\mu) \tag{7.43}$$

Since the functions F_2, F_3 and F_4 are approximately constant, the normalized crack length, C/a, is primarily a function of $K_c\phi/H\sqrt{a}$. Figure 7.60 shows

Figure 7.60 Normalized crack length C plotted as a function of the normalized fracture toughness, $K_c\phi/H\sqrt{a}$, for lateral (a) and radial (b) cracks. (From Evans and Wilshaw, 1976.)

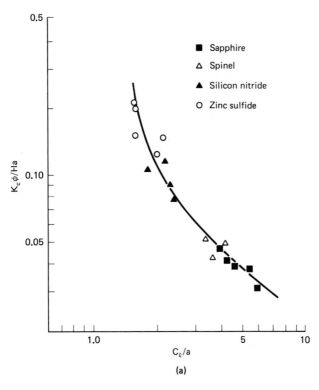

(a)

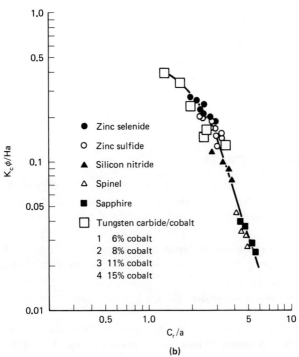

(b)

experimental results for radial and lateral cracks, respectively, as a function of the dimensionless quantity of $K_c\phi/H\sqrt{a}$, which represents a ratio of toughness to hardness. From Fig. 7.60, the length of the lateral crack at $C/a \geq 2$ is related to the applied load (which is related to hardness) and the stress intensity factor as

$$C_l \propto \left(\frac{L}{K_c}\right)^{3/4} \quad (7.44)$$

The depth of the lateral crack is approximately proportional to the radius of the contact zone, a.

Substituting Eq. (7.44) and making use of the fact that $h \propto a$, Eq. (7.38) may be expressed as

$$V_p \propto \left(\frac{L}{K_D}\right)^{3/2} a \quad (7.45)$$

where K_D is the dynamic stress intensity factor.

Eliminating a with $(L/H)^{1/2}$, Eq. (7.45) may be written as

$$V_p \propto \frac{L^2}{K_D^{3/2} H^{1/2}} \quad (7.46)$$

The total volume eroded by N particles is simply

$$V \propto \frac{NL^2}{K_D^{3/2} H^{1/2}} \quad (7.47)$$

L is related to the impact velocity as given by Eq. (7.33). At low impact velocities, Eq. (7.43) may be expressed as

$$V \propto \frac{NR^4 G^{4/5} U^{12/5} \rho^{6/5}}{K_D^{3/2} H_D^{1/2}} \quad (7.48)$$

Sargent et al. (1979) found that the experimental results obtained with small abrasive particles correlate with this Evans–Wilshaw model.

7.5 CONCLUDING REMARKS

In this chapter the material removal processes by hard particles and hard surfaces have been analyzed in detail. It was shown that abrasive wear of elastoplastic solids can create wear particles by cutting chips, but also that substantial subsurface plastic deformation occurs. The degree of subsurface plastic deformation at a given impingement angle increases as the negative rake angle of abrasive grains increases and as the interfacial shear strength between the abrasive particles and the material being removed increases. The abrasive wear coefficient is thus a sensitive function of both of these parameters.

This chapter also examined the erosive wear due to solid particle impingement. It was shown that erosive behavior of ductile materials normally involves plastic deformation and subsurface crack nucleation and propagation, unless the impingement angle is so small that chips are cut. The erosion of brittle materials exhibits two different types of behavior, depending on the size of the contact area. When the area is very small, the contact stress is so high that plastic indentation can occur from which lateral cracks can propagate, which cause erosive wear, whereas when the contact area is large, plastic deformation cannot occur and ring cracks form around the indenter from preexisting flaws. Theoretical models for these erosive phenomena have been presented in this chapter.

REFERENCES

ABEBE, M., and APPL, F. C., "A Slip-Line Solution for Negative Rake Angle Cutting," *Proceedings of the 9th North American Manufacturing Research Conference,* 1981, pp. 341–348.

AVIENT, B. W. E., and WILMAN, H., "New Features of the Abrasion Process Shown by Soft Metals; The Nature of Mechanical Polishing," *British Journal of Applied Physics,* Vol. 13, 1962, pp. 521–526.

AVIENT, B. W. E., GODDARD, J., and WILMAN, H., "An Experimental Study of Friction and Wear during Abrasion of Metals," *Proceedings of the Royal Society of London,* Vol. A258, 1960, pp. 159–179.

BELLMAN, R., JR., and LEVY, A., "Erosion Mechanism in Ductile Metals," *Wear,* Vol. 70, 1981, pp. 1–27.

BITTER, J. G. A., "A Study of Erosion Phenomena, Part I and Part II," *Wear,* Vol. 6, 1963, pp. 5–21, 169–190.

EVANS, A. G., and WILSHAW, T. R., "Quasi-static Solid Particle Damage in Brittle Solids: I. Observations, Analysis and Implications," *Acta Metallurgica,* Vol. 24, 1976, p. 939.

EVANS, A. G., GULDEN, M. E., EGGUM, G. E., and ROSENBLATT, M., "Impact-Damage in Brittle Materials in the Plastic Response Regime," Contract N0001475-C-0069, Report SC5023.9 TR, Rockwell International Science Center, Thousand Oaks, Calif., 1976 (Technical Report to the Office of Naval Research, 1976).

DATE, S. W., and MALKIN, S., "Effect of Grit Size on Abrasion with Coated Abrasives," *Wear,* Vol. 40, 1976, pp. 223–35.

FINNIE, I., "The Mechanism of Erosion of Ductile Metals," *Proceedings of the 3rd National Congress of Applied Mechanics,* ASME, New York, 1958, pp. 527–532.

FINNIE, I., WOLAK, J., and KABIL, Y., "Erosion of Metals by Solid Particles," *Journal of Materials,* Vol. 2, 1967, pp. 682–702.

GODDARD, J., HARKER, H. J., and WILMAN, H., "A Theory of the Abrasion of Solids Such as Metals," *Nature,* Vol. 184, 1959, pp. 333–335.

GULDEN, M. E., "Solid Particle Erosion of High-Technology Ceramics (Si_3N_4, Glass Bonded Al_2O_3, and MgF_2)," in *Erosion: Prevention and Useful Applications,* ASTM STP 664, W. F. Adler, ed., American Society for Testing and Materials, Philadelphia, 1979, pp. 101–122.

HANSEN, J. S., "Relative Erosion Resistance of Several Materials," in *Erosion: Prevention and Useful Applications,* ASTM STP 664, W. F. Adler, ed., American Society for Testing and Materials, Philadelphia, 1979, pp. 148–162.

HUTCHINGS, I. M., "Mechanisms of the Erosions of Metals by Solid Particles," *Erosion: Prevention and Useful Applications,* ASTM STP 664, W. F. Adler, ed., American Society for Testing and Materials, Philadelphia, 1979, pp. 59–76.

HUTCHINGS, I. M., WINTER, R. E., and FIELD, J. E., "Solid Particle Erosion of Metals: The Removal of Surface Material by Spherical Projectiles," *Proceedings of the Royal Society of London,* Vol. A348, 1976, pp. 379–392.

IVES, L. K., and RUFF, A. W., "Electron Microscopy Study of Erosion Damage in Copper," *Erosion: Prevention and Useful Applications,* ASTM STP 664, W. F. Adler, ed., American Society for Testing and Materials, Philadelphia, 1979, pp. 5–35.

JAHANMIR, S., "The Mechanics of Subsurface Damage in Solid Particle Erosion," *Wear,* Vol. 69, 1980, pp. 309–324.

JOHNSON, R. W., "A Study of the Pickup of Abrasive Particles during Abrasion of Annealed Aluminum on Silicon Carbide Abrasive Papers," *Wear,* Vol. 16, 1970, pp. 351–358.

KHRUSCHOV, M. M., "Resistance of Metals to Wear by Abrasion As Related to Hardness," *Institute of Mechanical Engineers Proceedings of Conference on Lubrication and Wear,* 1957, pp. 655–659.

KOMVOPOULOS, K., SAKA, N., and SUH, N. P., "Plowing Friction of Dry and Lubricated Metal Sliding," presented at the *ASLE/ASME Tribology Conference,* Atlanta, October, 1985, and for publication in the *Journal of Tribology,* Trans. ASME, 1986.

KOMVOPOULOS, K., SUH, N. P., and SAKA, N., "Wear of Boundary Lubricated Metal Surfaces," *Wear,* 1986.

LARSEN-BASSE, J., "Influence of Grit Diameter and Specimen Size on Wear during Sliding Abrasion," *Wear,* Vol. 12, 1968, pp. 35–53.

LARSEN-BASSE, J., "Some Effects of Specimen Size on Abrasive Wear," *Wear,* Vol. 19, 1972, pp. 27–35.

LARSEN-BASSE, J., and TANOUYE, P. A., "Abrasion of WC-Co Alloys by Loose Hard Abrasives," *Advances in Hard Material Tool Technology* (Proc. Int. Conf. Hard Material Tool Technology), Pittsburgh, Pa., 1976, pp. 188–199.

LEVY, A. V., "The Solid Particle Erosion Behavior of Steel as a Function of Microstructure," *Wear,* Vol. 68, 1981, pp. 269–287.

MOORE, M. A., and Douthwaite, R. M., "Plastic Deformation below Worn Surface," *Metallurgical Transactions,* Vol. 7A, 1976, pp. 1833–1839.

MULHEARN, T. O., and SAMUELS, L. E., "The Abrasion of Metals: A Model of the Process," *Wear,* Vol. 5, 1962, pp. 478–498.

NATHAN, G. K., and JONES, W. J. D., "The Empirical Relationship between Abrasive Wear and the Applied Conditions," *Wear,* Vol. 9, 1966, pp. 300–309.

NIELSON, J. H., and GILCHRIST, A., "Erosion by Stream of Solid Particles," *Wear,* Vol. 11, 1968, pp. 111–122.

OH, H. L., OH, K. D. L., VAIDYANATHAN, S., and FINNIE, I., "On the Shaping of Brittle Solids by Erosion and Ultrasonic Cutting," NBS Special Publication 348, National Bureau of Standards, Washington, D.C., 1972, p. 119.

RABINOWICZ, E., "Partial Use of the Surface Energy Criterion," *Wear,* Vol. 7, 1964, pp. 9–22.

RABINOWICZ, E., and MUTIS, A., "Effect of Abrasive Particle Size on Wear," *Wear,* Vol. 8, 1965, pp. 381–390.

RABINOWICZ, E., DUNN, L. A., and RUSSEL, P. G., "A Study of Abrasive Wear under Three-Body Conditions," *Wear,* Vol. 4, 1961, pp. 345–355.

RAMALINGAM, S., and WATSON, J. D., "Tool Life Distribution, Part 4: Minor Phase in Work Material and Multiple-Injury Tool Failure," *Journal of Engineering for Industry,* Vol. 100, 1978, pp. 201–209.

RICHARDSON, R. D. C., "The Abrasive Wear of Metals and Alloys," *Proceedings of the Institute of Mechanical Engineers,* Vol. 182, 1967, pp. 410–414.

SARGENT, G. A., MEHROTRA, P. K., and CONRAD, J., "Multiple Erosion of Pyrex Glass," in *Erosion: Prevention and Useful Application,* ASTM STP 664, W. F. Adler, ed., American Society for Testing and Materials, Philadelphia, 1979, pp. 77–135.

SHELDON, G. L., "Similarities and Differences in the Erosion Behavior of Materials," *Journal of Basic Engineering,* Vol. 92D, 1970, pp. 619–626.

SHELDON, G. L., "Effects of Surface Hardness and Other Material Properties on Erosive Wear of Metals by Solid Particles," *Journal of Engineering Materials and Technology,* Vol. 99H, 1977, pp. 133–137.

SHELDON, G. L., and FINNIE, I., "On the Ductile Behavior of Nominally Brittle Materials during Erosive Cutting," *Journal of Engineering for Industry,* Vol. 88B, 1966a, pp. 387–392.

SHELDON, G. L., and FINNIE, I., "The Mechanism of Material Removal in the Erosive Cutting of Brittle Materials," *Journal of Engineering for Industry,* Vol. 88B, 1966b, pp. 393–400.

SIN, H.-C., "Effect of Abrasive Grit Size on Abrasive Wear," S.M. thesis, MIT, 1978.

SIN, H.-C., SAKA, N., and SUH, N. P., "Abrasive Wear Mechanisms and the Grit Size Effect," *Wear,* Vol. 55, 1979, pp. 163–190.

SMELTZER, C. E., GULDEN, M. E., and COMPTON, W. A., "Mechanisms of Metal Removal by Impacting Dust Particles," *Journal of Basic Engineering,* Transactions of the ASME, Vol. 92, 1970, pp. 639–654.

SUH, N. P., "Microstructural Effects in Wear of Metals," in *Fundamental Aspects of Structural Alloy Design,* R. I. Jaffe and B. A. Wilcox, eds., 1977, pp. 565–593.

SUH, N. P., and TURNER, A. P. L., *Elements of the Mechanical Behavior of Materials,* McGraw-Hill, New York/Scripta, Silver Spring, Md., 1975.

TILLY, G. P., and SAGE, W., "The Interaction of Particle and Material Behavior in Erosion Processes," *Wear,* Vol. 16, 1970, pp. 447–465.

TIMOSHENKO, S. P., and GOODIER, J. N., *Theory of Elasticity,* McGraw-Hill, (3rd Ed.), New York, 1970.

WAHL, H., and HARTSTEIN, F., Strahlverschleiss, Franckh'sche Verlagshandlung, Stuttgart, 1946.

WESTBROOK, J. H., "Temperature Dependence of Hardness of Some Oxides," *Revue des Hautes Températures et des Réfractaires,* Vol. 3, 1966, p. 41.

WOOD, C. D., and ESPENSCHADE, P. W., *Transactions of the Society of Automotive Engineers,* Vol. 73, 1973, pp. 515–523.

YOUNG, J. P., and RUFF, A. W., "Particle Erosion Measurements on Metals," *Journal of Engineering Materials and Technology,* Vol. 99H, 1977, pp. 121–125.

APPENDIX 7.A

SLIP-LINE ANALYSIS FOR PLOWING MECHANISMS

Figure 7.A1 is a slip-line field for an asperity or a hard particle entrapped at the interface, plowing a softer surface (Abebe and Appl, 1981; Komvopoulos et al., 1985a). The material is assumed to be rigid–perfectly plastic. A plane strain condition is assumed. The stresses along any slip line can be obtained from Hencky's relations, starting from the boundary where the stresses are known. Then the stresses distribution at each point can be found using Mohr's circle. The stresses (normal and tangential) along the interface *ABJC* are given by the assumed friction coefficients along these boundaries, which can be related to the known stresses along the α and β lines of the slip-line field.

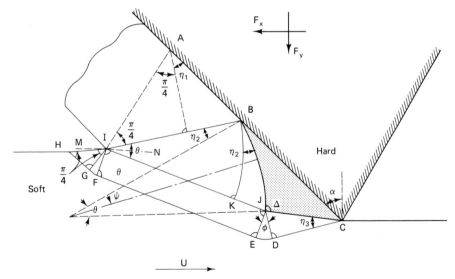

Figure 7.A1 Slip-line field for a hard particle or an asperity plowing the surface of a softer material. (From Abebe and Appl, 1981, and Komvopoulos et al., 1985.)

Along each slip line the hydrostatic pressure and the shear angle must satisfy the equilibrium and yield conditions. The characteristic equations for the α and β lines are (see Appendix 2.C for details)

$$p + 2k\phi = C_1 \quad (\alpha \text{ line})$$
$$p - 2k\phi = C_2 \quad (\beta \text{ line}) \qquad (2.C16)$$

where p is the hydrostatic pressure on a slip line, k is the shear strength of the material, ϕ is the shear angle, and C_1 and C_2 are constants. Using the characteristic equations and Mohr's circle, the stress distributions in each region and the resulting vertical and horizontal forces acting on the interface $ABJC$ can be calculated.

The slip-line field consists of a segment HI (ridge) and a rigid zone BJC which represents a dead zone of the soft surface material adhering to the hard asperity. The triangular fields AIB, HIG, and JDC and the rectangular field $IFEJ$ are networks of straight α and β lines. Since there is no curvature, the hydrostatic pressure (mean stress) remains the same throughout these fields. The centered fan fields IGF, IBK, JED and the field BKJ are networks of straight and circular orthogonal slip lines. In these fields the hydrostatic pressure is the same along a radial line, but changes from one radial line to another.

Since HI and IA are free surfaces, the α and β lines must intersect these surfaces at 45°. The angles η_1, η_2, and η_3 that α and β lines make with the interface $ABJC$ depend on the interfacial shear stress τ_i at the surfaces AB, BJ, and JC. The three angles η_1, η_2, and η_3 are directly related to the interfacial friction conditions through the relation

$$\eta_j = \frac{1}{2} \cos^{-1} \frac{\tau_j}{k} \qquad j = 1, 2, 3 \qquad (7.\text{A}1)$$

The pressure p and the shear stress τ along different segments of the slip line are as follows:

Along AI (stress free):

$$\sigma_{\text{I}} = 0 \qquad \sigma_{\text{II}} = -k \qquad \sigma_{\text{III}} = -2k$$

Along boundary AB:

$$\sigma_n^{AB} = k(1 + \sin 2\eta_1)$$
$$\sigma_t^{AB} = k \cos 2\eta_1$$

Along IB (β line):

$$p_{IB} = k \qquad \tau_{IB} = k$$

Along arc Bk (α line):

$$p_{BK} = k(1 + 2\psi) \qquad \tau_{BK} = k$$

Along boundary BJ:
$$\sigma_n^{BJ} = k(1 + 2\psi + \sin 2\eta_2)$$
$$\sigma_t^{BJ} = k \cos 2\eta_2 \qquad (7.A2)$$

Along HI (free surface):
$$\sigma_I = 0 \qquad \sigma_{II} = -k \qquad \sigma_{III} = -2k$$

Along GI (α line):
$$p_{GI} = k \qquad \tau_{GI} = k$$

Along IF (α line):
$$p_{IF} = k(1 + 2\theta) \qquad \tau_{IF} = k$$

Along JD (α line):
$$p_{JD} = k(1 + 2\theta + 2\phi) \qquad \tau_{JD} = k$$

Along boundary JC:
$$\sigma_n^{JC} = k(1 + 2\theta + 2\phi + \sin 2\eta_3)$$
$$\sigma_t^{JC} = -k \cos 2\eta_3$$

The lengths of OB and AB can be expressed in terms of the slip-line field angles and the length of JC as

$$(OB) = (JC) \frac{\cos(\theta/2) \sin(\eta_1 - \eta_2 + \theta + \Delta)}{\sin \theta \sin(\eta_1 - \eta_2 + \theta/2)} \qquad (7.A3)$$

$$(AB) = (JC) \frac{\cos(\eta_2 + \theta/2) \sin(\eta_1 - \eta_2 + \theta + \Delta)}{\sqrt{2} \cos[(\pi/4) - \eta_1] \sin \theta \sin(\eta_1 - \eta_2 + \theta/2)}$$

Then the resultant forces on the interface $ABJC$ are expressed as
$$F_x = F_x^{AB} + F_x^{BJ} + F_x^{JC}$$
$$= (AB)(\sigma_n^{AB} \cos \gamma - \sigma_t^{AB} \sin \gamma)$$
$$+ (OB) \left\{ \int_0^\theta [\sigma_n^{BJ} \cos(\eta_2 - \eta_1 + \gamma - \psi) \right.$$
$$\left. - \sigma_t^{BJ} \sin(\eta_2 - \eta_1 + \gamma - \psi)] \, d\psi \right\} \qquad (7.A4)$$
$$+ (JC)[\sigma_n^{JC} \cos(\pi - \Delta - \eta_1 + \eta_2 - \theta + \gamma)$$
$$+ \sigma_t^{JC} \sin(\pi - \Delta - \eta_1 + \eta_2 - \theta + \gamma)]$$
$$F_y = F_y^{AB} + F_y^{BJ} + F_y^{JC}$$
$$= (AB)(\sigma_n^{AB} \sin \gamma + \sigma_t^{AB} \cos \gamma)$$
$$+ (OB) \left\{ \int_0^\theta [\sigma_n^{BJ} \sin(\eta_2 - \eta_1 + \gamma - \psi) \right.$$

$$\qquad (7.A5)$$

$$+ \sigma_t^{BJ} \cos(\eta_2 - \eta_1 + \gamma - \psi)] \, d\psi \Big\}$$
$$+ (JC)[\sigma_n^{JC} \sin(\pi - \Delta - \eta_1 + \eta_2 - \theta + \gamma)$$
$$- \sigma_t^{JC} \cos(\pi - \Delta - \eta_1 + \eta_2 - \theta + \gamma)]$$

Substituting Eqs. (7.A2) and (7.A3) into Eqs. (7.A4) and (7.A5) and integrating, the horizontal and vertical forces acting on $ABJC$ may be expressed as

$$F_x = (JC)k \Bigg[\frac{\sin(\beta + \Delta - \gamma)}{\sin(\beta - \gamma - \theta/2)} \\
\left(\frac{\sqrt{2} \cos(\eta_2 + \theta/2) \sin[(\pi/4) + \eta_1 - \gamma]}{\sin \theta} \right. \\
+ (1 + \sin 2\eta_2) \cos\left(\beta - 2\gamma - \frac{\theta}{2}\right) \\
+ \frac{\theta \sin(\beta - 2\gamma)}{\sin(\theta/2)} \\
+ (\cos 2\eta_2 - 2) \sin\left(\beta - 2\gamma - \frac{\theta}{2}\right) \Bigg) \\
- (1 + 2\theta + 2\phi) \cos(\beta + \Delta - 2\gamma) \\
+ \sin(\beta + \Delta - 2\gamma - 2\eta_3) \Bigg]$$
(7.A6)

$$F_y = (JC)k \Bigg[\frac{\sin(\beta + \Delta - \gamma)}{\sin(\beta - \gamma - \theta/2)} \\
\left(\frac{\sqrt{2} \cos(\eta_2 + \theta/2) \cos[(\pi/4) + \eta_1 - \gamma]}{\sin \theta} \right. \\
- (1 + \sin 2\eta_2) \sin\left(\beta - 2\gamma - \frac{\theta}{2}\right) \\
+ \frac{\theta \cos(\beta - 2\gamma)}{\sin(\theta/2)} \\
+ (\cos 2\eta_2 - 2) \cos\left(\beta - 2\gamma - \frac{\theta}{2}\right) \Bigg) \\
+ (1 + 2\theta + 2\phi) \sin(\beta + \Delta - 2\gamma) \\
+ \cos(\beta + \Delta - 2\gamma - 2\eta_3) \Bigg]$$
(7.A7)

where
$$\beta = \eta_1 - \eta_2 + \theta + \gamma$$

At point J, the summation of all angles must be equal to 2π; thus

$$\Delta = \frac{\pi}{2} + \eta_2 + \eta_3 - \phi \tag{7.A8}$$

For the angles *HIM*, *BIN*, and *JIN* to be nonzero as shown in Fig. 7.A1, the following relations must be satisfied, respectively:

$$\theta \leq \frac{1}{2}\left(\frac{\pi}{4} + \gamma - \eta_1\right) \tag{7.A9}$$

$$\gamma - \eta_1 \geq 0 \tag{7.A10}$$

$$\theta + \eta_1 - \gamma \geq 0 \tag{7.A11}$$

Combining equations (7.A9), (7.A10), and (7.A11), the range for θ may be written as

$$\gamma - \eta_1 \leq \theta \leq \frac{1}{2}\left(\frac{\pi}{4} + \gamma - \eta_1\right)$$

$$\theta \leq \frac{\pi}{4} \tag{7.A12}$$

For the velocity discontinuities to be the same along the β line *HGFEDC*, the following relation obtained from the hodograph (Fig. 7.A2) must be satisfied:

$$\phi = \eta_3 + \eta_1 + \theta - \gamma - \sin^{-1}\left[\sqrt{2}\sin\left(\frac{\pi}{4} + \gamma - \eta_1 - 2\theta\right)\sin\eta_3\right] \tag{7.A13}$$

Finally, the coefficient of friction is given by

$$\mu = \frac{F_x}{F_y} \tag{7.A14}$$

where F_x and F_y are given by Eqs. (7.A6) and (7.A7), respectively. F_x and F_y must be evaluated subjected to the limitations given by Eqs. (7.A12) and (7.A13). Solutions for the coefficient of friction are shown in Figs. 7.34 and 7.36 as a function of the semiasperity angle γ and the penetration depth $w/2r$.

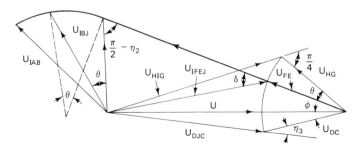

Figure 7.A2 Hodograph for the slip-line field shown in Fig. 7.A1. (From Abebe and Appl, 1981, and Komvopoulos et al., 1985.)

APPENDIX 7.B

PROOF OF THEOREMS OF LIMIT ANALYSIS[1]

The limit analysis is a powerful technique for determining the upper and lower bounds on a plastic deformation problem. This appendix derives the theorems of the limit analysis, starting from Drucker's postulate for the *principle of maximum plastic work*.[2]

Drucker's postulate is that "if a load increment is slowly applied to, and removed from, an element which is in some state of stress, the work done by the added stress, during the application and removal of the added stress, is positive, if plastic deformation has occurred during the cycle." If the initial state of stress is denoted by σ_{ij}^*, which is inside the yield locus, and the final state of stress, corresponding to the plastic strain rate $\dot{\varepsilon}_{ij}^p$, is denoted by σ_{ij}^f, the principle of maximum plastic work may be stated as

$$W^p = \int_{t_1}^{t_2} \sum_{i=1}^{3} \sum_{j=1}^{3} (\sigma_{ij}^f - \sigma_{ij}^*) \dot{\varepsilon}_{ij}^p \, dt \geq 0 \tag{7.B1}$$

where t_1 is the moment at which the state of stress satisfies the yield condition for the first time, $(t_2 - t_1)$ is the time increment during which the load is continuously applied, and $t = t_2$ is the point at which the load is finally removed. At $t > t_2$, the state of stress is finally brought back to σ_{ij}^*. Since unloading takes place elastically and since the elastic work done is recovered, Eq. (7.B1) represents only the work done during plastic deformation by the added increment of stress. Equation (7.B1) is satisfied if

$$\sum_{i=1}^{3} \sum_{j=1}^{3} (\sigma_{ij}^f - \sigma_{ij}^*) \dot{\varepsilon}_{ij}^p \geq 0 \tag{7.B2}$$

Equation (7.B2) will be used to derive the limit theorems.

The Lower-Bound Theorem (Statically Admissible Solution)

Consider an element enclosed by a surface S. The boundary conditions on the surface are specified in terms of the velocity components on a portion of the surface, S_V and in terms of the surface stress vector F_i on the remainder of the surface, S_F (see Fig. 7.B1). The surface stress vector F_i is defined as the sum of the three components of $\sigma_{ij}\nu_j$, where ν_j is the direction vector normal to the surface and pointing outward.

---

[1] From Suh and Turner, 1975.

[2] Drucker, D. C., "Stress-Strain Relations in Plastic Range of Metals—Experiments and Basic Concepts," *Rheology, Theory and Applications*, Vol. 1, Ed. by F. R. Eirich, Academic Press, 1956.

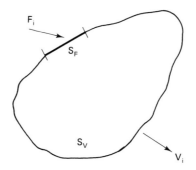

Figure 7.B1 Boundary conditions on a body, expressed in terms of the displacements of the boundary on the surface S_v and the tractions applied to the boundary on the surface S_F.

The actual stress field will be denoted by σ_{ij}, and any arbitrary stress field that satisfies the equilibrium condition and the boundary condition on S_F is denoted by σ_{ij}^*. The rate of work done by the external load is

$$\dot{W} = \int_S \sum_{i=1}^{3} F_i v_i \, dS = \int_S \sum_{i=1}^{3} \sum_{j=1}^{3} \sigma_{ij} v_j v_i \, dS \qquad (7.\text{B3})$$

where v_i is the actual velocity field associated with the stress field. Application of the divergence theorem to Eq. (7.B3) yields[3]

$$\dot{W} = \int_v \sum_{i=1}^{3} \sum_{j=1}^{3} (\sigma_{ij} v_{i,j} + \sigma_{ij,j} v_j) \, dV \qquad (7.\text{B4})$$

Since the equilibrium condition requires that $\sigma_{ij,j} = 0$, Eq. (7.B4) may be simplified to

$$\dot{W} = \int_v \sum_{i=1}^{3} \sum_{j=1}^{3} \sigma_{ij} v_{i,j} \, dV \qquad (7.\text{B5})$$

In terms of the deviator and hydrostatic stress components, Eq. (7.B5) may be rewritten as

$$\dot{W} = \int_v \sum_{i=1}^{3} \sum_{j=1}^{3} \left(\sigma'_{ij} + \frac{1}{3} \delta_{ij} \sum_{k=1}^{3} \sigma_{kk} \right) v_{i,j} \, dV$$

$$= \int_v \sum_{i=1}^{3} \sum_{j=1}^{3} \left(\sigma'_{ij} v_{i,j} + \frac{1}{3} \sum_{k=1}^{3} \sigma_{kk} v_{i,i} \right) dV \qquad (7.\text{B6})$$

The conservation of mass condition requires that $v_{i,i} = 0$, since the material is incompressible. The velocity field $v_{i,j}$ may be written in terms of the

---

[3] The notation $\sigma_{ij,j} = \partial \sigma_{ij}/\partial x_j$, while $v_{i,j} = \partial v_i/\partial x_j$. Commas denote partial differentiation with respect to the coordinate axis indicated by the subscript following the comma.

strain rate tensor $\dot{\varepsilon}_{ij}$ and rotation rate tensors $\dot{\omega}_{ij}$ as

$$v_{i,j} = \frac{1}{2}(v_{i,j} + v_{j,i}) + \frac{1}{2}(v_{i,j} - v_{j,i}) \qquad (7.B7)$$
$$= \dot{\varepsilon}_{ij} + \dot{\omega}_{ij}$$

Note that $\dot{\omega}_{ij}$ is a skew-symmetric tensor, whereas $\dot{\varepsilon}_{ij}$ and σ_{ij} are symmetric tensors. Note also that the product of a symmetric tensor and a skew tensor equals zero. Upon substitution of Eq. (7.B7), Eq. (7.B6) may be rewritten as

$$\dot{W} = \int_v \sum_{i=1}^{3} \sum_{j=1}^{3} \sigma'_{ij} \dot{\varepsilon}_{ij} \, dV = \int_S \sum_{i=1}^{3} F_i v_i \, dS \qquad (7.B8)$$

A similar expression may be obtained using an assumed stress:

$$\dot{W}^* = \int_v \sum_{i=1}^{3} \sum_{i=1}^{3} \sigma'^*_{ij} \dot{\varepsilon}_{ij} \, dV = \int_S \sum_{i=1}^{3} F^*_i v_i \, dS \qquad (7.B9)$$

where F^*_i is the unknown boundary condition in equilibrium with the assumed stress field σ^*_{ij}. Since F_i is known on S_F, Eq. (7.B9) may be written as

$$\dot{W}^* = \int_{S_V} \sum_{i=1}^{3} F^*_i v_i \, dS + \int_{S_F} \sum_{i=1}^{3} F_i v_i \, dS = \int_v \sum_{i=1}^{3} \sum_{j=1}^{3} \sigma'^*_{ij} \dot{\varepsilon}_{ij} \, dV \qquad (7.B10)$$

Similarly, Eq. (7.B8) may be written as

$$\dot{W} = \int_{S_V} \sum_{i=1}^{3} F_i v_i \, dS + \int_{S_F} \sum_{i=1}^{3} F_i v_i \, dS = \int_v \sum_{i=1}^{3} \sum_{j=1}^{3} \sigma'_{ij} \dot{\varepsilon}_{ij} \, dV \qquad (7.B11)$$

Subtracting Eq. (7.B10) from Eq. (7.B11), we obtain

$$\dot{W} - \dot{W}^* = \int_S \sum_{i=1}^{3} (F_i - F^*_i) v_i \, dS = \int_v \sum_{i=1}^{3} \sum_{j=1}^{3} (\sigma'_{ij} - \sigma'^*_{ij}) \dot{\varepsilon}_{ij} \, dV \qquad (7.B12)$$

Applying the principle of maximum plastic work, we have

$$W \geq W^* \qquad (7.B13)$$

Equation (7.B13) states that the actual work done is always greater than that produced by any assumed stress field. It should be noted that the statically admissible lower-bound solution must satisfy (1) equilibrium conditions, (2) boundary conditions, and (3) the yield condition.

The Upper-Bound Theorem (Kinematically Admissible Solution)

Consider again the same element discussed in the preceding section. This time the equilibrium condition will be disregarded and consideration given only to the deformation, including slip along an internal slip surface S_D (see Figure 7.B2). Let v_i be the true velocity field and v_i^* the assumed velocity field.

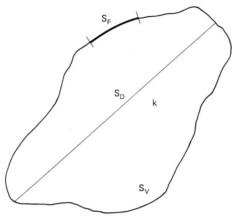

Figure 7.B2 Boundary conditions on a body, specified in terms of the displacements of the boundary on the surface S_v and the tractions applied to the boundary on the surface S_F. Slip is allowed to occur along some internal surface S_D with a shear stress k acting across the slip surface.

The rate of work done by the true stress and velocity field is derived as before:

$$\dot{W} = \int_S \sum_{i=1}^{3} F_i v_i \, dS = \int_S \sum_{i=1}^{3} \sum_{j=1}^{3} \sigma_{ij} \nu_j v_i \, dS \qquad (7.B14)$$

$$= \int_v \sigma'_{ij} \dot{\varepsilon}_{ij} \, dV + \int_{S_D} k |v_D| \, dS_D$$

where v_D is the absolute value of the relative velocity on S_D, and k is the critical shear stress. Expression (7.B14) is similar to Eq. (7.B8) except in the very last term, which incorporates the work done on internal slip planes. Similarly, the rate of work done by the assumed velocity field is

$$\dot{W}^* = \int_S \sum_{i=1}^{3} F_i v_i^* \, dS = \int_v \sum_{i=1}^{3} \sum_{j=1}^{3} \sigma'_{ij} \varepsilon_{ij}^* \, dV + \int_{S_D} k v_D^* \, dS_D \qquad (7.B15)$$

The principle of maximum plastic work states that if the deviator stress field, corresponding to an assumed strain rate field ε_{ij}^* (for the assumed velocity field v_i^*), is $\sigma'_{ij}{}^*$, then

$$\int_V \sum_{i=1}^{3} \sum_{j=1}^{3} (\sigma'_{ij}{}^* - \sigma'_{ij}) \dot{\varepsilon}_{ij}^* \, dV \geq 0 \qquad (7.B16)$$

Substituting Eq. (7.B16) into Eq. (7.B15), we obtain

$$\dot{W}^* = \int_S \sum_{i=1}^{3} F_i v_i^* \, dS \leq \int_V \sum_{i=1}^{3} \sum_{j=1}^{3} \sigma_{ij}^* \dot{\varepsilon}_{ij}^* \, dV + \int_{S_D} k|v_D^*| \, dS_D \qquad (7.\text{B}17)$$

Since $v_i^* = v_i$ on S_V, Eq. (7.B17) may be rewritten as

$$\int_{S_V} \sum_{i=1}^{3} F_i v_i \, dS + \int_{S_F} \sum_{i=1}^{3} F_i v_i^* \, dS \leq \int_V \sum_{i=1}^{3} \sum_{j=1}^{3} \sigma_{ij}^* \dot{\varepsilon}_{ij}^* \, dV$$
$$+ \int_{S_D} k|v_D^*| \, dS_D \qquad (7.\text{B}18)$$

Equation (7.B18) can be transformed to

$$\int_{S_V} \sum_{i=1}^{3} F_i v_i \, dS \leq \int_V \sum_{i=1}^{3} \sum_{j=1}^{3} \sigma_{ij}^* \dot{\varepsilon}_{ij}^* \, dV + \int_{S_D} k|v_D^*| \, dS_D$$
$$- \int_{S_F} \sum_{i=1}^{3} F_i v_i^* \, dS = \int_{S_V} \sum_{i=1}^{3} F_i^* v_i^* \, dS \qquad (7.\text{B}19)$$
$$= \int_{S_V} \sum_{i=1}^{3} F_i^* v_i \, dS$$

The right-hand side of the equation represents the work done by the unknown stress field which corresponds to the assumed velocity field on S_V. Equation (7.B19) states that the actual work done on S_V by the true stress and velocity fields is always less than that done by the assumed velocity field. In the limiting case of $S_F = 0$, it is clear from Eq. (7.B19) that the actual work done over the entire surface is also always less than the work done by the assumed velocity field and its corresponding stress field.

The upper-bound solutions can be obtained if the assumed displacement field satisfies the velocity boundary conditions and the condition of incompressability. The equilibrium conditions and the yield condition need not be satisfied.

8
WEAR DUE TO CHEMICAL INSTABILITY

8.1 INTRODUCTION

The friction and wear behavior of materials examined in Chapters 3 through 7 was controlled primarily by their mechanical behavior. In these chapters two basic assumptions were made regarding the geometric nature of the asperity interaction: (1) the distance between the asperity contacts is so large that the stress and temperature fields at an asperity are not affected by those due to the neighboring asperities, and (2) there is space available at the interface to accommodate and remove wear particles during the sliding motion. These assumptions are not valid when the normal load acting at the interface is so large that the real area of contact approaches the apparent area. This situation arises in metal cutting.

In metal cutting the interfacial temperature is high due to the large nominal stresses and high cutting speeds involved in generating metal chips. Under these conditions, the cutting tools can wear by three different mechanisms: chemical reaction, physical dissolution, and by delamination due to softening of the tool material. Because of the high interfacial temperature, chemical reactions between the chip and the tool material can occur, such as when titanium is cut with most cutting tools. On the other hand, when the contacting materials are soluble in each other, physical dissolution can occur, which is the primary mode of tungsten–carbide tool wear in cutting steel. When complex carbide tools (e.g., titanium–tungsten–carbide) are used, less soluble carbides are removed in chunks after the easily soluble tungsten carbide is gradually dissolved away. The high temperature also softens tool materials, which fail when the hardness of the tool is less than at least 4.5 times that of the workpiece (Suh, 1977). High-speed-steel (HSS) tools fail due to softening.

In this chapter the wear of tool materials due to chemical instability is examined, following a brief introduction to metal cutting and a review of the phenomenological aspects of tool wear.

8.2 BRIEF INTRODUCTION TO METAL CUTTING

Figure 8.1 shows schematically the two-dimensional chip-forming process. In a highly simplified form, the chip-forming process is similar to the plowing mechanism discussed in Chapter 7, except that the tools are sharper than the abrasive particles (i.e., the rake angle is either positive or slightly negative). As the cutting tool moves from the right to the left, the workpiece shears along the primary shear plane and the chip slides over the rake face of the tool. In addition to the primary shear zone, there is also a secondary shear zone in the chip near the chip–tool interface. The mechanics of the chip-forming process were first analyzed by Merchant (1945).

Deformation of the chip in the primary shear zone does not take place homogeneously, as shown in the micrograph of Fig. 8.2. The chip shows heterogeneous deformation, consisting of lamella structure and intense shear bands where the bulk of shear deformation has taken place. Highly cold-worked metals exhibit discrete shear bands of highly deformed metal separated by regions of little or no shear deformation. The distance between the shear deformation bands is large at very high cutting speeds. The heterogeneous nature of deformation and the localized shear deformation zone are also indicated by the rough nature of the free surface of the chip, as shown in Fig. 8.3. One of the unique aspects of chip deformation is that the slip takes place acrystallographically—the slip band propagates right across the grain boundaries—whereas in normal metal deformation plastic deformation occurs along the easy slip planes (Turley and Doyle, 1982).

The chips can be discontinuous (type I), continuous (type II), and continuous with a built-up edge (BUE) (type III). In cutting cast iron, discontinuous chips are generated due to the microcracks developed in the chip in the primary shear zone propagating across the brittle graphite phase. Discontinuous chips are also generated when the microcracks that are nucleated propagate along the shear bands. Microcracks are not normally generated in pure elements and single-phase metals, and therefore chips tend to be continuous. However, the shape of the chip is a function not only of the microstructure of materials but also of the tool geometry, as shown in Fig. 8.4. Normally, discontinuous chips are formed at low speeds and small or negative rake angles. The formation of discontinuous chips is believed to occur when the natural strain in the deformation band exceeds 1 (Turley and Doyle, 1982).

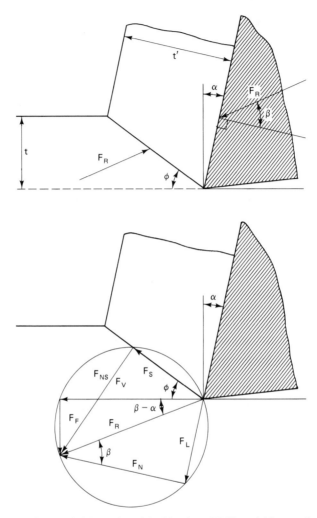

Figure 8.1 Schematic illustration of the Merchant (1945) model for metal cutting and the diagram of forces.

At intermediate cutting speeds, a built-up edge (BUE) forms near the tip of the cutting edge, as shown in Fig. 8.5 (Cook, 1966). When a BUE is present, the actual cutting is done by a BUE. However, a BUE is not stable and breaks away intermittently. It can be removed with the chip or remains attached to the workpiece. Consequently, when cutting conditions are suitable for BUE formation, the surface finish is poor. A BUE does not form when cutting pure and single-phase metals, whereas a BUE is continuously observed in cutting two-phase metals (Williams and Rollason, 1970). This is attributed to the ease of crack nucleation in these two-phase metals. At higher cutting speeds a BUE does not form.

Sec. 8.2   Brief Introduction to Metal Cutting

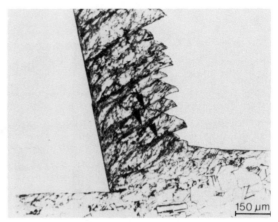

Figure 8.2 Optical micrograph of the midsection through a titanium chip produced by interrupting an orthagonal cut. Note that deformation in the chip is heterogeneous (some shear bands are arrowed). (From Turley and Doyle, 1982.)

The chip formation process in high-speed machining of AISI 4340 steel shows that the chips are discontinuous below 30 m/min, but continuous at 30 to 60 m/min (Komanduri et al., 1982). At cutting speeds above 60 m/min, cyclic instabilities in the cutting process occur due to the formation of fully developed intense shear bands which are separated by large areas (segments) of relatively less deformed material. As the cutting speed increases, the extent of contact between the segments is forced to decrease rapidly, separating completely at speeds greater than 1000 m/min. The speed at which the decohesion occurs between the chip segments depends on its microstructure and hardness. Figure 8.6 shows the chip shapes obtained at various cutting speeds.

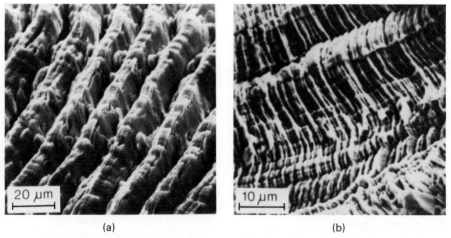

(a)   (b)

Figure 8.3 Scanning electron micrographs of the free surface of chips: (a) aluminum-alloy machining chip; (b) hardened-steel grinding chip. (From Turley and Doyle, 1982.)

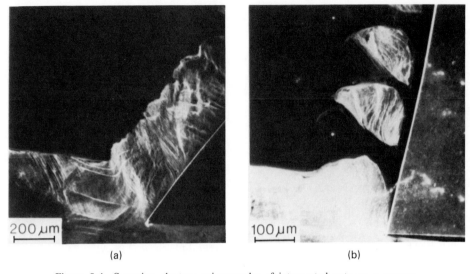

Figure 8.4 Scanning electron micrographs of interrupted cuts on a copper–silicon (8:) alloy machined inside the scannning electron microscope showing (a) continuous chip formation at $+30°$ rake angle; (b) discontinuous chip formation at $+5°$ rake angle. (From Turley and Doyle, 1982.)

The mechanics of chip formation and the machinability of materials have been investigated during the last four decades (Merchant, 1945; Shaw and Finnie, 1955; Dewhurst, 1979; Usui and Shirakashi, 1982; Oxley, 1982; Turley and Doyle, 1982). Machinability is measured in terms of such variables as the dimensional accuracy, surface finish, metal removal rate, duration of tool life, and the ease of chip removal (Tipnis, 1982). Although great

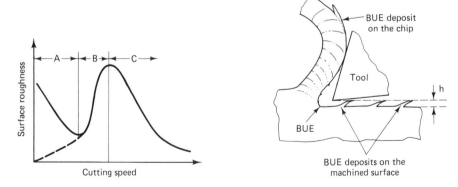

Figure 8.5 Built-up-edge (BUE) formation near the cutting edge of the tools at intermediate cutting speeds and its influence on surface finish. (Cook, 1966.)

Sec. 8.2 Brief Introduction to Metal Cutting

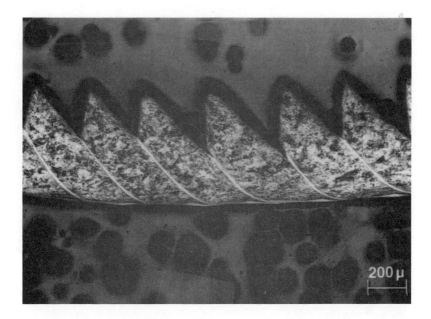

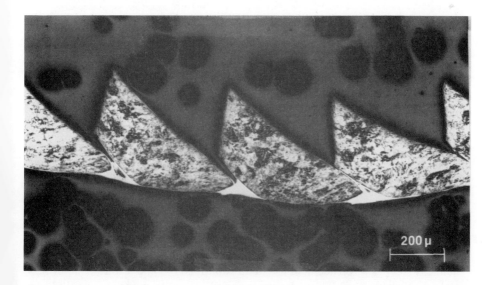

(b)

Figure 8.6 Shape of the chips generated at high-speed cutting of AISI 4340 steel: (a) 250 m/min; (b) 1000 m/min;

continued

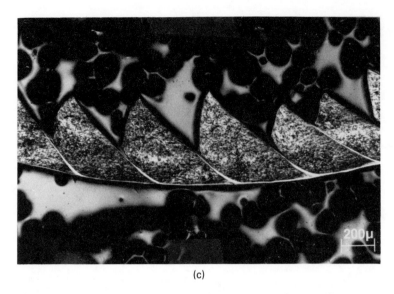

(c)

Figure 8.6 (*cont.*) (c) 2000 m/min. The chips show cyclic instability. (From Komanduri et al., 1982.)

strides have been made in understanding the mechanisms of chip formation both qualitatively and quantitatively, there is still much to be done to predict the machinability. Nevertheless, the existing knowledge can be applied qualitatively to solve many practical problems involving machinability.

Machinability of the workpiece is affected by the crystal structure, chemical composition, and microstructure (including the morphology of inclusions) of the workpiece. To improve machinability, such nonmetallic inclusions as manganese sulfides, selenides, tellurides, lead, and bismuth have been incorporated in steel (Tipnis, 1982). When sulfur is added to steel containing manganese in amounts ranging from 0.06% in plain carbon steel to 0.4% for low-alloy resulfurized steel, manganese sulfide is formed. The presence of manganese in steel allows the cutting speed to be increased by 25% or more, decreases the wear rate of high-speed-steel (HSS) tools, improves surface finish, lowers the friction between the chip–tool interface, and makes thinner chips. These inclusions act as crack nucleation sites, making it easy for chips to segment and fracture. Selenium is added to stainless steel as a replacement for sulfur (up to 0.03%) and also to resulfurized carbon steel (up to 0.07%). Selenium combines with manganese sulfide to form a more stable solution of manganese sulfide–selenide. Tellurium also combines with manganese to form MnTe. Lead is extensively used as a free machining additive, which exists as randomly occurring particles in steel because of its extremely low solubility. Bismuth is similar to lead in its role as a free machining additive.

Sec. 8.2   Brief Introduction to Metal Cutting

In metal cutting, the temperature rise at the interface is quite significant. In general, the temperature rise is most sensitive to the speed. With the rise of the interfacial temperature the wear of cutting tools increases exponentially, since the solubility, diffusivity, and the reciprocal of yield strength of materials follow an Arrhenius type of relationship with temperature. The chip–tool interface temperature can be determined using the interface as a thermocouple junction, or by employing detectors through small holes drilled through the tool, or by infrared photography. One such result is shown in Fig. 8.7. Various analytical techniques of estimating the interfacial temperature rise are given in Appendix 8.C.

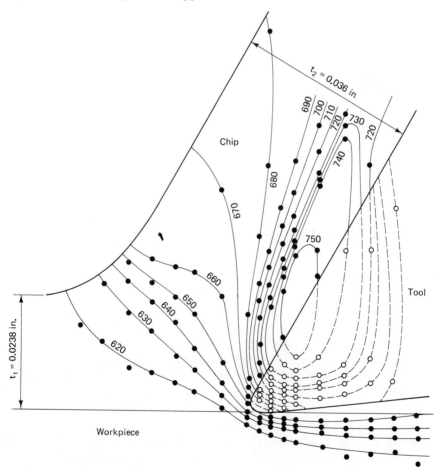

Figure 8.7 Temperature distribution in cutting zone as determined from infrared photograph. The material is free-cutting mild steel, and the cutting conditions are: speed, 75 ft/min (22.86 m/min); width of cut, $b = 0.25$ in. (6.3 mm); rake angle, $\alpha = 30°$; relief angle $= 7°$; and initial workpiece temperature, $T_0 = 611°C$. (After Boothroyd, 1963.)

8.3 CUTTING TOOL MATERIALS AND PHENOMENOLOGICAL ASPECTS OF TOOL WEAR

In the early nineteenth century plain carbon steel was used as the cutting tool material by forming martensite. However, these tools could not be used at high speeds because martensite is an unstable structure which decomposes into cemenite and ferrite at 100°C. In the mid-nineteenth century high-carbon steel was developed with the discovery of iron carbides. In the late nineteenth and early twentieth centuries high-speed-steel (HSS) tools were introduced with the discovery of the Taylor–White process of treating self-hardening tool steels containing chromium and tungsten (Taylor, 1907). HSS tools are still one of the most widely used cutting tool materials in industry. A major conceptual breakthrough that enabled the cutting of steel at much higher speeds was the invention of cemented tungsten carbide tools by Schroter (1926). He ball-milled tungsten carbide powder with cobalt powder, compacted the mixture, and sintered it to make a hard and tough compact that could cut steel at speeds up to 200 ft/min. Since then many attempts have been made to improve on the basic theme of Schroter, which led to the development of cemented complex carbides in 1929 (Schwarzkopf and Kieffer, 1960) and cemented titanium carbide tools in the mid-1950s (Humenik and Parikh, 1956). The development of these carbide tools increased cutting speed to 800 ft/min. Until recently, the basic hypothesis underscoring these tool developments has been that the harder the tool material, the better is its wear resistance—but that, of course, is not quite true.

During the past two decades, other tools, such as aluminum oxide, cubic boron nitride (CBN), and various coated tools have been introduced. Aluminum oxide tools are extremely wear resistant in cutting steel but are very brittle, rendering them useful only in limited applications. To overcome this lack of toughness, cemented alumina with metal (e.g., nickel–zirconium) binder was made (Suh and Fillion, 1980). The introduction of alumina tools has enabled cutting speeds to increase to as high as 3000 ft/min (914.4 m/min) in cutting steel. Cubic boron nitride (CBN) tools have been found to be superior to carbides in machining nickel-based superalloys and glass-fiber-reinforced composites. Diamond tools are used in machining nonferrous metals, especially in precision machining of optical surfaces.

Since the early 1970s various coated tools have been introduced to prolong tool life (Suh, 1977). The carbide tools and HSS tools have been coated with chemically stable compounds such as titanium nitride (TiN), titanium carbide (TiC), hafnium carbide (HfC), and aluminum oxide (Al_2O_3). Coating is normally accomplished by chemical vapor deposition of various kinds, and by sputtering, which is a physical deposition process. These deposition techniques are described further in Chapter 9.

Sec. 8.3 Cutting Tool Materials and Phenomenological Aspects of Tool Wear

In general, cutting tools must satisfy three basic requirements: (1) wear resistance, (2) fracture resistance, and (3) resistance to gross plastic deformation of the entire tool. HSS tools have excellent fracture resistance, but cannot cut steel at high speeds because of its limited resistance to gross plastic deformation and wear. Alumina tools have good wear resistance, but they have very low toughness. Cemented carbide tools have an optimum combination of all these properties, although their low toughness is still the limiting factor in intermittent cutting operations.

Figure 8.8 schematically illustrates the various forms of tool wear. The wear of the rake face, or crater wear, is normally observed when cutting steel with cemented carbides, cubic boron nitride (CBN), and HSS tools at high cutting speeds, whereas the flank wear is normally the mode of wear at low cutting speeds. The tool fails when the crater depth is typically 0.005 in. (0.12 mm), and catastrophic failure of the tool occurs when the wear land on the flank side becomes about 0.030 to 0.050 in. (0.75 to 1.25 mm). Groove wear can also limit tool life, although in many cases crater wear and flank wear are two major causes of tool failure. Flank wear is believed to be caused by abrasion of the tool by hard particles of the workpiece, whereas crater wear is caused primarily by the dissolution of tool materials.

Tools can also fail by two other mechanisms. One commonly occurring failure mode is the fracture of the tool; a less common failure is the plastic deformation of the cutting edge. The fracture of the tool is a particularly critical problem for alumina tools without binders, coated cemented carbide tools, and cemented tools, especially when they are used in intermittent cutting operations. In cemented carbides the metallic phase acts as a crack

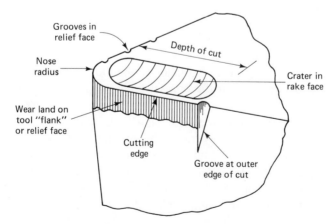

Figure 8.8 Schematic representation of the geometry of a worn tool used in turning. (From Cook, 1973.)

arrester by undergoing plastic deformation when the crack tip reaches the binder phase. In alumina tools the resistance to crack propagation is controlled by using small grains and monitoring the purity of the powder. Grain growth inhibitors such as manganese oxide are often used to increase toughness in these ceramic tools (Hasselman et al., 1973).

The fracture resistance is normally characterized by means of transverse rupture or bend test. For the cutting tools to be useful in intermittent cutting operations, the transverse rupture strength (TRS) must exceed 300,000 psi (2.07×10^9 Pa). Aluminum oxide tools have a TRS of about 100,000 psi (6.9×10^8 Pa) and cemented carbide tools have a strength of about 80,000 to 300,000 psi (1.24×10^9 to 2.7×10^9 Pa). Most of the coated cemented carbide tools have 5 to 20% less TRS, depending on the thickness of the coating; that is, the thicker the coating, the less the TRS (Schintlmeister et al., 1975). When a cutting tool is not sufficiently hard, the nose of the tool deforms plastically, making the tool unusable as a cutting tool. This is the case when cemented carbide tools are used to cut hard steel at speeds greater than 800 ft/min (24.38 m/min), whereas alumina tools can easily withstand the loads when cutting steel at a speed in excess of 2000 ft/min (609.6 m/min).

8.4 WEAR MECHANISMS OF CUTTING TOOLS

Cutting tools wear by the following mechanisms:

1. Abrasive wear of all tool materials due to the presence of hard particles in the workpiece (Ramalingam and Watson, 1973; Ramalingam et al., 1978).
2. Delamination wear of HSS and alumina tools when the tool undergoes plastic deformation leading to subsurface crack nucleation and propagation. This can occur when HSS tools soften due to annealing during machining (Cook, 1973; Suh, 1977).
3. Solution wear due to the dissolution of the tool materials (e.g., carbides, CBN, and diamond) in the chip at high cutting speeds (Kramer and Suh, 1980).
4. Chemical reaction of the tool material with the workpiece and subsequent removal of the reacted layer, which occurs when titanium is machined with, for example, a diamond tool (Hartung and Kramer, 1982).
5. Wear due to the diffusion of elements in the tool material into the chip, leading to weakening and ultimate failure of the cutting tool material (Cook and Nayak, 1969; Loladze, 1962).

The mechanism of abrasive wear was discussed extensively in Chapter 7 and delamination wear in Chapter 5. Therefore, these two wear mechanisms that lead to wear of cutting tools will not be discussed further in this chapter.

However, the wear due to chemical instability of the cutting tools, which is responsible for both the solution wear and the wear by chemical reaction, is a unique mechanism that has not been considered thus far and warrants further consideration in this chapter. Although this type of wear is normally observed when the interfacial temperature reaches a high value, it can also occur at low temperatures in sliding contact, depending on the chemical stability of the materials. Another high-temperature phenomenon considered in the past is diffusion wear, but there is little quantitative information that supports the existence of diffusion-controlled wear in practice, although theoretically it is a possible mechanism.

8.5 SOLUTION WEAR

8.5.1 Competing Rate Mechanisms

When a pair of sliding materials come in contact at the interface, the materials dissolve in each other if the free energy of the material pair decreases by the formation of a solution. The rate of dissolution increases with the interfacial temperature rise. The dissolved material can be transported by many different mechanisms:

1. Diffusion away from the interface
2. Convective transport by the macroscopic flow away from the interface
3. Subsurface failure below the dissolved layer, creating loose wear particles of the dissolved materials by delamination wear

This type of wear mechanism occurs when a hard cemented tungsten carbide or CBN tool is used to cut steel. Obviously, the concern here is the wear of the carbide tool. The wear rate is controlled either by the rate of dissolution or by one of the transport mechanisms cited above. In many practical situations, the wear rate due to these transport mechanisms is expected to increase in the order listed, when the transport rate, rather than the dissolution rate, is the rate-determining process. Therefore, when convective transport is present, diffusion is not an important mechanism, since the material transport by convection is much faster. Similarly, when subsurface cracks readily propagate and can create large wear particles, the wear rate due to delamination is so much greater than that due to convection that transport by convection can be ignored. This was the case in normal sliding wear discussed in Chapter 5.

When the wear rate is controlled by the dissolution rate rather than by convective transport or diffusion it is called *solution wear* (Kramer and Suh, 1980; Kramer, 1979). When the rate is controlled by mass diffusion rate, the wear is called *diffusion wear*. It should be noted that all these wear processes are caused by the fact that the materials are not chemically stable enough to resist dissolution or decomposition. Hence, these wear processes may be called *wear due to chemical instability*.

Convective transport of dissolved materials from the interface can occur due to both relative tangential motion (i.e., the velocity component parallel to the interface) of the sliding surface and the velocity component normal to the interface. The tangential motion may, however, have a very limited capacity to remove the dissolved material since the chip material at the interface can very quickly reach a fully saturated state while sliding along the tool surface. Therefore, in metal cutting the convective transport of the dissolved material from the interface is done primarily by the velocity component normal to the surface. Figure 8.9 shows a plausible velocity gradient of the tangential component of the chip velocity, v_x, very near the chip–tool interface. The material very near the interface slows down because of the interfacial friction (i.e., $\dfrac{\partial v_x}{\partial x}$ is negative). Then, from the continuity relation for the plane strain case,

$$\frac{\partial v_x}{\partial x} + \frac{\partial v_y}{\partial y} = 0 \tag{8.1}$$

It can be seen that $\partial v_y/\partial y$ must be greater than zero. This positive velocity component normal to the interface, v_y, can transport away the material dissolved at the interface. If v_y can remove all the dissolved material, the wear rate will be controlled by the dissolution rate. Then the transport rate from the interface is

$$\frac{1}{A}\frac{\partial m}{\partial t} = \left(cv_y - D\frac{\partial c}{\partial y}\right)_{y=0} \tag{8.2}$$

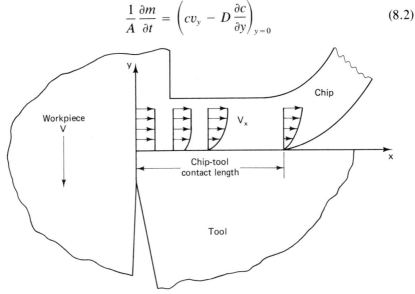

Figure 8.9 Plausible velocity gradient of the tangential component of the chip veocity. (From Kramer and Suh, 1980.)

where A = interfacial area
          c = concentration of the dissolved material at the interface
          D = diffusivity of the dissolved material
         m = mass of the element leaving the interface
         v_y = y component of the velocity
         y = coordinate axis perpendicular to the interface

8.5.2 Rate of Dissolution

When two solids A and B come into contact, the materials may dissolve in each other if the free energy decreases due to the formation of a solution. The rate of dissolution $\partial c/\partial t$ of species A right at the interface may be written, assuming that the dissolved material is not transported away, as

$$\dot{c} = \frac{\partial c}{\partial t} = N_0 \nu \beta \exp\left(-\frac{\Delta G}{kT}\right) \tag{8.3}$$

where ΔG is the activation energy for dissolution, which may be taken to be equal to the change in free energy due to the formation of solid solution; T the absolute temperature at the interface; k the Boltzmann constant; N_0 the number of available substitutional sites for A in B per unit volume; ν the frequency of atomic oscillation; and β the difference between the number of atoms leaving the solution and joining the solution. The product $N_0\nu\beta$ is sometimes called the frequency factor. If we denote the maximum solubility as c_s, then N_0 may be related to c_s and c as

$$N_0 = \alpha(c_s - c) \tag{8.4}$$

where α is a proportionality constant. Substituting Eq. (8.4) into Eq. (8.3) and rearranging the terms, we obtain

$$\frac{dc}{c_s - c} = \nu\beta' \exp\left(-\frac{\Delta G}{kT}\right) dt \tag{8.5}$$

where $\beta' = \alpha\beta$. Integration of Eq. (8.5) from $t = 0$ to $t = t$ yields

$$c = c_s \left\{1 - \exp\left[-\nu\beta' \exp\left(-\frac{\Delta G}{kT}\right)t\right]\right\} \tag{8.6}$$

At large values of the exponential term (i.e., large t and T) the amount of material A dissolved in B is approximately equal to c_s, the equilibrium concentration level. It can be seen that the effect of the Gibbs free-energy change on the rate of dissolution is very large, as indicated by the double exponential function. The Gibbs free-energy change associated with dissociation of most carbides, oxides, and nitrides, and their dissolution in the chip material, are discussed later in this section.

If the material is transported away from the interface rapidly, the number of available sites for dissolution will not decrease as rapidly as indicated by Eq. (8.4). Then N_0 may be expressed as

$$N_0 = \alpha'(c_s - c)^n \tag{8.7}$$

where n is a constant which is less than or equal to 1, and α' is a proportionality constant. In this case it will take longer to reach the steady-state value.

8.5.3 Steady-State Wear by Dissolution

When the solution at the interface is at the equilibrium concentration c_s and when the wear process has reached a steady state, the wear rate of the material being dissolved away may be written as (Kramer and Suh, 1980; Kramer, 1979)

$$\dot{V} = \frac{\partial V}{\partial t} = \int K\left(-D\frac{\partial c}{\partial y} + c_s v_y\right) dA \tag{8.8}$$

where $\dot{V}$ = wear rate of the material being dissolved
  K = ratio of the molar volumes of the tool material being dissolved and the deforming workpiece material, respectively
  D = diffusivity of the slowest diffusing solute constituent in the deforming material (i.e., the material being cut in the case of metal cutting)
  c = concentration of the dissolving material (i.e., the tool) in the deforming material (i.e., the material being cut)
  c_s = equilibrium concentration of the tool material (i.e., the dissolving material) in the material being cut
  v_y = bulk velocity of the chip material at the sliding interface perpendicular to the interface
  A = interfacial area

The diffusion flux term [i.e., the first term on the right-hand side of Eq. (8.8)] is much smaller than the second term when the dissolving material deforms a great deal, establishing a sharp velocity gradient at the sliding surface. In this case the wear rate may be expressed as

$$\dot{V} = \int K c_s v_y \, dA \tag{8.9}$$

Equation (8.9) may be expressed in terms of local crater wear rate, v_c, as

$$v_c = K c_s v_y \tag{8.10}$$

In most cases, the equilibrium concentration can be computed for a given pair of materials. In theory, v_y can also be determined from knowledge of the interfacial friction, the temperature field, and the deformation behavior of the deforming body (i.e., the chip). In practice, however, it is difficult

Sec. 8.5 Solution Wear

because the frictional force at the interface cannot be predicted a priori. It is expected that v_y is the same when a given material (i.e., deforming solvent material) is slid against various solute materials with similar characteristics, such as group IVB carbides. In this case Eq. (8.10) may be used to determine the relative wear rates.

The equilibrium concentration c_s of the dissolving material in chip material can be calculated from thermodynamic considerations. At equilibrium, the free energy of formation of the dissolving material must be equal to the free-energy change at constant pressure and temperature associated with the formation of solution, which may be expressed as (see Appendix 8.A for a brief introduction to thermodynamics)

$$\Delta G_{ij} = \Delta \overline{G}_i + \Delta \overline{G}_j = \Delta \mu_i + \Delta \mu_j \tag{8.11}$$

where ΔG_{ij} = free energy of formation of the compound ij (i.e., the tool material)
$\Delta \overline{G}_i = \Delta(\partial G/\partial n_i)_{T,P,n_j}$ = relative partial molar free energy of the solution due to the change in the amount of component i in the solvent material (i.e., the chip material)
$\Delta \overline{G}_j = \Delta(\partial G/\partial n_j)_{T,P,n_i}$ = relative partial molar free energy of the solution due to the change in the amount of component j in the solvent material (i.e., the chip material)
μ_i = chemical potential of component i = $(\partial G/\partial n_i)_{T,P,n_j}$
μ_j = chemical potential of component j = $\left(\dfrac{\partial G}{\partial n_j}\right)_{T,P,n_i}$

The relative partial molar free energy of solution of component i can be expressed in terms of the relative partial molar excess free energy of solution [i.e., Eq. (8.A38)] as

$$\Delta \overline{G}_i = \Delta \overline{G}_i^{xs} + RT \ln c_i \tag{8.12}$$

where c_i is the concentration (i.e., the mole fraction) of component i and ΔG^{xs} is the relative partial molar excess free energy of solution of component i. If the solution obeys Henry's law, the excess free energy will remain constant with solute concentration at a given temperature.

Substituting Eq. (8.12) for $\Delta \overline{G}_i$ and $\Delta \overline{G}_j$ into Eq. (8.11) and solving for equilibrium concentrations, we obtain

$$\ln c_i + \ln c_j = \frac{1}{RT}(\Delta G_{ij} - \Delta \overline{G}_i^{xs} - \Delta \overline{G}_j^{xs}) \tag{8.13}$$

From stoichiometry the relationship between the concentrations c_i and c_j can be found. When c_i and c_j are the same, Eq. (8.13) may be written as

$$c_i = c_j = \exp\left[\frac{1}{2RT}(\Delta G_{ij} - \Delta \overline{G}_i^{xs} - \Delta \overline{G}_j^{xs})\right] \tag{8.14}$$

By estimating the values of ΔG_{ij}, $\Delta \overline{G}_i^{xs}$, and $\Delta \overline{G}_j^{xs}$, the concentrations c_i and c_j can be found. Appendix 8.B gives various estimated values of these thermodynamic properties.

The relative wear rates of various carbides of the group IVB and VB metals in cutting steel can be estimated by taking the ratio of Eq. (8.10) for different tool materials as

$$\frac{(v_c)_1}{(v_c)_2} = \frac{(Kc_s)_1}{(Kc_s)_2} \tag{8.15}$$

For example, the relative wear ratio of tools coated with TiC and HfC can be estimated as follows:

1. The solubility of TiC in α-iron, at equilibrium, can be calculated from

$$\Delta G_{\text{TiC}} = \Delta \overline{G}_{\text{Ti}} + \Delta \overline{G}_{\text{C}} \tag{8.16}$$

The partial molar excess free energy of Ti in iron and C in iron are -6.9 and 7.6 kcal/mol, respectively. The free energy of formation of TiC is -39.52 kcal/mol. The concentration of Ti and C are equal. The substitution of these numbers into Eq. (8.14) gives concentrations at $T = 1600$ K of

$$C_{\text{TiC}} = C_{\text{Ti}} = C_{\text{C}} = 1.86 \times 10^{-3} \quad (\text{mol/mol}) \tag{8.17}$$

2. By similar calculation, the concentration of dissolved Hf and C at 1600 K is determined to be

$$C_{\text{HfC}} = C_{\text{Hf}} = C_{\text{C}} = 1.97 \times 10^{-4} \quad (\text{mol/mol}) \tag{8.18}$$

3. The relative wear rates can be calculated using Eq. (8.15) as

$$\frac{\text{wear rate of TiC}}{\text{wear rate of Hfc}} = \frac{(1.86 \times 10^{-3})(12.20)}{(1.97 \times 10^{-4})(15.04)} = 7.65 \tag{8.19}$$

Table 8.1 presents the calculated values of the relative wear rates of various group IVB and VB carbides and experimental results. The experimental results are obtained using CVD coated carbide tools, which is described in greater detail in Chapter 9. The agreement is excellent, except for the wear rate of TiC. These values are obtained by Kramer (1979) assuming that the iron remains as body-centered-cubic iron, although the temperatures are high enough to transform it to γ-iron (i.e., face-centered cubic) under normal steady-state conditions. The assumption is reasonable in that the $\alpha \rightarrow \gamma$ transition takes a finite period of time, whereas the temperature in the chip rises rapidly during cutting, which may not allow sufficient time for the $\alpha \rightarrow \gamma$ transition.

The discrepancy between the theoretically predicted and experimentally determined wear rate of TiC is due to the reaction of TiC with oxygen which comes from the steel workpiece. The number of oxygen atoms in

Sec. 8.5 Solution Wear

TABLE 8.1 Comparison of Theoretical Predictions with Test Results: Predicted Relative Wear Rates at Various Temperatures[a]
$[v_{wear}(HfC) = 1]$

	Temperature				
Carbide	1600 K	1500 K	1400 K	1300 K	Test Result
HfC	1	1	1	1	1
TiC	7.65	8.82	10.6	12.8	2.75
$TiC_{0.75}O_{0.25}$	2.26	2.41	2.61	2.86	—
ZrC	4.44	4.87	5.47	6.20	5.51
TaC	6.33	7.19	8.39	9.98	10.6
	(9.43)	(10.7)	(12.5)	(14.9)	
NbC	9.13	10.6	12.7	15.6	23.7
	(31.0)	(36.0)	(43.1)	(52.8)	
WC	(107.)	(153.)	(215.)	(332.)	(237.)

[a] Terms without parentheses are calculated on the basis of solubility estimated from thermodynamic properties. Terms in parentheses are calculated using the reported solubilities (see Table 8.3).
Source: Kramer and Suh (1980).

the steel is believed to be sufficient to convert the titanium carbide into titanium oxycarbide TiC_xO_y. Cutting tests with coatings of TiC and $TiC_{0.75}O_{0.25}$ in machining steel showed that both tools have similar wear rates (Carson et al., 1976), supporting the hypothesis that TiC is indeed converted into titanium oxycarbide during the machining process. The agreement between the predicted wear rate for $TiC_{0.75}O_{0.25}$ and the experimentally obtained wear rate for TiC in Table 8.1 is excellent. While all the carbides listed in Table 8.2 can substitute carbon with oxygen, only titanium and niobium can form stable monoxides.

The estimated and reported solubilities of potential tool materials in α-iron are given in Table 8.3, which shows that oxides are the most chemically stable materials, their solubility being several orders of magnitude lower than those of carbides. Table 8.4 shows the wear rates and the tool life

TABLE 8.2 Solubilities of the Carbides in the Iron-Group Metals at 1523 K

	Solubility [wt % (mol %)]		
Carbide	Cobalt	Nickel	Iron
WC	22 (7.9)	12 (3.9)	7 (2.2)
TiC	1 (1.0)	5 (4.9)	<0.5 (<0.5)
VC	6 (5.6)	7 (6.6)	3 (2.7)
NbC	5 (2.9)	3 (1.7)	1 (0.53)
TaC	3 (.93)	5 (1.6)	0.5 (0.15)

TABLE 8.3 Estimated and Reported Solubilities of Potential Tool Materials in α-Fe at 1600 K

Potential Tool Material	Free Energy of Formation (cal/mol)	Estimated Equilibrium Concentration (solubility)	Exptl. Results[a] Extrapolated to 1600 K
ZrO_2	$-190,300$[b]	3.60×10^{-8}	
Al_2O_3	$-278,300$[b]	5.55×10^{-7}	
Ti_2O_3	$-260,800$[b]	8.22	
TiO_2	$-156,300$[b]	1.52×10^{-6}	
TiO	$-91,020$[c]	1.40×10^{-5}	
HfN	$-52,604$[c]	1.53×10^{-4}	
HfC	$-49,122$[d]	1.97	
ZrH	$-51,356$[c]	2.93	
$TiC_{0.75}O_{0.25}$	$-52,395$[e]	5.42	
ZrC	$-42,714$[d]	8.42	
TiN	$-45,150$[d]	1.04×10^{-3}	
TaC	$-34,604$[d]	1.41	2.1×10^{-3}
TiC (iron)	$-39,520$[d]	1.86	6.1×10^{-3}
NbC	$-32,236$[d]	2.01	6.8×10^{-3}
BN (graphitic)	$-26,100$[f]	9.65	
WC	$-8,144$[d]	—	2.6×10^{-2}
VC	$-23,416$[d]	—	3.2×10^{-2}
TiC (nickel)	$-39,520$[d]	2.24×10^{-2}	6.3×10^{-2}
Diamond	—	9.30	
Si_3N_4	$-51,850$[b]	9.50	
β-SiC	$-14,548$[b]	4.30×10^{-1}	

[a] From Edwards and Raine (1953).
[b] From Stull and Prophet (1971).
[c] From Schick (1966).
[d] From Hultgren et al. (1973).
[e] From Caron et al. (1976).
[f] From Elliott et al. (1963).
Source: Kramer and Suh (1980).

for these materials, assuming that the wear rates of these oxide tools are also controlled by the solubility (Kramer, 1979). According to this calculation the tool made of alumina should last more than 23 days when the interfacial temperature is at 1300 K (see Table 8.4). Clearly, this kind of tool life has not been obtained with alumina tools. This indicates that these tools fail by other mechanisms, such as abrasive wear and delamination wear. As shown in Tables 8.5 and 8.6, oxides have much lower melting points and have lower flow strength at a given temperature than carbides, making them susceptible to deformation-induced mechanical wear processes. Delamination wear of alumina has been observed (Swain, 1975).

Sec. 8.5 Solution Wear

TABLE 8.4 Predicted Relative Solution Wear Rates of Potential Steel-Cutting Materials at 1300 K

Potential Tool Material	Predicted Relative Solution Wear Rate [v_{wear}(HfC) = 1]	Estimated Time for 25 μm of Wear	
ZrO_2	0.0000367	26	months
Al_2O_3	0.00124	23	days
Ti_2O_3	0.00245	12	days
TiO_2	0.00313	9.1	days
TiO	0.0333	21	hr
HfN	0.680	60	min
HfC	1.	41	min
ZrN	1.56	26	min
$TiC_{0.75}O_{0.25}$	2.86	14	min
TiN	5.92	6.9	min
ZrC	6.20	6.6	min
TaC	9.98	4.1	min
TiC (iron)	12.8	3.2	min
NbC	15.6	2.6	min
BN	57.0	43	sec
WC	332	7.4	sec
VC	381	6.5	sec
Diamond	445	5.5	sec
TiC (nickel)	998	2.5	sec
Si_3N_4	5,440	0.45	sec
β-SiC	10,700	0.23	sec

Source: Kramer (1979).

TABLE 8.5 Melting Points of Refractory Carbides and Oxides

Carbides	T_m (°F)	Oxides	T_m (°F)	Nitrides	T_m (°F)
WC	5030	V_2O_3	3578	HfN	5990
Al_4C_3	4900 ± 100	TiO_2 (brookite)	3340 ± 30	TaN	5590
VC	4950 ± 150	Al_2O_3	3720 ± 40	ZrN	5400
NbC	6330 ± 135	MgO	5117 ± 35	TiN	5310
TaC	7080 ± 100	ZrO_2	5010 ± 150	NbN	3720
ZrC	6225 ± 40	HfO_2	5140 ± 90	VN	3720
TiC	5550 ± 30	NbO	3533	CrN	2730
HfC	7100 ± 100	—	—	—	—

Sources: Data from Schwarzkopf and Kieffer (1954) and American Ceramic Society (1966).

TABLE 8.6 Hardness at Specified Temperatures of the Phases That May Be Present in a Steel

Phase	Hardness, H_v (kgf/mm^2)		
	400°C	600°C	800°C
Iron	45	27	10
Iron and interstitials	90	27	10
TiO	1300	1000	650
FeO	350	210	50
MgO	320	220	130
NiO	200	140	100
MnO	120	60	45
Al_2O_3	1300	1000	650
SiO_2	700	2100	300
ZrO_2	650	400	350
TiO_2	380	250	160
$MgAl_2O_4$	1250	1200	1050
$ZrSiO_4$	400	290	140

Sources: Data from Ramalingam and Watson (1973) and Westbrook (1966).

8.6 DIFFUSION-CONTROLLED WEAR

It was stated in the preceding section that the diffusion rate is normally slower than the convective transport of materials in typical metal-cutting situations. However, there can be a special situation where the tool material is shielded from the workpiece by a barrier of another substance that prevents direct contact and thus the formation of solid solutions. Such an observation was reported by Hartung and Kramer (1982) in machining a titanium alloy (Ti–6Al–4V) with a polycrystalline diamond tool. They found that certain regions of the diamond tool wore rapidly due to the high chemical reactivity of titanium, but adjacent regions did not wear much during the same cutting operation (Fig. 8.10). They postulated that under certain conditions a thin layer of titanium from the chip adhered to the tool and reacted with the tool to form titanium carbide, which prevented further contact between the chip and the tool. Once such a layer is formed, the wear rate is controlled by the diffusion rate of the tool material through the "reaction layer."

A polycrystalline diamond tool that had turned titanium for 20 minutes at 61 m/min (200 ft/min) was analyzed using a scanning electron microscope (SEM) with energy-dispersive X-ray analysis (EDX) before and after etching the worn tool with hydrofluoric acid for 20 minutes to remove unreacted metallic titanium (Hartung and Kramer, 1982). Figure 8.11 shows a backscatter image of the rake face and its titanium map of the unetched tool by EDX. It shows that the crater is covered with titanium and/or titanium compounds.

Sec. 8.6   Diffusion-Controlled Wear

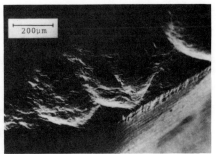

Figure 8.10 Comparison of the wear of cemented tungsten carbide (left) and polycrystalline cubic boron nitride (right) in the machining of Ti 6A1-4V at 200 sfpm (61.0 m/min). 0.005 in./rev (0.125 mm/rev) feed and 0.050 in. (1.25) mm) depth. Cutting times: WC, 10 minutes; CBN, 0.5 minute. Adherent titanium has been removed by etching HF. (From Hartung and Kramer, 1982.)

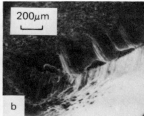

Figure 8.11 Energy dispersive x-ray analysis of a polycrystalline diamond tool after machining Ti 6A1–4V for 20 minutes. (a) electron backscatter image; (b) associated titanium map of the worn crater area; (c) secondary image of the worn tool surface. (From Hartung and Kramer, 1982.)

Figure 8.12 shows the same diamond tool after etching with hydrofluoric acid. Secondary images show the worn surface after free metallic titanium has been removed. The X-ray spectra of the etched surface show that a titanium compound is present on the crater area, the greatest concentration being present on the unworn regions of the surface. The surfaces were then subjected to the Auger electron spectroscopy (AES), which showed that the titanium compound on the unworn region of the surface consisted of titanium oxycarbide at the surface and titanium carbide at a depth of about 100 nm below the titanium oxycarbide surface.

Adelsberg and Cadoff (1967) have shown that the formation of a TiC layer on graphite from molten titanium follows a parabolic growth law of the following form:

$$x = (K_p t)^{1/2}$$
$$K_p = 0.2 \exp\left(-\frac{61{,}800}{KT}\right) \frac{\text{cm}^2}{\text{sec}} \tag{8.20}$$

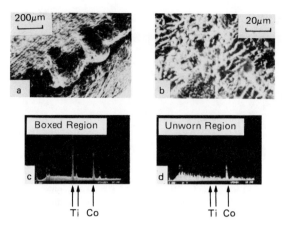

Figure 8.12 Energy dispersive x-ray analysis of the polycrystalline diamond insert of Figure 8.11 after etching in HP for 20 minutes to remove any unbound titanium from the tool surface; (a) secondary image of the etched tool surface; (b) magnification of the boxed area in the tool crater; (c) boxed region, (d) unworn region. (From Hartung and Kramer, 1982.)

where x is the layer thickness (cm), K_p the parabolic growth constant (cm^2/sec), t the carburization time (sec), T the temperature (K), and R the universal gas constant = 2 cal/mol-K. According to Eq. (8.20), it takes less than 0.5 second to form the experimentally observed 100-nm-thick TiC layer on the surface of the diamond tool after 20 minutes of cutting. Assuming that the 100-nm-thick layer is the steady-state thickness of TiC, the wear rate due to diffusion of carbon from the diamond tool to titanium chip can be computed using Fick's law as (Hartung and Kramer, 1982)

$$V = -\frac{v_t}{c_t} D \frac{\partial(c_i/v_i)}{\partial y} = -\frac{v_t D}{c_t \delta}\left(\frac{c_b}{v_b} - \frac{c_0}{v_0}\right) \quad (8.21)$$

where V is the wear rate of the tool (cm/sec), v_t the molar volume of the tool material (cm^3/mol), c_t the concentration of carbon in the tool material (mol/mol), v_i the molar volume of titanium carbide at joint i in the titanium carbide reaction layer (cm^3/mol), c_i the concentration of carbon at point i in the TiC$_s$ reaction layer, D the diffusion coefficient of carbon in the TiC reaction layer (cm^2/sec), and δ the thickness of TiC reaction layer (cm). The subscript b refers to the reaction layer–chip boundary, and subscript 0 to the reaction layer–tool boundary (cm). The term v_t/c_t is required to convert the concentration gradient into volume occupied by a mole of carbon in TiC$_x$. The carbon concentration at the reaction layer–chip boundary is assumed to be the equilibrium concentration of carbon in titanium; and that at the reaction layer–tool boundary is assumed to be the carbon concentration in diamond. Using the value for D of 9.26×10^{-11} cm^2/sec (Adelsberg and Cardoff, 1967), v_t of 3.42 cm^3/mol for diamond, c_t of 1 for diamond, v_b of 12.02 cm^3/mol, v_0 of 12.21 cm^3/mol, c_b of 0.67, c_0 of 0.98, and t of 100 nm, the predicted wear rate of diamond at 1400 K was found to be 0.47 μm/min. At 1300 and 1500 K it was calculated to be 0.1 and 1.85 μm/min, which agreed well with the experimental results 0.6 to 1.6 μm/min when machining titanium with diamond at 200 ft/min (61 m/min).

These experimental results and the theoretical prediction show that a diffusion barrier can control the wear rates under special conditions.

8.7 CONCLUDING REMARKS

In this chapter it was shown that when a pair of materials sliding against each other form a solid solution, wear of the dissolving material (i.e., solute) can occur. The dissolved material can be transported away from the interface by the following mechanisms: the mass diffusion of the dissolving material in the solvent material due to the concentration gradient, the convective transport of the dissolved material due to gross motion of the solvent material, and the generation of loose wear particles through subsurface crack propagation. It was shown that in the case of carbides of group IVB and VB metals, it is the solubility of the carbides in iron that determines the wear rates of these carbide tools in machining steel. The dissolved material is transported away by convection. The mass diffusion rate is too slow to account for the observed wear rate of these carbides in cutting steel. These wear processes, which are caused by chemical instability of materials, are important when the interfacial temperature is high such as in metal cutting.

A means of minimizing the wear due to chemical instability is to eliminate the direct contact of sliding materials by introducing a stable third material. It was shown that this may occur in some actual sliding situations such as when part of the chip material adheres to the tool material in machining titanium.

REFERENCES

ADELSBERG, L. M., and CADOFF, L. H., "The Reaction of Liquid Titanium and Hafnium with Carbon," *Transactions of the Metallurgical Society of ASME*, Vol. 239, 1967, pp. 932–935.

AMERICAN CERAMIC SOCIETY, *Engineering Properties of Selected Ceramic Materials*, American Ceramic Society, Columbus, Ohio, 1966.

BLOK, H., "Theoretical Study of Temperature Rise at Surfaces of Actual Contact under Oiliness Lubricating Conditions," *Proceedings of General Discussion on Lubrication and Lubricants*, Institution of Mechanical Engineers, London, 1938, p. 222.

BOOTHROYD, G., "Temperatures in Orthogonal Metal Cutting," *Proceedings of the Institute of Mechanical Engineers*, Vol. 177, 1963, pp. 789–810.

CARSON, W. W., LEUNG, C. L., and SUH, N. P., "Metal Oxycarbides as Cutting Tool Materials," *Journal of Engineering for Industry*, Transactions of the ASME, Vol. 98, 1976, pp. 279–286.

COOK, N. H., *Manufacturing Analysis*, Addison-Wesley, Reading, Mass., 1966.

COOK, N. H., "Temperature of Sliding Surfaces," unpublished lecture notes, MIT, Cambridge, Mass., April 1970.

COOK, N. H., "Tool Wear and Tool Life," *Journal of Engineering for Industry*, Transactions of the ASME, Vol. 95, 1973, pp. 931–938.

COOK, N. H., and BHUSHAN, B., "Sliding Surface Interface Temperatures," *Journal of Lubrication Technology*, Vol. 95, No. 1, 1973, pp. 59–64.

COOK, N. H., and NAYAK, R. N., "Development of Improved Cutting Tool Materials," Technical Report of AFML-TR-69-185, U.S. Air Force Materials Laboratory, June 15, 1969.

DEWHURST, P., "The Effect of Chip Breaker Constraints on the Mechanics of the Machining Process," *Annals of the CIRP*, Vol. 28, 1979, pp. 1–5.

EDWARDS, R., and RAINES, T., "The Solid Solubilities of Some Stable Carbides in Cobalt, Nickel, and Iron at 1250°C," *Plansee Proceedings*, 1952. F. Benesousky, ed., Mettalwerk Plansee Ges., MBH, Reutte/Tyrol, 1953, p. 22.

ELLIOT, R., *Constitution of the Binary Alloys, First Supplement*, McGraw-Hill, New York, 1965.

ELLIOTT, J., GLEISER, M., and RAMAKRISHNA, V., *Thermochemistry for Steelmaking*, Addison-Wesley, Reading, Mass., 1963.

HANSEN, M., *Constitution of the Binary Alloys*, McGraw-Hill, New York, 1958.

HARTUNG, P., and KRAMER, B. M., "Tool Wear in Titanium Machining," *CIRP Annals*, International Institution for Production Engineering Research, Vol. 31, 1982, pp. 78–80.

HASSELMAN, D. P., KANE, G. E., SCLAR, C. B., KIM, C. H., GHATE, B., SMITH, W. C., and BASU, D., "Synthesis and Evaluation of Multi-component Ceramic Oxide Cutting Tool Materials," National Science Foundation Hard Materials Research, Vol. 2, NTIS Publ. PB-221908, 1973.

HILLARD, J., "Iron–Carbide Phase Diagrams: Isobaric Sections of the Eutectoid Region at 35, 50, and 65 Kilobars," *Transactions of the Metallurgical Society of ASME*, Vol. 27, 1963, p. 249.

HULTGREN, R., et al., *Selected Values of the Thermodynamic Properties of Binary Alloys*, American Society for Metals, Metals Park, Ohio, 1973.

HUMENIK, M., and PARIKH, N. M., "Fundamental Concepts Related to Microstructure and Physical Properties of Cermet Systems," *Journal of the American Ceramic Society*, Vol. 39, 1956, p. 60.

JAEGER, J. C., "Moving Sources of Heat and the Temperature at Sliding Contacts," *Proceedings of the Royal Society*, N.S.W., Vol. 56, 1942, pp. 203–224.

KAUFMAN, L., "Condensed State Reactions at High Pressures," in *Energetics in Metallurgical Phenomena*, Vol. 3, W. Mueller, ed., Gordon and Breach, New York, 1967, p. 53.

KOMANDURI, R., SCHROEDER, T., HAZRA, J., VON TURKOVICH, B. F., and FLOM, D. G., "On the Catastrophic Shear Instability in High-Speed Machining of an AISI 4340 Steel," ASME Paper 82, PROD-7, 1982.

KRAMER, B. M., Ph.D. thesis, MIT, Cambridge, Mass., 1979.

KRAMER, B. M., and SUH, N. P., "Tool Wear by Solution: A Quantitative Understanding," *Journal of Engineering for Industry*, Transactions of the ASME, Vol. 102, 1980, pp. 303–339.

KUBASCHEWSKI, O., and DENCH, W., "The Heats of Formation in the Systems Titanium Aluminum and Titanium-Iron," *Acta Metallurgica*, Vol. 3, 1955, p. 339.

KUBASCHEWSKI, O., and EVANS, E., *Metallurgical Thermochemistry*, Pergamon Press, New York, 1958.

LOLADZE, T. N., "Adhesion and Diffusion Wear in Metal Cutting," *Journal of the Institute of Engineers (India), Part MEZ*, Vol. 43, 1962.

LOEWEN, E. G., and SHAW, M. C., "On the Analysis of Cutting Tool Temperatures," *Transactions of the ASME*, Vol. 76, 1954, pp. 217–231.

MERCHANT, M. E., "Mechanics of the Metal Cutting Process: II. Plasticity Conditions in Orthogonal Cutting," *Journal of Applied Physics*, Vol. 16, 1945, p. 318.

MURAKA, P. D., BARROW, G., and HINDUJA, S., "Influence of the Process Variables on the Temperature Distribution in Orthogonal Machining Using the Finite Element Method," *International Journal of Mechanical Science*, Vol. 21, 1979, pp. 445–456.

OXLEY, P. L. B., "Machinability: A Mechanics Machining Approach," *On the Art of Cutting Metals—75 Years Later*, ASME Publ. PED-Vol. 7, 1982, pp. 37–83.

RAMALINGAM, S., and WATSON, J. D., "Tool Life Distribution, Part 4: Minor Phases in Work Material and Multiple-Injury Tool Failure," *Journal of Engineering for Industry*, Transactions of the ASME, Vol. 100, 1973, pp. 201–209.

RAMALINGAM, S., PENG, Y. I., and WATSON, J. D., "Tool Life Distribution: Part 3. Mechanism of Single Injury Tool Failure and Tool Life Distribution in Interrupted Cutting," *Journal of Engineering for Industry*, Transactions of the ASME, Vol. 100, 1978, pp. 193–200.

SCHICK, H., *Thermodynamics of Certain Refractory Compounds*, Academic Press, New York, 1966.

SCHINTLMEISTER, W., PACHER, O., PFAFFINGER, K., AND RAINE T., "Structure and Strength Effects in CVD Titanium Carbide and Titanium Nitride Coatings," *Proceedings of the 5th International Conference on CVD*, American Nuclear Society, Hinsdale, Illinois, pp. 523–539.

SCHROTER, K., "Hard Metal Composition," U.S. Patent 1,721,416, July 26, 1926.

SCHWARZKOPF, P., and KIEFFER, R., *Refractory Hard Metals*, Macmillan, New York, 1954.

SCHWARZKOPF, P., and KIEFFER, R., *Cemented Carbide*, Macmillan, New York, 1960.

SHAW, M. C., and FINNIE, I., "The Shear Stress in Metal Cutting," *Transactions of the ASME*, Vol. 77, No. 2, 1955, p. 115.

SHUNK, F., *Constitution of the Binary Alloys, Second Supplement*, McGraw-Hill, New York, 1969.

STULL, D., and PROPHET, H., *JANAF Thermochemical Tables*, 2nd ed., U.S. Superintendent of Documents Number C13.48:37, 1971.

SUH, N. P., "Coated Carbide—Past, Present, and Future," *Carbide Journal*, Vol. 9, 1977, pp. 3–9.

SUH, N. P., and FILLION, P., "Optimization of Cutting Tool Properties through the Development of Alumina Cermet," *Wear*, Vol. 62, 1980, pp. 123–137.

SWAIN, M. V., "Microscopic Observations of Abrasive Wear of Polycrystalline Alumina," *Wear*, Vol. 35, 1975, pp. 185–189.

TAY, A. O., STEVENSON, M. G., and DAVIS, G. DE VAHL, "Using the Finite Element Method to Determine Temperature Distributions in Orthonormal Machining," *Proceedings of the Institution of Mechanical Engineers*, Vol. 188, 1974, pp. 627–638.

TAYLOR, F. W., "On the Art of Cutting Metals," *Transactions of the ASME*, Vol. 28, 1907, p. 31.

TIPNIS, V., "Influence of Metallurgy on Machinability—Free Machining Steels," *On the Art of Cutting Metals—75 Years Later*, ASME Publ. PED-Vol. 7., 1982, pp. 119–131.

TURLEY, D. M., and DOYLE, E. D., "Microstructural Behavior—Its Influence on Machining," *On the Art of Cutting Metals—75 Years Later*, ASME Publ. PED-Vol. 7, 1982, pp. 99–118.

USUI, E., and SHIRAKASHI, T., "Mechanics of Machining—From Descriptive to Predictive Theory," *On the Art of Cutting Metals—75 Years Later*, ASME Publ. PED-Vol. 7, 1982, pp. 13–35.

WAGMAN, et al., *Selected Values of the Chemical Thermodynamic Properties*, NBS Technical Note 270-3, U. S. Government Printing Office, Washington, D.C., 1968.

WAGMAN, et al., NBS Technical Note 270-4, 1969, and NBS Technical Note 270-5, 1971.

WESTBROOK, J. H., "Temperature Dependence of Hardness of Some Oxides," *Revue des Hautes Températures et des Réfractaires*, Vol. 3, 1966, p. 47.

WILLIAMS, J. E., and ROLLASON, E. C., "Metallurgical and Practical Machining Parameters Affecting Build-up Edge Formation in Metal Cutting," *Journal of the Institute of Metals*, Vol. 98, 1970, p. 144.

APPENDIX 8.A

BRIEF INTRODUCTION TO THE THERMODYNAMICS OF SOLIDS

Consider the closed thermodynamic system shown in Fig. 8.A1. The *first law of thermodynamics* states that the internal energy change of the system, dE, is equal to the difference between the heat transferred to the system, δq, and the work done by the system, δw:

$$dE = \delta q - \delta w \quad (8.\text{A1})$$

where the differential sign δ implies that δq and δw are not exact differentials. If the work done by the closed system is only volumetric expansion work, $P\,dV$, then

$$\delta w = P\,dV \quad (8.\text{A2})$$

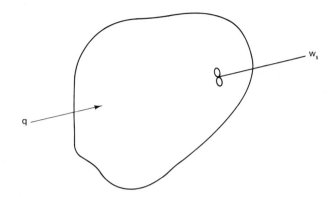

Figure 8.A1 Closed thermodynamic system into which heat q is transferred and the shaft work w_s is done on the system.

where P is the pressure and V is the volume. If the shaft work, δw_s, is also done on the system, then

$$\delta w = P\,dV - \delta w_s \tag{8.A3}$$

The *second law of thermodynamics* states that the entropy change of the system is greater than or equal to the heat transferred from the surroundings to the system as

$$dS \geq \frac{\delta q}{T} \tag{8.A4}$$

where the equal sign applies to a reversible process and the unequal sign to an irreversible process. For the closed system at equilibrium without any shaft work, Eqs. (8.A2) and (8.A4) may be substituted into Eq. (8.A1) to yield

$$dE = T\,dS - P\,dV \tag{8.A5}$$

T, S, P, and V are state functions (i.e., thermodynamic properties). An equilibrium state represents a minimum internal change.

The Gibbs free energy, G, which is a very useful thermodynamic function in chemistry and physics, is defined as

$$G = H - TS = E + PV - TS \tag{8.A6}$$

where $H = E + PV$ is also a state function called *enthalpy*. For the formation of a solution at constant pressure and temperature, through a reversible process, or for a solution at equilibrium,

$$dG = 0 \quad\quad dT = 0 \quad\quad dP = 0 \tag{8.A7}$$

For an irreversible, spontaneous process, at constant temperature and pressure,

$$dG < 0 \tag{8.A8}$$

since $dS > \delta q/T$.

When the process occurs at constant temperature and volume, the Helmholtz free energy, F, is a useful thermodynamic function, which is defined as

$$F = E - TS \tag{8.A9}$$

For a reversible process at constant temperature and volume,

$$dF = 0 \quad dT = 0 \quad dV = 0 \tag{8.A10}$$

For an irreversible process

$$dF < 0 \tag{8.A11}$$

The change in the Gibbs free energy for a process can be obtained by differentiating Eq. (8.A6) as

$$dG = dH - S\, dT - T\, dS \tag{8.A12}$$

For a process that occurs at constant temperature, dG becomes

$$dG = dH - T\, dS \tag{8A.13}$$

The physical significance of the Gibbs free energy is that it is equal to the maximum useful work that can be done by a system at constant pressure and temperature. This can be seen by expressing the Gibbs free energy for a system on which external shaft work is done. For this purpose substitute Eq. (8.A3) into Eq. (8.A1) to obtain

$$dE \leq T\, dS - P\, dV + \delta w_s \tag{8.A14}$$

which may be substituted into the expression for the Gibbs free energy, Eq. (8.A6), to obtain

$$dG \leq V\, dP - S\, dT + \delta w_s \tag{8.A15}$$

At constant pressure and temperature, Eq. (8.A15) reduces to

$$-dG \geq -\delta w_s \tag{8.A16}$$

which states that the maximum work that can be done by a system at constant temperature and pressure is equal to the negative of the Gibbs free-energy change.

Components can be added to an open system by crossing the system boundary (i.e., control volume). Therefore, the internal energy and the Gibbs free energy of an open system are functions of the thermodynamic properties and the concentration of chemical species in the system:

$$E = E(S, V, n_1, n_2, n_3, \ldots, n_k)$$
$$G = G(T, P, n_1, n_2, n_3, \ldots, n_k) \tag{8.A17}$$

The change in E and G may be expressed as

$$dE = \left(\frac{\partial E}{\partial S}\right)_{V,n_i} dS + \left(\frac{\partial E}{\partial V}\right)_{S,n_i} dV + \sum_i \left(\frac{\partial E}{\partial n_i}\right)_{S,v,n_j} dn_i \quad (8.A18)$$

$$dG = \left(\frac{\partial G}{\partial P}\right)_{T,n_i} dP + \left(\frac{\partial G}{\partial T}\right)_{P,n_i} dT + \sum_i \left(\frac{\partial G}{\partial n_i}\right)_{P,T,n_j} dn_i$$

From the definition of the Gibbs free energy and the first law of thermodynamics, we obtain

$$\left(\frac{\partial E}{\partial S}\right)_{V,n_i} = T \quad \left(\frac{\partial E}{\partial V}\right)_{S,n_i} = -P$$

$$\left(\frac{\partial G}{\partial P}\right)_{T,n_i} = V \quad \left(\frac{\partial G}{\partial T}\right)_{P,n_i} = -S \quad (8.A19)$$

The rate of change in the Gibbs free energy due to the addition of components is called the *partial molar Gibbs free energy*, $\overline{G}_i$, or the chemical potential of component i, μ_i:

$$\mu_i = \left(\frac{\partial G}{\partial n_i}\right)_{P,T,n_j} = \overline{G}_i \quad (8.A20)$$

Chemical potential may be interpreted as the additional capability of a phase to do work as a result of an infinitesimal increase in the amount of component i. At constant temperature and pressure, the variation in the Gibbs free energy is due only to variations in the number of moles of the phase:

$$dG = \sum_i \mu_i \, dn_i \quad (8.A21)$$

For an ideal gas at constant temperature, the Gibbs free energy is

$$dG = V \, dP \quad (8.A22)$$

Using the ideal gas law, $PV = nRT$, Eq. (8.A22) may be integrated to yield

$$G - G^\circ = nRT \ln \frac{P}{P^\circ}$$

or $\quad (8.A23)$

$$\mu(g) - \mu^\circ(g) = RT \ln \frac{P}{P^\circ}$$

where the superscript $^\circ$ refers to the standard state of 1 atm. Then Eq. (8.A23) can be written as

$$\mu(g) = \mu^\circ(g) + RT \ln P \quad (8.A24)$$

$\mu°(g)$ is the chemical potential of the gas, which is a function of only temperature. For a mixture of gases, the chemical potential of component i is

$$\mu_i(g) = \mu_i°(g) + RT \ln P_i \tag{8.A25}$$

where P_i is the partial pressure of the gas component. In terms of the mole fraction of component i, $c_i = P_i/P$, Eq. (8.A25) may be written as

$$\mu_i(g) = \mu_i°(g) + RT \ln P + RT \ln c_i \tag{8.A26}$$

Equation (8.A25) applies only to ideal gases. However, the expression is so convenient that the form of the equation is preserved even for nonideal gases. For nonideal gases, P_i in Eq. (8.A25) is replaced by f_i, which is called *fugacity*.

The chemical potential of an ideal solution can be similarly written as

$$\mu_i(\text{sol}) = \mu_i°(\text{sol}) + RT \ln c_i \tag{8.A27}$$

If the component i in the solution is in equilibrium with the vapor phase of the same component,

$$\mu_i(\text{sol}) = \mu_i(g) \tag{8.A28}$$

Combining Eqs. (8.A25) and (8.A27), we obtain

$$P_i = c_i \exp\left[\frac{\mu_i°(\text{sol}) - \mu_i°(g)}{RT}\right] \tag{8.A29}$$

Henry's law for a dilute solution can be written as

$$P_i = K_i c_i \tag{8.A30}$$

which states that the partial pressure of an ideal volatile solute is proportional to the mole fraction in the solution.

The expression for chemical potential given by Eq. (8.A27) was for an ideal solution. For real solutions the actual concentration c_i is replaced by an idealized one, called activities a_i, which is the ratio of the fugacity of the substance in the state in which it happens to be to its fugacity in its standard state (i.e., $a_i = f_i/f_i°$). For real solutions, the chemical potential may be expressed as

$$\mu_i = \mu_i° + RT \ln a_i \tag{8.A31}$$

The relationship between the actual concentration c_i and the activity a_i is expressed by an activity coefficient γ_i as

$$a_i = \gamma_i c_i \tag{8.A32}$$

Substituting Eq. (8.A32) into Eq. (8.A31), the expression for chemical potential may be written as

$$\mu_i = \mu_i^\circ + RT \ln \gamma_i + RT \ln c_i$$

or (8.A33)

$$\Delta\mu_i = \mu_i - \mu_i^\circ = RT \ln \gamma_i + RT \ln c_i$$

The concept of excess free energy and regular solution was introduced by Hildebrand (1929), which is useful in dealing with nonideal solutions. A regular solution is defined as one for which the entropy of formation and hence the partial molar[1] entropies are the same as for an ideal solution. Then the partial molar Gibbs free energy of component i of solution, as well as other thermodynamic properties, may be written as the sum of the ideal part, $\overline{G}^{id}$, and the excess part, $\overline{G}^{xs}$, as

$$\overline{G}_i = \overline{G}_i^{xs} + \overline{G}_i^{id} \tag{8.A34}$$

Subtracting the corresponding thermodynamic function for the pure component, $n_i \overline{G}_i^\circ$, Eq. (8.A34) may be written as

$$\Delta\overline{G}_i = \Delta\overline{G}_i^{xs} + \Delta\overline{G}_i^{id} \tag{8.A35}$$

At constant temperature, the relative partial molar free energy of an ideal solution is

$$\Delta\overline{G}_i^{id} = \Delta\overline{H}_i^{id} - T \Delta\overline{S}_i^{id} \tag{8.A36}$$

The partial molar enthalpy for the ideal solution, $\Delta\overline{H}^{id}$, is equal to zero at constant temperature. The partial molar entropy change for an ideal solution, $\Delta\overline{S}^{id} = \Sigma_i^n c_i \Delta\overline{S}_i^{id}$, is

$$\Delta\overline{S}^{id} = R \sum_i^n (c_i \ln c_i) \tag{8.A37}$$

Substituting Eqs. (8.A37) and (8.A36) into Eq. (8.A35), we obtain

$$\Delta\overline{G}_i = \Delta\overline{G}_i^{xs} + RT \ln c_i \tag{8.A38}$$

where $\overline{G}_i^{xs}$ is called the partial molar excess free energy of solution of component i. The first term on the right-hand side of Eq. (8.A34) represents the excess free energy associated with enthalpy of mixing, whereas the second term is associated with entropy of mixing. When two elements being mixed have the same affinity, the excess free energy is equal to zero and is called a *perfect solution*. When they have stronger affinity to each other than to their own species, the free energy decreases more than that

---

[1] The name *partial molal* property is also used, mostly by chemists.

given by the entropic term. Conversely, when they have weaker affinity, the free energy change is less negative.

APPENDIX 8.B

ESTIMATION OF THERMODYNAMIC PROPERTIES[2]

Estimation of the Relative Partial Molar Free Energies of Solution of the Tool Constituents in the Chip

The purpose of this appendix is to estimate the free energies of solution from the limited data that are available. The relative partial molar free energies of solution of the tool constituents in the chip must be known before the solubility of the tool material can be estimated.

All tool constituents are assumed to dissolve in the chip in accordance with Henry's law. Therefore, the relative partial molar free energy of solution of component i may be expressed as

$$\Delta \overline{G}_i = RT \ln a_i = RT \ln (\gamma_i c_i) \qquad (8.B1)$$

where a_i, γ_i, and c_i are the activity, activity coefficient, and the concentration of component i, respectively, and γ_i is a constant in the solution range. Alternatively,

$$\Delta \overline{G}_i = RT \ln \gamma_i + RT \ln c_i = \Delta \overline{G}_i^{xs} + RT \ln c_i \qquad (8.B2)$$

where $\Delta \overline{G}_i^{xs}$ is a constant within the solution range at a given temperature.

At the solubility limit, the solution will be in equilibrium with a second phase. Therefore, the free energy of solution of component i at the solubility limit must equal the free energy of formation of the second phase per gram atom of i. Many free energies of solution are estimated herein using this technique. The solubility limits are estimated from phase diagrams and are summarized in Table 8.B1.

In all cases of steel cutting, tool constituents are assumed to be dissolving in body-centered-cubic iron. The free-energy change for transformation to the face-centered-cubic modification at 1600K is very small, -6 cal/mol at atmospheric pressure (Elliott et al., 1963). The magnitude of the free-energy change will be slightly larger at the high pressures encountered in metal cutting, due to the volume contraction that occurs on transformation (Hillard, 1963; Kaufman, 1967). In addition, the severe deformation in the cutting zone may provide significant opportunities for mechanical activation of the transformation. However, given the observed sluggishness of the

---

[2] From Kramer and Suh, 1980.

TABLE 8.B1 Estimated Solubility Limits of the Elements of Interest in Body-Centered-Cubic Iron at 1600 K

Element	Solubility (at %)	Phase in Equilibrium	Note
Ti	8	Fe$_2$Ti	Sol. in α-Fe at 1600 K
Zr	3	Fe$_2$Zr	Sol. in δ-Fe at the Fe–Fe$_2$Zr eutectic temperature (1603 K)
Hf	0.9	Fe$_2$Hf	Sol. in δ-Fe at 1603 K
Nb	2.6	Fe$_2$Nb	Sol. in δ-Fe at 1600 K
Ta	1.9	Fe$_2$Ta	Sol. in δ-Fe at 1600 K
Ti (Ni)	12.5	Ni$_3$Ti	Sol. at the Ni–Ni$_3$Ti eutectic temperature (1560 K)
Si	19	Liquid	Sol. in α-Fe at 1600 K
C (Ni)	2.7	Graphite	Sol. at the Ni–graphite eutectic temperature (1591 K)
O	0.032	Liq. FeO	Sol. in δ-Fe at 1 atm and 1673 K

Sources: Elliott (1965), Hansen (1958), and Shunk (1969).

$\gamma \rightarrow \alpha$ transformation in steel and realizing that the total time of heating of the chip is on the order of 0.1 msec, the assumption of a metastable α phase in the γ stability field seems the most reasonable one.

The relative partial molar excess free energies of formation of interest are summarized in Table 8.B2. The techniques used to estimate these values are detailed in the following sections.

Estimation of the Free Energies of Solution of the Group IVB and VB Metals

Kubaschewski and Dench (1955) have investigated the system Fe–Ti by heating cold-pressed powder compacts until reaction occurred. They were unable to produce Fe$_2$Ti by this means. Instead, mixtures of FeTi and Fe were obtained. However, specimens showing X-ray powder diffraction peaks for Fe$_2$Ti have been produced by cooling FeTi melts in vacuo. Kubaschewski suggests that when more than one intermetallic compound forms between elements, melting point may be indicative of the relative enthalpies of formation, those compounds with similar melting points having similar enthalpies of formation per gram atom. FeTi decomposes peritectically to Fe$_2$Ti and liquid at 1,590 K. Fe$_2$Ti melts congruently at 1700 K. These temperatures are quite similar. Therefore, it is expected that Fe$_2$Ti and FeTi have similar enthalpies of formation on a gram-atom basis.

The enthalpy of formation of FeTi is -4850 ± 150 cal/g-atom. Taking the enthalpy of formation of Fe$_2$Ti to be equal, on a gram-atom basis, to the enthalpy of formation of the compound, Fe$_2$Ti is $-14,550 \pm 450$ cal/mol.

TABLE 8.B2 Estimated
Excess Free Energies of
Solution of Tool
Constituents in α-Iron

Tool Constituent	$\Delta \overline{G}_i^{xs}$ (cal/mol)
Ti	−6,900
Ti (Ni)	−26,800
Zr	−5,000
Hf	−2,100
V	−9,100
Nb	−100
Ta	−200
Al	−10,700
Si	−16,700
B	−2,100
C	+7,600
C (Ni)	+11,600
O	−12,600
N	+5,700

Source: Kramer and Suh (1980).

The entropy of formation of Fe_2Ti is unknown. The fact that Fe_2Ti can be produced on cooling from the melt but not on heating suggests positive entropy of formation and increased stability at high temperatures. $[\Delta H(298 \text{ K}) - \Delta G(298 \text{ K})]$ of Fe_2Nb is reported as 700 cal/mol (Wagman et al., 1968; 1969; 1971), indicating a $\Delta S(298 \text{ K})$ of 2.35 cal/mol-K for Fe_2Nb.

Rather than introducing significant terms on the basis of one determination of unknown accuracy, the enthalpies of formation of the intermetallic compounds have been taken as a measure of the free energies of formation (i.e., $\Delta S = 0$). It should be kept in mind that the free energies is overestimated in this way and that the error may be significant. In the case of Fe_2Nb, if $\Delta S(1600K) = \Delta S(298K)$, $-T \Delta S(1600K)$ is equal to $-3,760$ cal/mol. When more reliable data concerning the entropies of formation and the variation of the enthalpies of formation with temperature for the compounds of interest become available, a determination can be made.

As pointed out by (Kubaschewski and Evans, 1958), the methods of estimating enthalpies of formation are not very reliable and their uncertainty is such that as many methods as possible should be used to estimate a single value.

The structures and melting points of the group IVB and group VB intermetallic compounds of the type Fe_2X are summarized in Table 8.B3. All five compounds are of the Laves type and have very closely related

TABLE 8.B3 Melting Points and Volume Changes on Formation of Group IVB and VB Intermetallic Compounds with Iron

Formula	Structure Type	Melting Point (K)	Effective Volume Change on Formation (%)	Note
FeTi	CsCl	1590	−13.28	Peritectic temp.
Fe$_2$Ti	MgZn$_2$	1700	−14.05	Melts congruently
Fe$_2$Zr	MgCu$_2$	1918	−14.62	Melts congruently
Fe$_2$Hf	MgNi$_2$	2083	−13.75	Melts congruently
Fe$_2$Nb	MgZn$_2$	1928	−10.13	Melts congruently
Fe$_2$Ta	MgZn$_2$	2048	−11.40	Melts congruently

Source: Elliott (1965), Hansen (1958), and Shunk (1969).

structures. All five compounds melt congruently and have melting point maxima corresponding to the composition Fe$_2$X. Within a given group it may be reasonable to estimate relative enthalpies of formation by means of absolute melting-point ratios. Free energies of formation of Fe$_2$Zr and Fe$_2$Hf are estimated from that of Fe$_2$Ti in this way. The free energy of formation of Fe$_2$Ta is estimated from the known values for Fe$_2$Nb. Results are summarized in Table 8.B4.

Determinations from melting-point ratios show regular variation in stability with increase in atomic number within a group in keeping with observed homologous relationships in other systems (Kubaschewski and Evans, 1958). However, it should be noted that the quantitative determination of enthalpies of formation from melting-point ratios is somewhat novel. Therefore, independent checks are required.

Enthalpies of formation can also be calculated from the volume change on formation of a compound. Effective volume changes on formation are summarized in Table 8.B3. The details of the techniques used in calculation

TABLE 8.B4 Estimated Free Energies of Formation of Group IVB and VB Intermetallic Compounds (cal/mol)

Formula	Δ from Melting Point	Δ from Volume Change	ΔG Average	Note
FeTi	—	—	−9,700	ΔH(298 K)
Fe$_2$Ti	−14,550	−15,390	−14,970	See the text
Fe$_2$Zr	−16,420	−16,010	−16,215	See the text
Fe$_2$Hf	−17,830	−16,630	−17,230	See the text
Fe$_2$Nb	—	—	−11,800	ΔG(298 K)
Fe$_2$Ta	−12,530	−13,280	−12,905	See the text
Ni$_3$Ti	—	—	−33,500	ΔH(298 K)

Source: Kramer and Suh (1980).

are given by Kramer (1979). These calculations include a geometric factor that compensates for lack of close packing in non-close-packed structures. The factor is 0.95 for the CsCl-type structure. This factor has also been applied to the body-centered-cubic metals, which have identical packing. The geometric factor for the Laves phases has been estimated by reconciling the volume change of Fe_2Nb to the known enthalpies of formation of FeTi and Fe_2Nb and the volume change of FeTi, which has CsCl-type structure. The factor so determined is 0.89.

Calculations based on the volume change of formation of Fe_2Hf indicate an enthalpy of formation of $-15{,}060$ cal/mol. This value runs contrary to the trend toward increasing stability with atomic number in the series Ti,Zr and Nb,Ta. According to Kubaschewski and Evans (1958), calculations of the enthalpies of formation from the volume changes on formation are particularly unreliable for elements with atomic numbers greater than 57 (i.e., those subject to the lanthanide contraction). Therefore, two additional estimates of the enthalpy of formation of Fe_2Hf are obtained by assuming that homologous relationships exist between Fe_2Hf and the intermetallic compounds formed from the neighboring elements. Extension of the series Ti,Zr yields an estimated heat of formation of $-16{,}630$ cal/mol. An analogy between Nb-Ta and Zr-Hf yields $-18{,}190$ cal/mol. The average of the directly determined value and the two values determined from homologous series is $-16{,}630$ cal/mol. This value is entered in Table 8.B4.

The average estimated free energies of Table 8.B4 can be employed with the estimated solubility limits of Table 8.B1 to determine relationships for the free energies of solution of the group IVB and group VB metals. An example will suffice.

With reference to Eq. (8.B2) at the solubility limit of titanium in α-iron at 1600K:

$$\Delta \overline{G}_{Ti} = \Delta \overline{G}_{Fe_2Ti} = -14{,}970 = \Delta \overline{G}_{Ti}^{xs} + 3200 \ln 0.08 \qquad (8.B3)$$

Therefore, the excess free energy of solution of Ti in α-Fe at the solubility limit is -6890 cal/mol. If in accordance with Henry's law, the excess free energy of solution is assumed to remain constant within the solubility range, the free energy of solution of titanium in α-iron at 1600K can be calculated as

$$\Delta \overline{G}_{Ti} = -6890 + RT \ln c_{Ti} \qquad (8.B4)$$

The calculated values of $\Delta \overline{G}_i^{xs}$ for the group IVB and VB metals are entered in Table 8.B2.

Estimation of the Free Energies of Solution of Aluminum, Silicon, and Boron

The data of Kubaschewski and Dench (1955) show the enthalpy of solution of aluminum in solid iron is $-12{,}500$ cal/mol of aluminum. The enthalpy of solution is the average enthalpy of solution calculated from experimental determinations of the enthalpy of formation of iron–aluminum alloys at five concentrations ranging from 5 to 25 at %. Since 1.5 at % aluminum stabilizes α-iron, all determinations are for the solution of aluminum in a α-iron.

The increase in the integral enthalpy of formation of iron–aluminum alloys with aluminum concentration is quite linear in the range 5 to 25 at % aluminum. However, the solubility of aluminum oxide is so low that even a determination at 5 at % differs in concentration by several orders of magnitude from expected concentrations. Significant changes in the enthalpy of solution of aluminum may occur at the low aluminum concentrations of interest herein. These are not reflected in the present calculation. Assuming regularity, the free energy of solution of aluminum in iron is

$$\Delta \overline{G}_{Al} = -12,500 + RT \ln c_{Al} \quad \text{cal/mol} \quad (8.B5)$$

In the temperature range of interest, the free energy of fusion of aluminum must be subtracted to obtain the free energy of solution of molten aluminum in solid iron. At 1600K, the free energy of fusion of aluminum is -1800 cal/mol. Therefore,

$$\Delta \overline{G}_{Al}^{M} = -10,700 + RT \ln c_{Al} \quad \text{cal/mol} \quad (8.B6)$$

Iron dissolves 19 at % silicon at 1600 K. The solution is in equilibrium with liquid composed of 21 at % silicon. Elliot et al. (1963) gives the variation in the temperature range 1693 to 1973K of the activity coefficient of silicon in a 21 at % silicon–iron liquid solution. Extrapolating to 1600K, the partial molar free energy of solution of silicon in a 21 at % silicon–iron liquid at 1600K is $-22,000$ cal per mole of silicon. The partial molar free energy of solution of silicon in the 19 at % solid solution will be equal. Therefore, the free energy of solution of silicon in iron at 1600K is

$$\Delta \overline{G}_{Si} = -16,700 + RT \ln c_{Si} \quad \text{calories per mole} \quad (8.B7)$$

Essentially no data are available concerning solutions of boron in iron. Hansen (1958), Elliot (1965), and Shunk (1969) give the solubility of boron in α-iron at 1173K as 0.03 at %. This solution is in equilibrium with Fe_2B. Shaffer (1964) gives the activation energy for the reaction $B + \gamma - Fe \rightarrow Fe_2B$ as 21,160 cal/mol. Taking this as the free energy of formation in the absence of other thermodynamic data,

$$\Delta \overline{G}_{B} = -2,100 + RT \ln c_{B} \quad \text{cal/mol} \quad (8.B8)$$

There is no reason to believe that this estimate is accurate.

Estimation of the Free Energies of Solution of Carbon, Oxygen, and Nitrogen

The free energy of solution of carbon in α-iron is best estimated from the solubility data for VC in iron of Edwards and Raine (1952) and the thermodynamic data concerning solid solutions of vanadium and iron of Hultgren et al. (1973). Edwards and Raine (1952) give the solubility of VC in iron at 1523K as 3 wt %, equivalent to 2.67 mol %. 2.67 at % vanadium is sufficient to stabilize α-iron at 1523K (Hansen, 1958; Elliot, 1965; Shunk, 1969).

The excess partial molar free energy of formation of alloys of 2.67 at % vanadium in solid iron at 1600K may be interpolated from the data of Hultgren et al. (1973) to be -9051 cal/mol. Using this value at 1523K, the partial molar free energy of solution is obtained as

$$\Delta \overline{G}_v = -9{,}051 + RT \ln 0.0267 = -20{,}100 \text{ cal/mol} \tag{8.B9}$$

The free energy of formation of $V_{0.53}C_{0.47}$ at 1523K is $-23{,}560$ cal/mol (Hultgren et al., 1973).

For the reaction $VC \rightarrow [V]_{Fe} + [C]_{Fe}$, equilibrium will be established when [Eq. (8.11)]

$$\Delta G_{VC} = \Delta \overline{G}_V + \Delta \overline{G}_C \tag{8.B10}$$

Therefore

$$\Delta \overline{G}_C^{xs} = +7600 \text{ cal/mol at } 1523K \tag{8.B11}$$

The solubility of carbon in nickel at 1600K is 2.7 at %. This solution is in equilibrium with graphite. Therefore,

$$0 = \Delta \overline{G}_C^{xs}(\text{nickel}) + RT \ln 0.027 \tag{8.B12}$$
$$\Delta \overline{G}_C^{xs}(\text{nickel}) = +11{,}600 \text{ cal/mol}$$

The solubility of oxygen in δ-iron at atmospheric pressure is given as $\log (\text{wt } \% \text{ 0}) = 5.51 - (12{,}630/T)$ (Hansen, 1958; Elliot, 1965; Shunk, 1969). The solubility at 1673K, the $\gamma \rightarrow \delta$ transition temperature, is 0.032 at %. The solution will be in equilibrium with molten FeO. The free energy of formation of FeO at 1673K is $-39{,}510$ cal/mol (Stull and Prophet, 1971). Therefore

$$\overline{G}_0^{xs} = -12{,}600 \text{ cal/mol} \tag{8.B13}$$

Elliot et al. (1963) give the solubility of nitrogen in iron at 1 atm. In the temperature range of interest, the solubility of nitrogen in body-centered-cubic iron is approximated by

$$c_N = 3.28 \times 10^{-4} + 4.706 \times 10^{-7} T \tag{8.B14}$$

The partial heat of solution of nitrogen in iron at 1600K can be determined from the relationship (Kubaschewski and Evans, 1958):

$$\Delta \overline{H}_N = RT^2 (d \ln c_{sat}/dT) = +5670 \text{ cal/mol} \tag{8.B15}$$

The phase in equilibrium with the N–Fe solid solution at 1600K is unknown. Examination of existing phase diagrams (Hansen, 1958; Elliot,

1965; Shunk, 1969) which are not constructed above 1000K suggests that compounds of iron and nitrogen will not form at high temperatures and that the α-solid solution will be in equilibrium with either γ-iron or ε-iron solid solutions. The heats of transformation from α-iron to these phases may be quite small and are taken as zero. Therefore,

$$\Delta \overline{G}_N^{xs} = +5{,}670 \text{ cal/mol} \qquad (8.B16)$$

This estimate must be regarded as extremely approximate. Experimental determination at 1600K of the solubility of VN in iron and the free energy of formation of VN might yield a greatly improved estimate.

Calculated values of the excess free energies for the solution of the elements of interest in body-centered-cubic iron are listed in Table 8.B2.

APPENDIX 8.C

TEMPERATURE DISTRIBUTION AT THE SLIDING INTERFACE

The temperature at the sliding interface rises due to the work done by the external force. The temperature distribution is a function of the load, speed, the surface topography, material properties, and the environment. When the normal load acting at the interface is large, the real area of contact is nearly equal to the apparent area of contact and the heat transfer is one-dimensional. This situation exists at the chip–tool interface in metal cutting. When the applied load is low, the sliding contact occurs at a number of asperities which are distributed throughout the contact area. In this case, heat transfer from these asperity contacts into the bulk is three-dimensional. The determination of the temperature distribution in this lightly loaded case requires accurate information on the size and distribution of asperity contacts, which is not normally available. This type of interfacial contact occurs at many sliding interfaces, such as gears, cams, and so on. When a liquid lubricant is present at the sliding interface, heat is also transferred by the lubricant. At high speeds and loads, the lubricant also affects the load distribution at the interface (i.e., elastohydrodynamic lubrication).

In this appendix, the temperature distribution in two different kinds of dry sliding contacts is analyzed and compared with experimental results. The key references for materials presented in this appendix are Blok (1938), Jaeger (1942), Loewen and Shaw (1953), Cook (1970), and Cook and Bhushan (1973). For a detailed mathematical derivation, readers are referred to Jaeger (1942).

Governing Equations

The energy equation for heat conduction in solids may be written as

$$\frac{\partial^2 \Theta}{\partial x^2} + \frac{\partial^2 \Theta}{\partial y^2} + \frac{\partial^2 \Theta}{\partial z^2} + \frac{\dot{W}}{k} = \frac{1}{\alpha}\frac{\partial \Theta}{\partial t} \qquad (8.C1)$$

where Θ is temperature, (x, y, z) are the Cartesian coordinates, α is the thermal diffusivity equal to $k/\rho c$, t is time, and $\dot{W}$ is the rate of heat generated within the solid per unit volume (i.e., cal/m^3-sec). k is the thermal conductivity in cal/°C-sec-m and ρc is the volumetric specific heat in cal/°C-m^3. Equation (8.C1) is valid for an isotropic solid whose thermal properties do not change as a function of temperature.

At the sliding contacts, the heat source moves along the surface, and therefore the temperature at a given point changes with time. When there is more than one heat source and if they are close together, the temperature at a given point will be affected by many of these heat sources.

If the solid undergoes plastic deformation, the plastic work will be transformed into thermal energy. In this case $\dot{W}$ is equal to

$$\frac{d}{dt} \bar{\sigma} \, d\bar{\varepsilon}$$

where $\bar{\sigma}$ and $\bar{\varepsilon}$ are the effective stress and strain, respectively. $\dot{W}$ can be readily evaluated when numerical techniques are used to account for the bulk deformation of the sliding surfaces (e.g., Tay et al., 1974), but for the analysis presented in this appendix, it will be assumed to be zero.

Energy Partition

Blok (1938) developed the concept of energy partition to determine the fraction of heat generated at the sliding interface that goes into each one of the sliding bodies. It is done by simply solving for the temperature distribution in each body due to the heat conducted into the solid and by assuming that the interfacial temperature is the same for both bodies. If heat generation per unit time is q, the temperature distribution in each of the two sliding surfaces may be written as

$$\Theta_1(x, y, z, t) = R_1 q f_1(k_1, \rho_1, c_1, v, x, y, z, t)$$
$$\Theta_2(x, y, z, t) = (1 - R_1) q f_2(k_2, \rho_2, c_2, v, x, y, z, t) \qquad (8.C2)$$

If the contact occurs at $z = 0$, Θ_1 and Θ_2 may be assumed to be equal to each other at the contact point. Then, equating the two equations above at $z = 0$, the energy partition factor R_1 can be determined. When the heat transfer is one-dimensional, R_1 is simply a function of thermal conductivities.

Jaeger's Approach to a Moving-Heat-Source Problem

Starting from a general solution for temperature rise due to sudden release of heat Q in an infinite solid, Jaeger determined the temperature rise due to various heat sources. The temperature rise at the point (x, y, z)

at time t in an infinite solid due to a quantity of heat Q instantaneously released at (x', y', z') at $t = 0$ can be obtained by finding a solution that satisfies Eq. (8.C1) with $\dot{W} = 0$. The solution for the temperature rise due to this point source is

$$\Theta - \Theta_i = \frac{Q\alpha}{8k(\pi\alpha t)^{3/2}} \exp\left[-\frac{(x-x')^2 + (y-y')^2 + (z-z')^2}{4\alpha t}\right] \quad (8.C3)$$

where Θ_i is the initial temperature, which will be assumed to be equal to zero.

The temperature rise due to an instantaneous line source can be obtained by replacing Q with $Q\,dy'$ and by integrating the resulting equation with respect to y' from $-\infty$ to ∞. Here Q is the heat released per unit length. The solution for temperature rise at (x, y, z) due to the line heat source passing through $y = 0$ and parallel to the y axis may be expressed as

$$\Theta - \Theta_i = \frac{Q}{4\pi k t} \exp\left[-\frac{(x-x')^2 + (z-z')^2}{4\alpha t}\right] \quad (8.C4)$$

Point source. The steady-state temperature rise due to a *point heat source* moving at a constant velocity along the x axis on the surface of a semi-infinite half space $z < 0$ (see Fig. 8.C1) can be obtained from Eq. (8.C3). If we let the source be at the origin at $t = 0$, then at the time t ago, the heat source was at $x' = -Vt$. Then the temperature rise due to heat $(dQ = Q\,dt)$ liberated at $(x = -Vt)$ is

$$d\Theta_{x,y,z} = \frac{2Q\,dt}{8\rho c(\pi\alpha t)^{3/2}} \exp\left[-\frac{(x+Vt)^2 + y^2 + z^2}{4\alpha t}\right] \quad (8.C5)$$

The factor 2 is the result of adapting a solution for an infinite solid to a semi-infinite solid. Integrating Eq. (8.C5) over all past time, the steady-state temperature rise is obtained as

$$\Theta = \frac{Q}{4\pi rk} \exp\left[-\frac{V(r+x)}{2\alpha}\right] \quad (8.C6)$$

where $r = (x^2 + y^2 + z^2)^{1/2}$.

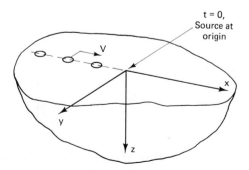

Figure 8.C1 Geometry of a moving-point heat source. (From Cook, 1970.)

Band source. The steady-state temperature due to a band source of width $2l$ moving with velocity V in the plane $z = 0$ of the semi-infinite solid $z < 0$ with no loss of heat from the plane $z = 0$ may be obtained by following a procedure similar to that used for a point source. Figure 8.C2 shows a band heat source which is parallel to the y axis and of length $2l$ parallel to the x axis. Heat is liberated at the rate of q per unit time per unit area over the area of the source. We again assume a steady state (i.e., that the motion has gone on infinitely long). We calculate the temperature at the instant when the center of the band heat source is at the origin, at $t = 0$. At time t earlier, the center of the band was at $-Vt$. The temperature at (x, y, z) at $t = 0$ due to a line heat source of $2q\,dx'\,dt'$ per unit length, parallel to the y axis and through the point $(x' - VT, 0, 0)$, is obtained from Eq. (8.C4) as

$$\Theta = \frac{q\,dx'\,dt}{2\pi kt} \exp\left[-\frac{(x - x' + Vt)^2 + z^2}{4\alpha t}\right] \tag{8.C7}$$

To find the temperature at zero time for a band of length $2l$ which has been moving for an infinite time, Eq. (8.C7) may be integrated with respect to x' from $-l$ to l and with respect to t from $-\infty$ to 0. The solution may be written as

$$\Theta = \frac{2q\alpha}{\pi k V} \int_{X-L}^{X+L} K_0(Z^2 + u^2)^{1/2} \exp(-u)\,du \tag{8.C8}$$

where $K_0(s)$ is the modified Bessel function of the second kind and the dimensionless quantities are defined as

$$X = \frac{Vx}{2\alpha} \qquad Y = \frac{Vy}{2\alpha} \qquad Z = \frac{Vz}{2\alpha} \qquad L = \frac{Vl}{2\alpha} \tag{8.C9}$$

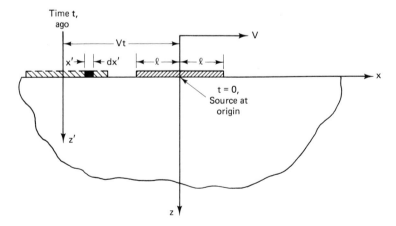

Figure 8.C2 Geometry of band source. (From Cook, 1970.)

The solution given by Eq. (8.C8) for the temperature rise is plotted in the sliding plane (i.e., $z = 0$) in terms of dimensionless quantities in Fig. 8.C3. It shows that the temperature is maximum near the trailing edge of the contact at high sliding speeds, whereas at low sliding speeds the temperature distribution is nearly symmetrical about $x = 0$. At an intermediate sliding speed the maximum lies at some point between the center and the rear of the source. Approximate equations for both maximum and mean temperature rise at very high sliding speeds ($L > 10$) and very low sliding speeds ($L < 0.5$) are given as follows:

$L > 10$ (high sliding speed):

$$\Theta_{max} \simeq 1.6 \frac{ql}{k} \left(\frac{Vl}{\alpha}\right)^{-1/2} \qquad (8.C10)$$

$$\overline{\Theta} \simeq \frac{2}{3} \Theta_{max}$$

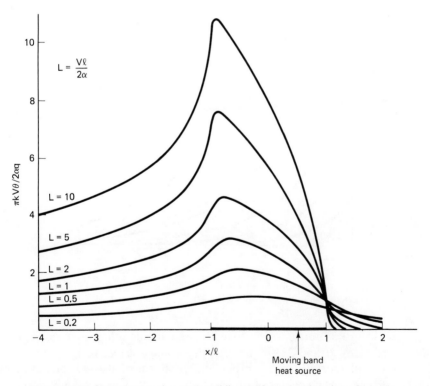

Figure 8.C3 Temperature rise at the sliding surface as a function of position and sliding speed. (From Jaeger, 1942.)

$L < 0.5$ (low speeds):

$$\Theta_{max} \simeq 0.64 \frac{ql}{k} \ln \frac{6.1\alpha}{Vl} \qquad (8.C11)$$

$$\bar{\Theta} \simeq 0.64 \frac{ql}{k} \ln \frac{5\alpha}{Vl}$$

The temperature gradient can be computed by differentiating Eq. (8.C8) with respect to z. Figure 8.C4 is a plot of temperature rise at $x = -L$ (i.e., the temperature at the trailing edge of the heat source) as a function of depth for various values of the sliding speed.

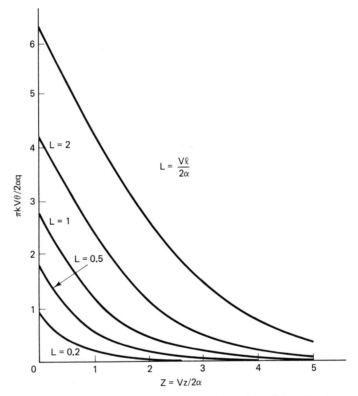

Figure 8.C4 Temperature distribution at the trailing edge of a band heat source ($z = -l$) sliding at velocity V along the x axis. (From Jaeger, 1942.)

Rectangular source. The steady-state temperature for a rectangular source of sides $2l$ and $2b$, parallel to the x and y axes, respectively, can also be calculated following a procedure similar to that used for the band source problem. The center of the heat source moves with velocity V along the x axis in the plane $z = 0$ of a semi-infinite solid $z < 0$ with no loss of heat from the plane $z = 0$. The results of the analysis are similar to those of a band source. The temperature distribution is plotted in Fig. 8.C5. The maximum and the mean temperatures may be approximated by the following equations:

$L > 10$ (high sliding speed):

$$\Theta_{max} \simeq 1.6 \frac{ql}{k} \left(\frac{Vl}{\alpha} \right)^{-1/2}$$

$$\overline{\Theta} \simeq \frac{2}{3} \Theta_{max} \frac{ql}{k} \left(\frac{Vl}{\alpha} \right)^{-1/2}$$

(8.C12)

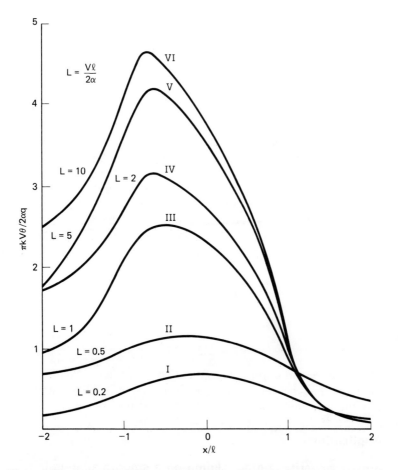

Figure 8.C5 Temperature rise at the surface of a semi-infinite solid due to a rectangular heat source of $2l \times 2b$ moving with velocity V along the x axis. The center of the heat source at $x/l = 0$. The length of heat source along the x axis is $2l$. (From Jaeger, 1942.)

$L < 0.5$ (low sliding speed):

$$\Theta_{max} \simeq 1.1 \frac{ql}{k} \qquad (8.\text{C}13)$$

$$\bar{\Theta} \simeq 0.95 \frac{ql}{k}$$

Stationary heat sources. The temperature rise due to a stationary heat source of $(-l < x < l; -l < y < l, z = 0)$ can be determined following a procedure similar to those discussed earlier. The maximum temperature occurs at $x = y = z = 0$, which is given by

$$\Theta_{max} = 1.1 \frac{ql}{k} \qquad (8.\text{C}14)$$

the mean temperature is

$$\bar{\Theta} = 0.95 \frac{ql}{k} \qquad (8.\text{C}15)$$

Transient heat source. When the heat source has moved only a finite time T, the temperature field may not be at a steady state. In this case the basic equation can be numerically integrated from 0 to T with respect to time to obtain the solution for the transient case. The temperature rise at the center of a square source is shown in Fig. 8.C6 in terms of the dimensionless temperature $\pi k V\Theta/2q\alpha$ versus $V^2T/2\alpha$ for various values of L. In Fig. 8.C7 the time T required to reach half the final value is plotted in terms of $V^2T'/2\alpha$ versus L. The relationship between the time taken to reach the final temperature and the sliding velocity may approximately be given as

$$\frac{V^2T}{\alpha} \simeq 2.5L \simeq 2.5 \frac{Vl}{\alpha}$$

or $\qquad (8.\text{C}16)$

$$VT \simeq 2.5l$$

VT is the distance slid during time T, and l is $\frac{1}{2}$ the slider length. Therefore, (8.C16) states that the center of a square resource reaches "steady state" after moving a distance of only 1.25 slider lengths.

Applications

Square asperity contact sliding on a smooth semi-infinite solid. Consider the case of a semi-infinite slider with a square contact of a side $2l$ (solid 2 in Fig. 8.C8) moving with velocity V on a semi-infinite solid (solid 1 in Fig. 8.C8). The heat generated at the interface is q, of which

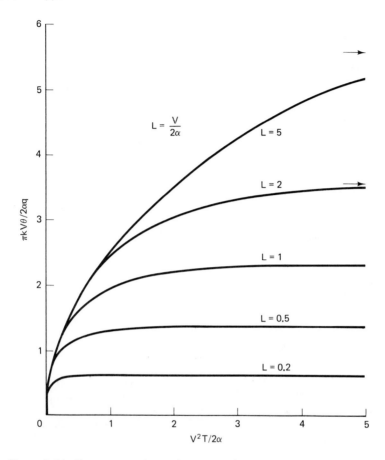

Figure 8.C6 Temperature rise at the center of a square heat source that has been moving for a finite period of time T. (From Jaeger, 1942.)

a fraction r goes to the stationary solid and the rest $(1 - r)$ goes in to the slider. The partition function r can be obtained by assuming that the surface temperatures of solid 1 and solid 2 are the same.

If $L = Vl/2\alpha_1$ is small, the temperature rise given by Eq. (8.C15) for a stationary heat source can be equated to that due to the moving heat source given by Eq. (8.C13) as

$$0.95 \frac{rql}{k_1} = 0.95 \frac{(1 - r)ql}{k_2} \tag{8.C17}$$

to obtain the partition function as

$$r = \frac{k_1}{k_1 + k_2} \tag{8.C18}$$

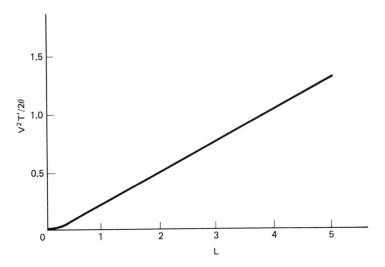

Figure 8.C7 Time T' required to reach half the final steady-state temperature versus L for a square heat source. $L = Vl/2\alpha$. (From Jaeger, 1942.)

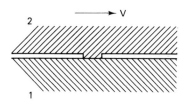

Figure 8.C8 Idealized asperity contact involving one asperity (2) moving over a smooth semi-infinite solid (1). (From Jaeger, 1947.)

For the case $L > 10$, Eq. (8.C14) may be equated to Eq. (8.C12) to obtain

$$\frac{rql}{k_1}\left(\frac{\alpha_1}{Vl}\right)^{1/2} = 0.95\frac{(1-r)\,ql}{k_2} \tag{8.C19}$$

The partition function is then

$$r = \frac{1}{1 + (k_2/k_1)(\alpha_1/Vl)^{1/2}} \tag{8.C20}$$

The heat generated at each contact, q, may be expressed as

$$q = \tau V \tag{8.C21a}$$

where τ is the shear stress at each junction.

Many square asperity contacts sliding on a smooth surface. When the normal load L is applied on two sliding surfaces, it is carried by a large number of asperity contacts, the number of contacts being nearly proportional to the normal load. In this case the temperature rise at a junction is the sum of that due to the temperature rise at an individual asperity contact, independent of any other contacts, and that due to thermal interaction with other asperities. The temperature rise at individual contacts was given in Eqs. (8.C10) through (8.C15), and the temperature rise due to thermal interaction with other asperities depends on the spacing and the number of asperities.

To simplify the problem, we will consider the case of uniformly distributed contacts of diameter d, all having the same heat generation. The heat generated at each contact may be approximated as (Cook and Bhushan, 1973)

$$q = \mu V H \tag{8.C21b}$$

where μ is the coefficient of friction and H is the indentation hardness. Then the individual temperature rise at a square source can be obtained from Eqs. (8.C12) and (8.C13), which may be expressed as

$\dfrac{Vd}{2\alpha} > 10$ (high speed):

$$\Theta = 0.71 \frac{\mu H}{\rho c} \left(\frac{Vd}{\alpha}\right)^{1/2} \tag{8.C22a}$$

$\dfrac{Vd}{2\alpha} < \dfrac{1}{2}$ (low speed):

$$\Theta = 0.48 \frac{\mu H}{\rho c} \left(\frac{Vd}{\alpha}\right) \tag{8.C22b}$$

where $d = 2l$. It should be noted that according to Eqs. (8.C22), the mean temperature rise at an individual contact does not depend on the load, but depends very strongly on the sliding speed.

Experimental results show that d may be of the order of 10 to 20 μm. It is also interesting to note that the product of material properties and the friction coefficient, $\mu H/\rho c$, is about 300°F for steel, indicating that in most sliding situations $Vd/2\alpha$ is less than $\frac{1}{2}$, since the interfacial temperature rise is only a few tens of degrees.

The temperature rise due to asperity thermal interactions may be very roughly approximated as follows (Cook, 1970). If the mean contact spacing is S and the diameter of the contact is d, the ratio $(d/s)^2$ may be related to the normal pressure σ and hardness H as

$$\left(\frac{d}{s}\right)^2 = \frac{A_r}{A} = \frac{\bar{\sigma}}{H} \tag{8.C23}$$

where A_r and A are the real and the apparent area of the contact, respectively. The temperature rise at an asperity contact due to the heat generated at all other asperities will not be greatly affected by the source geometry at other asperities when $s/d \gg 1$. Furthermore, if there are many such heat sources, the heat generated may be approximated as being uniformly distributed over the entire apparent area of contact with an average intensity $\bar{q}$. $\bar{q}$ is proportional to q as

$$\bar{q} = \frac{A_r}{A} q = \frac{\sigma}{H} q \qquad (8.\text{C}24)$$

Equations (8.C12) and (8.C13) give the temperature rise due to a heat source distributed over the entire area. The primary error in this approximation is that the contribution of heat generation at the asperity of interest is counted twice, once independently and once as part of the heat generated over the apparent area of contact, which includes the specific asperity. Therefore, the temperature rise due to the latter must be subtracted. Then the total temperature rise may be expressed as

$$\bar{\Theta} = \Theta_i + \Theta_a - \Theta_s \qquad (8.\text{C}25)$$

where Θ_i = temperature rise at individual contact
Θ_a = average temperature rise at the entire apparent area of contact (of slider length $2l$)
Θ_s = temperature rise which is taken into account twice (heat generated over the asperity spacing $s/2$)

Using appropriate dimensions for characteristic geometric factor in Eqs. (8.C12) and (8.C13), the temperature can be expressed as

High velocity:

$$\bar{\Theta} = \left(\frac{Vd}{2\alpha}\right)^{1/2} + \frac{\sigma}{H}\left(\frac{Vl}{\alpha}\right)^{1/2} - \frac{\sigma}{H}\left(\frac{Vs}{2\alpha}\right)^{1/2} \qquad (8.\text{C}26)$$

Low velocity:

$$\bar{\Theta} = 0.95 \left(\frac{Vd}{2\alpha} + \frac{\sigma}{H}\frac{Vl}{\alpha} - \frac{\sigma}{H}\frac{Vs}{2\alpha}\right) \qquad (8.\text{C}27)$$

Rough surface sliding over another rough surface. When a rough surface with many asperities slides over another rough surface with many asperities, the temperature rise depends on the sliding velocity. When the sliding velocity (i.e., Vd/α) is large, the temperature will continue to rise until the termination of the asperity contact since the entire process approaches an adiabatic condition. On the contrary, when the sliding velocity is low, the temperature initially rises, since the contact area increases, requiring more external work to deform the asperity junction, and then decreases with the decreases in the contact area. Cook and Bhushan (1973) derived very approximate expressions for peak temperatures as:

$L > 10$ (high speed):

$$\Theta_{max} = 0.59 \frac{\mu H}{\rho c} \left(\frac{Vd}{\alpha}\right)^{1/2} + \frac{\mu \sigma}{\rho c} \left(\frac{Vl}{\alpha}\right)^{1/2} \qquad (8.C28a)$$

$L < 1$ (low speed)

$$\Theta_{max} = 0.26 \frac{\mu H}{\rho c} \frac{Vd}{\alpha} + \frac{\mu \sigma}{\rho c} \frac{Vl}{\alpha} \qquad (8.C28b)$$

Equation (8.C28b) was obtained assuming that all of the heat generated by the frictional force goes into the asperity. However, in actual sliding situations part of the heat is also conducted into the counterface. Therefore, Eqs. (8.C28a) and (8.C28b) must be modified to include the partition of energy given by Eq. (8.C18). Furthermore, the junction area of a given asperity contact changes as a function of its relative contact position. Cook and Bhushan (1973) have shown that the area-averaged temperature rise at individual junction during slow sliding speeds is 0.85 times the peak temperature when the varying contact area is taken into account, while the interaction temperature rise due to other asperities is not affected by the varying contact area of the individual contact area.

The calculated temperature values for various combinations of materials (assuming that the surface hardness is twice the bulk hardness) are given in Table 8.C1. The calculated values for the temperature rise (in °F) are compared with experimentally obtained values in Fig. 8.C9 for various sliding speeds (given in in./sec) and in Fig. 8.C10 for various normal loads (given in psi). The agreement is reasonable.

Analysis of cutting-tool temperatures. In metal cutting the temperature rise may be assumed as a first-order approximation, to be due to the shear work done at the primary shear plane in the chip and also due to the frictional work done at the chip–tool interface (see Fig. 8.C11). If we denote the component of the force directed along the shear plane as F_s and the velocity of the chip relative to the workpiece as V_s, the heat that flows from the primary shear zone per unit time per unit area may be expressed as (see Loewen and Shaw, 1954)

$$q_1 = \frac{F_s V_s}{tb \csc \phi} = u_s V \sin \phi \qquad (8.C29)$$

where u_s is the shear energy per unit volume of metal cut and V the cutting speed. If we denote r_1 as the fraction of the energy that leaves shear zone with the chip, $(1 - r_1)q_1$ goes into the workpiece. Then the mean temperature rise of the chip due to the work done in the shear plane is

$$\overline{\Theta}_s = \frac{r_1 u_s}{\rho_1 c_1} + \Theta_0 \qquad (8.C30)$$

TABLE 8.C1 Calculated Values of Temperature Rise*

Material Pair		f	d_{max} $(10^{-6}$ in.$)$	$k_1 + k_2$ $\left(\dfrac{\text{lb}}{\text{sec-}°\text{F}}\right)$	H $(10^5$ psi$)$	$\partial\bar{\theta}/\partial V$ at $\bar{\sigma}$		$\partial\bar{\theta}/\partial\bar{\sigma}$ and $\bar{\theta}_0$ at $V = 20$ in./sec†	
1	2					$\dfrac{°F}{\text{in./sec}}$	psi	$\dfrac{°F}{\text{psi}}$	°F
1. Brass	1095	0.22	93	20.2	2.8	1.47	60	0.076	25.5
2. Bronze	1095	0.25	55	13.5	2.8	1.65	60	0.13	25
3. 4140	C.I.	0.62	90	11.2	2.8	7.3	60	0.39	122
4. 4140	Graphite	0.31	66	8.5	1.0	1.83	60	0.26	21
5. Graphite	Copper	0.41	56	51.0	1.0	0.37	60	0.056	3.9
6. Graphite	Al	0.45	53	23.0	1.0	0.86	60	0.137	9.1
7. 4140	Al	0.60	70	25.0	2.1	2.0	60	0.168	31
8. 4140	Al	0.25	60	18.0	3.5	1.57	60	0.097	26.4
9. Sint. bronze	1095	0.20	133	13.0	2.3	2.37	60	0.11	41
10. 4140	Babbit	0.75	75	8.3	0.3	2.8	60	0.63	17.8
11. Ti	Mg	0.32	82	21.0	1.1	0.96	175	0.0395	9.6
12. Ni	Mg	0.35	57	27.0	1.1	0.95	350	0.34	5.7

* From Cook, N. H., and Bhushan, B., "Sliding Surface Interface Temperatures," *Journal of Lubrication Technology*, Vol. 95, No. 1, 1973, pp. 59–64. Reprinted by permission of the American Society of Lubrication Engineers. All rights reserved.

† $\bar{\theta}_0$ is $\bar{\theta}$ at $\bar{\sigma} = 0$, i.e., $\bar{\theta}_0$ is the flash temperature at an asperity contact without any temperature rise due to heat conduction from other asperities.

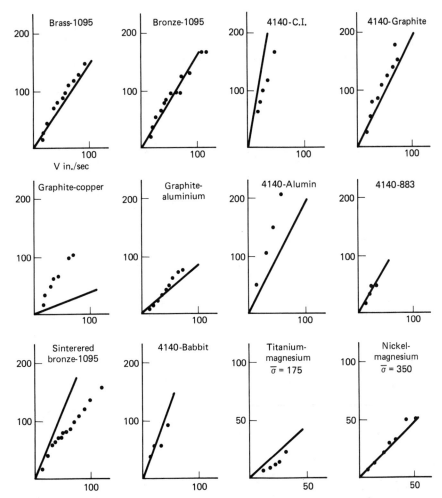

Figure 8.C9 Temperature versus sliding speed. Lines are calculated, points are experimental data. $\bar{\sigma} = 60$ psi except as noted. (From Cook, N. H., and Bhushan, B., "Sliding Surface Interface Temperatures," *Journal of Lubrication Technology*, Vol. 95, No. 1, 1973, pp. 59–64. Reprinted by permission of the American Society of Lubrication Engineers. All rights reserved.)

where Θ_0 is the ambient workpiece temperature and $c_1\rho_1$ is the volume specific heat at the mean temperature between Θ_s and Θ_0. The mean temperature rise of the workpiece due to the moving heat source may be expressed as

$$\Theta = 0.75 \frac{ql}{k}\left(\frac{Vl}{2\alpha}\right)^{-1/2} + \Theta_0 \qquad \text{for } \frac{Vl}{2\alpha} > 0.5 \qquad (8.\text{C}31)$$

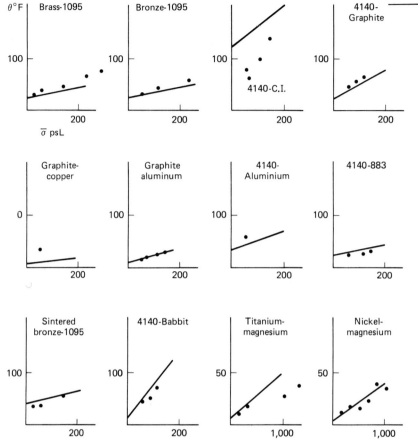

Figure 8.C10 Temperature versus normal stress. Lines are calculated, points are experimental data. $V = 20$ in./sec. (From Cook, N. H., and Bhushan, B., "Sliding Surface Interface Temperatures," *Journal of Lubrication Technology*, Vol. 95, No. 1, 1973, pp. 59–64. Reprinted by permission of the American Society of Lubrication Engineers. All rights reserved.)

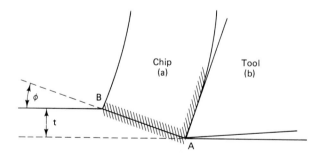

Figure 8.C11 Major sources of energy dissipated in cutting: (a) shear energy; (b) friction energy. (From Loewen, E. G., and Shaw, M. C., "On the Analysis of Cutting Tool Temperatures," *Transactions of the ASME*, Vol. 76, 1954, pp. 217–231. Reprinted by permission of the American Society of Mechanical Engineers.)

which may be expressed as

$$\Theta_s = 0.754 \frac{(1 - r_1)q_1(t \csc \phi /2)}{k_1} \left(\frac{Vl}{2\alpha_1}\right)^{-1/2} + \Theta_0 \qquad (8.C32)$$

$$L_1 = \frac{V_s(t \csc \phi /2)}{2\alpha_1} = \frac{V\gamma t}{4\alpha_1}$$

where t is the depth of cut and γ is the strain in the chip.

Equating Eqs. (8.C30) and (8.C32), the partition function is obtained as

$$r_1 = \frac{1}{1 + 1.328(\alpha_1 \gamma / Vt)^{1/2}} \qquad (8.C33)$$

Equation (8.C33) shows that at high cutting speeds, all the work done in the primary shear plane is carried by the chip and no heat is transferred into the workpiece.

The temperature rise at the chip–tool interface can be analyzed by treating the heat source to be moving relative to the chip and stationary relative to the tool. The heat generated at the interface, q_2, is

$$q_2 = \frac{FV_c}{ab} = \frac{u_f V_t}{a} \qquad (8.C34)$$

where F is the friction force along the tool face, V_c the chip velocity, a the contact length, b the chip width, and u_f the specific energy per unit volume of metal cut consumed to overcome the friction. If r_2 is the fraction of q_2 that flows into the chip, the temperature rise can be determined, using Eq. (8.C31), as

$$\Delta\Theta_f = \frac{0.754 r_2 q_2(a/2)}{k_2} \left[\frac{V_c(a/2)}{2\alpha_2}\right]^{-1/2} \qquad (8.C35)$$

The mean temperature of the chip surface along the tool face is the sum of Θ_s and $\Delta\Theta_f$ given by Eqs. (8.C32) and (8.C35).

The temperature field in the tool can be determined using an expression similar to Eq. (8.C15), which may be written as

$$\overline{\Theta}_t = \frac{(1 - r_2)q_2 a}{k_3} \overline{A} + \Theta'_0 \qquad (8.C36)$$

where $\overline{A}$ is a geometric factor which is equal to b/a for a lathe tool and to $b/2a$ for an orthogonal tool.

Equating the mean surface temperature of the chip and the tool surface temperature, the partition function r_2 is determined to be

$$r_2 = \left(\frac{u_f Vt\overline{A}}{k_3} - \Theta_s + \Theta_0'\right) \bigg/ \left[\frac{u_f Vt\overline{A}}{k_3} + \frac{0.754 u_f}{\rho_2 c_2}\left(\frac{Vt}{\dfrac{ar}{t}\alpha_2}\right)^{1/2}\right] \qquad (8.C37)$$

where r is the chip–thickness ratio, which is equal to V_c/V.

According to the Loewen–Shaw analysis, the shear stress on the shear plane is the most important variable in determining the temperature rise on the tool face; the temperature rise is directly proportional to the shear stress. The next set of important variables are the cutting speed V, the depth of cut t, the shear strain γ, and the product of the thermal properties (kpc) for the workpiece. The temperature rise depends on the one-half power of these variables. The coefficient of friction is the third important variable. Finally, the relative length of the chip–tool contact to the depth of cut plays a role in affecting the temperature rise. Figure 8.C12 shows the effect of the cutting speed on the temperature rise.

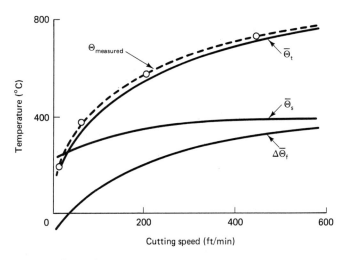

Figure 8.C12 Comparison between observed and computed cutting temperatures for SAE B1113 steel when cut with a K2S cemented carbide tool with a 20° rake angle: $t = 0.0023$ in., $b = 0.151$ in. (From Loewen, E. G., and Shaw, M. C., "On the Analysis of Cutting Tool Temperatures," *Transactions of the ASME,* Vol. 76, 1954, pp. 217–231. Reprinted by permission of the American Society of Mechanical Engineers.)

With the advent of the finite element method, the temperature field in the chip and the tool can be more accurately predicted considering the shear deformation that occurs in the chip (Muraka et al., 1979). However, the analytical results presented in this appendix are still valuable since they give much clearer physical insight and the relative importance of various parameters.

9

NOVEL METHODS OF IMPROVING TRIBOLOGICAL BEHAVIOR OF SLIDING SURFACES

9.1 INTRODUCTION

One of the important reasons for developing theories of friction and wear is to solve practical tribological problems based on the fundamental scientific understanding provided by the theories. If the theories are correct, it should be possible to devise a means of changing the friction and wear coefficients as required for a specific application based on the theoretical predictions. Even such problems as the electrical contact resistance between sliding surfaces should be solved based on the basic understanding of the processes involved in sliding contacts. In the past, many tribological problems have been solved empirically with a minimal input from fundamental theories. In this chapter some "new" techniques which have been devised based on the theories discussed in the preceding chapters will be presented.

The empirical approaches of the past involved the following techniques: lubrication of the interface to lower the surface traction; hardening the surface layer by such means as precipitation, nitriding, and carburizing to reduce wear; coating the surface either to harden or lubricate the surface; and the use of an "incompatible" set of materials to minimize friction and avoid seizure. All these techniques were based on the idea that lubrication of the interface, the use of hard materials, and the use of an "incompatible" pair of materials are good in lowering friction and wear. Basically these are sound ideas, but further improvement can be made only by proper use of the fundamental theories presented in this book. Well-known "old" techniques such as the use of boundary lubrication will not be treated in this chapter.

The theories presented in the preceding chapters imply that many things can be done to control the friction and wear of materials. The friction theory presented in Chapter 3 implies that friction can be reduced if the wear particles can be eliminated from the interface, since the energy is consumed in plowing of the surface with the wear particles and in some cases, in deforming the entrapped wear particles themselves. Another implication of the theory is that the adhesion component can be reduced by lubrication of the interface. Also, if the normal load can be reduced, the frictional force will be reduced, since it will decrease both plowing and adhesion.

The wear theories presented state that the wear rate of metals and polymers can be reduced by a number of different means. Some of the important implications are as follows:

1. *Lower surface traction:* The wear rate of metals and polymers can be reduced if the forces acting at the interface are decreased. Therefore, all techniques that reduce the frictional force should also decrease the wear rate.
2. *Prevent the plastic deformation of the surface and subsurface layer:* Since plastic deformation affects the crack nucleation and propagation rates, the hardness of the elastoplastic material should be as high as possible. Another possibility is to lubricate the surface or put a soft layer which will deform and prevent the subsurface from deformation.
3. *Eliminate crack nucleation sites:* Second-phase particles and impurities are potential crack nucleation sites. Hard, ductile materials without second-phase particles are highly desirable tribological materials. Certainly, materials with preexisting flaws are not desirable. If the second-phase particle must be present, it must be strongly bonded to the matrix.
4. *Reduce crack propagation rates:* Since the crack propagation in elastoplastic solids is due to crack tip displacement, hard materials are better than soft materials. Another desirable characteristic of the material is the self-healing ability of the cracks so that they reweld before the next asperity applies load. The judicious placing of strong fibers may also prevent crack propagation.
5. *Increase the chemical stability of materials:* In applications where the interfacial temperature is high, the adhesion and the solution wear of the interface can be eliminated if one of the materials is so chemically stable that a chemical reaction or dissolution cannot occur.
6. *Strengthen the outermost layer of the surface of crystalline polymers:* Since highly crystalline polymers wear by thin-film transfer when the

applied stress exceeds the bond strength between oriented molecules, it will be desirable to alter the structure of the polymers so as to prevent the shearing from occurring.

Several new techniques were motivated by the foregoing deductions. These techniques are described in this chapter.

Utilizing the observation that the wear particles increase the friction coefficient by plowing the surface (see Chapter 3), a means of maintaining a low coefficient in composites was invented (Suh and Burgess, 1983). Similarly, the understanding that the wear particles also increase the electrical contact resistance by decreasing the real contact area and by becoming oxide particles led to the invention of a method of having low and constant electrical contact resistance between sliding surfaces (Suh et al., 1983).

The delamination theory of wear presented in Chapter 5 indicates that the nucleation and propagation of subsurface cracks are responsible for wear. To prevent the crack nucleation through plastic deformation of the subsurface, a thin, soft metallic layer was plated on a hard substrate (Jahanmir et al., 1976). The important contribution of the theory to this technique was the idea that the thickness of the soft layer must be less than a critical value to prevent the generation of wear particles by the delamination of the soft layer itself.

Various methods of increasing the surface hardness have been tried. One technique is ion implantation, which is used extensively in semiconductor industry to change the electrical properties. A very thin surface layer can be hardened by driving in excess amount of interstitial atoms into the surface. Such a layer prevents plowing of the surface under light loads (Shepard and Suh, 1982). Ion implantation can also be used to alter the surface chemistry to minimize adhesion. Another technique, called ion plating, may be used to create a hard layer on a soft substrate (or vice versa) to prevent the delamination wear.

In the case of polymers the surface layer can be cross-linked to prevent deformation and wear particle generation as discussed in Chapter 6 (Youn and Suh, 1981). Such plastics as HDPE and POM can be subjected to low-temperature plasma treatment which cross-link the molecules near the surface and increase the flow strength of the surface layer. This retards the thin-film transfer and other wear particle generation process.

The life of cutting tools can be increased by depositing chemically more stable compounds on carbide tools (Suh, 1977) and even on high-speed-steel tools (Su and Cook, 1977). Although research on this concept was started less than 20 years ago, this technology is already well established in industry. Various carbide tools are now coated with TiC, HfC, TiN, Al_2O_3, and a combination of these compounds.

Each of these techniques will be reviewed without attempting to be comprehensive in reviewing all the work done on these subjects. The goal is to illustrate how the theories in this book can be used in solving practical problems rather than provide a thorough review of the literature. The reader may skip the sections describing the experimental details and read only the introductory and concluding sections of each case study if interested only in obtaining general ideas about these novel applications.

9.2 PREVENTION OF WEAR OF COMPOSITES BY THE CREATION OF MICROVOIDS TO TRAP WEAR PARTICLES

The composite used was made of graphite fibers and polyurethane. The microvoids were created by saturating the resin with nitrogen gas at high pressure and by instantaneously releasing the pressure to nucleate microvoids through the creation of thermodynamic instability. These microvoids trap wear particles and prevent the coefficient of friction from increasing with the sliding distance. These specimens were tested using the pin-on-disk type of testing arrangement (Suh and Burgess, 1984; Burgess, 1983). The details of the experimental conditions and results are given in this section.

9.2.1 Sample Preparation

Adiprene L-83 Urethane Rubber Prepolymer and Caytur 21 Urethane Curative were used as matrix material for the composite. The two components were mixed such that a 95% theoretical stoichiometric amount of curative was present. Say-5300 silicone-based surfactant was added (10 wt %) to the Adiprene/Caytur mixture, both foamed and unfoamed specimens. The surfactant performs two functions: (1) it reduces the energy required to form new surfaces, leading to an enhancement of bubble formation; and (2) it promotes bubble stability by equalizing the bubble surface tension.

The fibers used were Union Carbide T-300 High Strength Carbon Fibers, sized for epoxy resin. The fiber tows were wound around a flat winding board. After winding, the resulting fiber-wound boards were impregnated with the polyurethane resin mixture. One-inch-wide prepregs with fibers parallel to the long axis were cut and laid on top of one another to a thickness of approximately 1 in. A load of 100 lb was applied to the mold for 10 hours to allow the excess resin to leak out.

The uncured sample was then cut into 2-in.-long sections which were placed in shorter molds that could be adjusted for different amounts of volume expansion with respect to the original thickness. These adjustable molds were then set for the desired volume expansion and placed in a pressure chamber, which was pressurized with nitrogen at 500 and 1500 psi. After saturating the polyurethane resin with nitrogen gas for 24 hours, the pressure was quickly released (30 to 40 seconds) to allow nucleation

Sec. 9.2 Prevention of Wear of Composites

and growth of microvoids. The samples were then placed in a forced-air oven and cured at a temperature of 110°C for an hour.

Cured composite samples were sliced into 0.125-in.-thick wafers with the fibers aligned perpendicular to the flat wafer surface. They were then mounted in polyester resin and polished using 240-, 320-, and 600-grit silicon carbide paper and 0.3- and 0.05-μm alumina abrasives. The bottom surface of the polyester mount was ground so that the top and bottom surfaces were parallel. To remove the polishing debris from the surface, samples were brushed with distilled water and vacuum cleaned with a pipette connected to a water trap and vacuum pump. Finally, the samples were allowed to dry for at least 10 hours in a desiccator with calcium chloride crystals.

Figure 9.1 is an SEM micrograph of a cross section of the composite sample containing voids. This micrograph is a good representation of the morphology of voids generated by this technique. Optical micrographs of a void sample, cut and polished parallel to the fiber axis, are shown in Figs. 9.2 and 9.3. The broken fibers are a result of polishing. The voids are oblong or elliptical, and tend to grow parallel to the fiber axis.

Figure 9.1 SIM micrograph of voids in polished sample. (From Burgess, 1983.)

The void morphology of samples was characterized in terms of the multiples of the fiber cross-sectional area based on a fiber diameter of 7 μm. A histogram of the void sample is plotted in Fig. 9.4.

9.2.2 Friction and Wear Tests

The standard pin-on-disk type of apparatus was used to conduct friction and wear tests with a $\frac{1}{2}$-in. AISI 52100 steel bearing ball, as the slider. The composite sample was clamped to the rotating table and the steel ball was stationary. A wear track diameter was chosen such that a rotation of 200

Figure 9.2 Cross section of void sample. (From Burgess, 1983.)

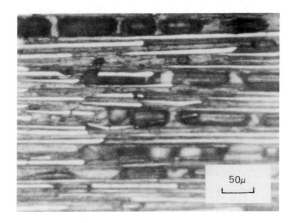

Figure 9.3 Higher magnification of voids; broken fibers are a result of simple preparation. (From Burgess, 1983.)

rev/min yielded a sliding velocity of 5 cm/sec. The normal load applied was 1 kg. The average and local friction coefficients were measured during the test, since the frictional force varied from location to location on the disk.

After testing, wear debris was collected for examination. Wear volume was measured using surface profiles after 100,000 revolutions at four equidistant locations on the circular wear tracks with a Dek-Tak surface profilometer. The cross-sectional areas of the wear track profiles were measured with a planimeter and averaged. Weight loss was very small and was difficult to measure due to moisture content variations and the accumulation of wear debris in the voids.

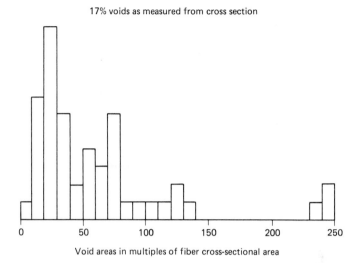

Figure 9.4 Histogram of void sizes. (From Burgess, 1983.)

9.2.3 Experimental Results

The friction coefficients for specimens with and without voids are plotted as a function of the number of rotations in Fig. 9.5. The friction coefficient at the onset of sliding for both types of samples was very high, approximately 0.35 to 0.45. It then dropped from these high values to the steady-state value of approximately 0.2 within the first 1000 revolutions.

As sliding proceeded the frictional behavior between the two samples differed significantly. The sample without voids exhibited a large increase in friction coefficient, reaching values as high as 0.43 in some tests. It also showed a large scatter in the friction values from sample to sample. The samples with voids exhibited a stable and very nearly constant friction coefficient, increasing only from approximately 0.18 to 0.20 during 50,000 revolutions. The scatter in the data was also significantly less when the voids were present.

To ensure that the low friction exhibited by the samples with voids was repeatable and not the result of contamination during sample preparation, a completely independent set of samples was processed, prepared, and tested. The test results were very similar to those shown in Fig. 9.5.

9.2.4 Micrographs of the Wear Tracks

Figures 9.6 and 9.7 are SEM micrographs of the wear tracks of samples with and without voids, respectively, after 50,000 revolutions. The sample without voids had a continuous film of wear debris on the wear track.

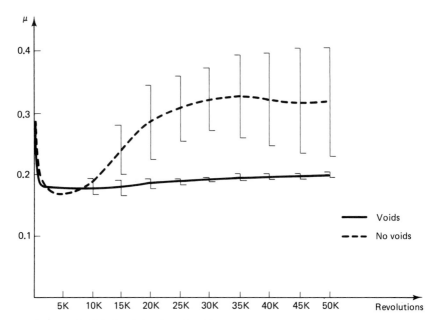

Figure 9.5 Friction behavior of void and nonvoid samples. (From Burgess, 1983.)

Figure 9.8 is a higher magnification of this film. The film has a very fine texture and is an agglomeration of fine wear particles. Figure 9.9 is an SEM micrograph of these fine wear particles. The sizes of these particles are much smaller than the fiber diameter. The samples with voids, on the other hand, show much less evidence of such a film on the wear track, and rarely in large continuous patches, as shown in Fig. 9.6.

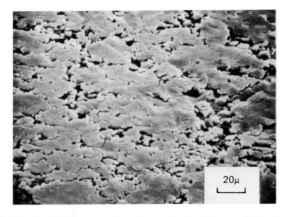

Figure 9.6 Wear track of nonvoid sample covered with wear debris film. (From Burgess, 1983.)

Sec. 9.2 Prevention of Wear of Composites 421

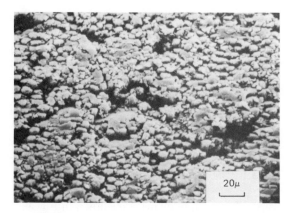

Figure 9.7 Clean wear track of void sample. (From Burgess, 1983.)

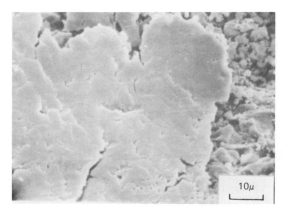

Figure 9.8 Wear debris film. (From Burgess, 1983.)

Figure 9.9 Fine wear particles on wear track. (From Burgess, 1983.)

Another common feature of the surface is the presence of cracks on the wear track (Fig. 9.10). These cracks tend to be oriented perpendicular to the sliding direction. They are more prevalent in the sample with voids, running from void to void. However, neither type of sample exhibited large scale fracture or fiber pull-out. All sample surfaces maintained their integrity as a fiber-reinforced composite with normal fiber orientation. These results suggest that the cracks generated at the surface do not cause wear. The cracks cannot penetrate the surface very deeply because the tensile region is very shallow, being confined to the surface layer. Furthermore, the crack propagation direction cannot change in the subsurface to a direction parallel to the surface due to the presence of the fibers, which are perpendicular to the surface.

A micrograph of a worn fiber tip was shown in Fig. 9.11. The smooth appearance of the fiber is typical of the majority of fibers on wear tracks of the samples with and without the voids. It indicates that gradual wear took place.

Profiles of wear tracks after 100,000 revolutions show that the wear tracks of the samples without voids were deeper and narrower than those of the samples with voids. The magnitude of the depth of the wear tracks was of the same order of fiber diameter (approximately 5 to 10 μm). There was no indication that the fibers broke off by buckling. Fibers wore off gradually.

Visual examination of wear tracks revealed an interesting difference between the samples with and without voids. The latter samples had a substantial amount of wear debris on the periphery of the wear track (Figs. 9.12 and 9.13). The particles were a black, chalky, carbon powder. The samples with voids, by contrast, exhibited a significant reduction in the

Figure 9.10 Transverse cracks in wear track. (From Burgess, 1983.)

Figure 9.11 Smooth appearance of a worn fiber tip. (From Burgess, 1983.)

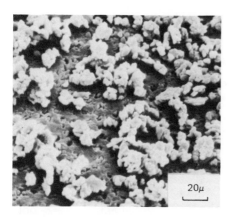

Figure 9.12 Peripheral wear debris. (From Burgess, 1983.)

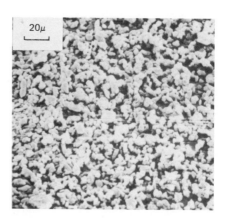

Figure 9.13 Peripheral wear debris closer to wear track. (From Burgess, 1983.)

amount of this powder present on the edges of the wear track. The aspect ratios of the fibrous wear debris were seldom greater than unity.

The bearing steel ball counterfaces slid against the samples with and without voids showed a noticeable polished spot in all cases. However, sliding against the samples with voids resulted in a larger and smoother wear spot. Also, plowing grooves were present over much of the polished spot. Sliding on the composites without voids also produced plowing grooves, but of a much less severe nature. The grooves appeared narrow and shallow, resulting in less alteration of the original surface features. The wear rates of both composite samples were about the same, 10^{-14} cm^3/cm. Metallic wear debris was not present on the wear track and hence it was assumed that the steel ball wear was negligible.

9.2.5 Discussion of the Role of Voids

The experimental results support the hypothesis that the introduction of the voids should eliminate the time-dependant behavior of the friction coefficient by trapping the wear debris from the sliding interface. It is interesting to note that the contribution of sticky wear debris to the friction coefficient was as large as the combined contribution by all other mechanisms to friction: plowing the metal counterface by fiber tips, and plowing the composite by asperities of the metal counterface. The incorporation of voids into the material microstructure eliminates a significant portion of the friction due to the deformation of wear particles at the sliding interface.

The wear rate of graphite-fiber-reinforced composite was analyzed in Chapter 6, assuming that the wear rate of the carbon fiber controls the wear rate of the composite. The predicted wear rate was on the order of

10^{-10} cm³/cm. The experimentally measured wear rate was also on the order of 10^{-10} cm³/cm. As stated in Chapter 6, agreement to within an order of magnitude between theoretical and experimental wear rates is good in view of the assumptions made in the analysis.

The voids in the materials reduced the friction coefficient, but did not have a significant effect on the wear rate. In metals, the wear rate is a power function of friction due to the power relation between crack growth and applied surface tractions. However, in these composites the predicted wear rate based on the gradual removal of graphite fibrils is linearly proportional to the coefficient of friction (see Chapter 6). Therefore, a large reduction in the wear rate is not expected with these continuous graphite fiber/polyurethane composites, especially when the wear debris protects the surface in addition to increasing the friction coefficient.

The effectiveness of voids in controlling the friction coefficient is not diminished by continuous sliding. The experimental results show that the voids are very effective even after 100,000 revolutions. When a proper number of voids of an optimum size is present on the surface, the voids are not completely filled with wear debris. Some of the wear particles that were entrapped in the voids apparently came out of the voids once the slider had moved over the voids. Furthermore, new voids are generated as the specimen gradually wears away.

9.3 CONTROL OF ELECTRICAL CONTACT RESISTANCE

Electrical contacts for electronic applications are made by first plating copper with nickel and then with a thin layer of soft metals such as gold, gold–cobalt alloys, and lead–tin alloys. The functional requirement of the contacts is that the resistance be less than 20 mΩ. However, it has been found in practice that even a good electrical contact gradually loses its low resistance with repeated use and eventually becomes ineffective as the contact resistance exceeds the specified value. This problem was solved by hypothesizing that the contact resistance increases for two reasons: due to the presence of wear particles at the interface, which limit the contact area by separating the interface; and due to the oxidation of the entrapped wear particles, thus becoming insulators. The proposed solution to solve the problem was by creating a modulated surface to trap wear particles, as shown in Fig. 9.14. (Suh et al., 1984; Liou, 1983).

9.3.1 Description of Experiments

The experimental arrangement is shown schematically in Fig. 9.15, which is similar to the fretting apparatus. During the test, two specimens were subjected to small-amplitude cyclic relative motion at the interface.

Sec. 9.3 Control of Electrical Contact Resistance

Figure 9.14 Geometry of a modulated surface. (From Saka, Liou, and Suh, 1984.)

One sample was in the form of metal plated on flat composite board used for printed circuits, and the other was a small pin to represent an edge-card connector. The flat sample board was clamped to the table driven by an electromagnetic shaker. The pin and its housing were clamped in a holder which was connected to a rigid arm through a piezoelectric force

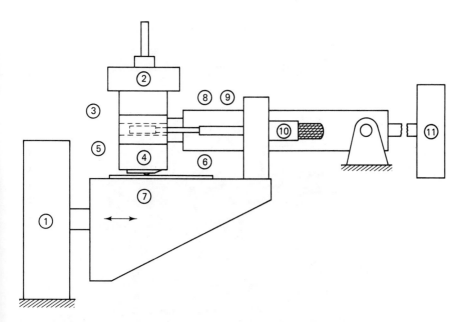

Figure 9.15 Schematic diagram of the fretting apparatus. 1, Shaker; 2, dead weight; 3, LVDT; 4, pin holder; 5, pin; 6, PC board; 7, vibrating table; 8, force transducer; 9, rigid arm; 10, micrometer; 11, counterweight.

transducer. The actual testing was done under a normal load of 1.96 N (200 g). The relative reciprocating motion was 200 μm, which was measured using a linear variable differential transformer (LVDT) attached to the pin holder and the shaker table. Figure 9.16 shows the sinusoidal displacement and the resulting frictional force measurement, which is close to a square wave. The electrical contact resistance was measured using a Wheatstone bridge circuit. The electric current through the contact was 20 mA. All the tests were done at room temperature (293 to 298 K) and the relative humidity was 55 to 65%. The total distance slid was 40 m (i.e., 100 kilocycles).

Both bulk materials and their coatings were tested. Electroplated Sn–34% Pb, Cu, and Ni; coatings of Sn–34% Pb and Au–1% Co on flat copper

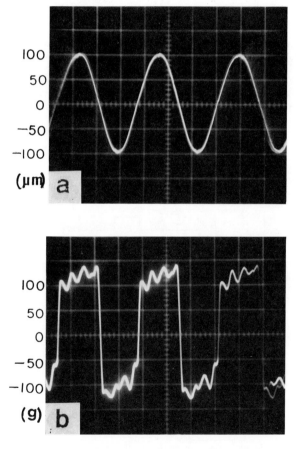

Figure 9.16 Displacement (a) and friction force (b) from an oscilloscope.

Sec. 9.3   Control of Electrical Contact Resistance

substrate; and also Sn–34% Pb coated on a modulated copper substrate were chosen for study because of their low resistivity and wide range of hardness (see Table 9.1). Gold–cobalt alloy, which is currently used in printed circuit boards, was used for comparison. All of the plating on copper substrate was done using nickel as an intermediate layer. The hardness and resistivity values are given in Table 9.2. The pin was a Au–30% Ag alloy of 0.7 mm diameter spot-welded on a phosphor–bronze pin.

Scanning electron microscopy and optical microscope were used to take micrographs of the wear track. An energy-dispersive analyzer (EDX) was also used to determine the composition of the contact area. Auger electron spectroscopy (AES) was also used to determine the surface composition and to detect if the substrate had been exposed, by examining before and after 3 minutes of argon sputtering.

TABLE 9.1   Material Combinations[a]

Base metals
  Sn–34 Pb (100 μm)
  Cu   (100 μm)
  Ni   (100 μm)
Coatings on Cu laminates
  Sn–34 Pb (1 μm)/Ni (3.75 μm)/Cu (35 μm)
  Au–1 Co (1–25 μm)/Ni (3.75 μm)/Cu (35 μm)
Coatings on modulated Cu surfaces
  Sn–34 Pb (1 μm)/Ni (3.75 μm)/Cu (modulated)
  Sn–34 Pb (1 μm)/Ni (30 μm)/Cu (modulated)

[a] Numbers in parentheses indicate the thickness of the material.

TABLE 9.2   Resistivity and Hardness of Test Materials[a]

Material	Resistivity ($\mu\Omega$-cm)	Hardness [MPa (kg/mm^2)]
Sn–34 Pb	9.6 ± 0.2	11 ± 2
Cu	1.74 ± 0.0	103 ± 9
Ni 7.3 ± 0.6	508 ± 32	
Au–1 Co	10.4 ± 2.0	163 ± 20
Au–30 Ag	13.8 ± 0.0	158 ± 8

[a] The composition values are in weight percent.

9.3.2 Experimental Results

Sn–Pb, Cu, and Ni samples. Figure 9.17 shows the friction coefficient as a function of the distance slid (i.e., cycles) for electroplated Sn–Pb, Cu, and Ni samples. All three materials have low friction coefficients (about 0.22 to 0.34) at the beginning of sliding, but the friction coefficient increases to high steady values (about 0.6 to 0.7) at about 20 to 40 × 10^3 cycles. The reason for this increase was discussed in great detail in Chapter 3. The electric contact resistance for Sn–Pb alloy is shown in Fig. 9.18. The initial static contact resistance is about 7 mΩ. The dynamic contact resistance starts to increase after only a few hundred cycles of oscillation, reaching over 100 mΩ after several thousand cycles. Figure 9.19(a) is the micrograph of the contact zone of Sn–Pb alloy after 10^5 cycles, which shows that there was a significant amount of plastic deformation. Figure 9.19(b) is a micrograph

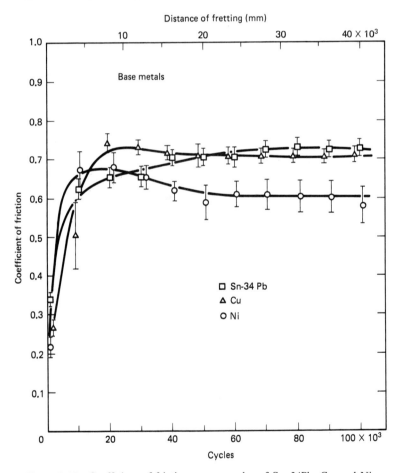

Figure 9.17 Coefficient of friction versus cycles of Sn–34Pb, Cu, and Ni.

Sec. 9.3 Control of Electrical Contact Resistance

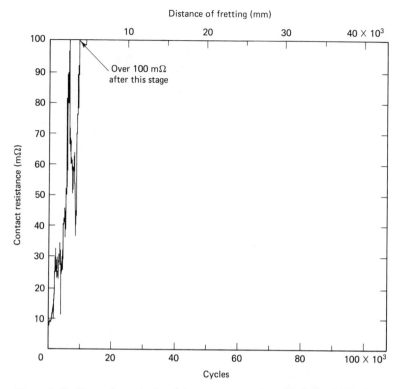

Figure 9.18 Dynamic contact resistance versus cycles of bulk Sn–34Pb alloy.

of the pin. EDX shows that Sn–Pb transferred from the flat surface to the gold–silver pin. Both surfaces were covered with black debris.

The copper specimen slid against the Au–Ag pin showed similar electrical contact resistance behavior to Sn–Pb specimens. However, the static contact resistance of the nickel specimen was high, being about hundreds of milliohms. The resistance then decreased to 9 mΩ after a few tens of cycles and stayed at this low value (Fig. 9.20). The examination of the nickel surface with SEM and EDX indicated the nickel surfaces were covered with the gold–silver alloy (Fig. 9.21) and nickel did not wear. Most of the wear debris was Au–Ag.

Sn–Pb and Au–Co coatings on flat Cu laminates. The friction coefficients of Au–Co and Sn–Pb coatings on the Ni–Cu substrate are plotted as a function of the number of cycles in Fig. 9.22. They start at low values of friction coefficient (0.25 for Sn–Pb and 0.35 for Au–Co) and reach a steady-state value (0.77 for Au–Co at 10^4 cycles and 0.72 at 5×10^4 cycles for Sn–Pb), which is consistent with the friction mechanisms postulated in Chapter 3.

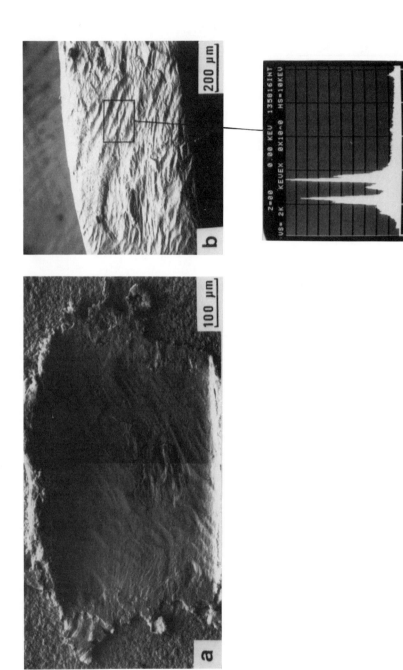

Figure 9.19 Bulk Sn–34Pb alloy (a) and the Au–30Ag pin (b) after 10^5 cycles. (From Saka, Liou, and Suh, 1984.)

Sec. 9.3   Control of Electrical Contact Resistance   431

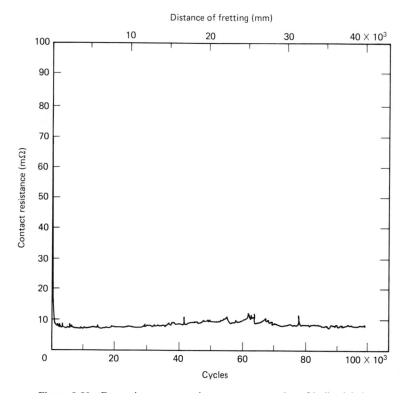

Figure 9.20 Dynamic contact resistance versus cycles of bulk nickel.

Figure 9.23 is a plot of the contact resistance of Au–Co alloy coating versus the number of cycles. The static contact resistance was 8 mΩ and the dynamic contact resistance was constant at 9 mΩ throughout the test. These results are quite different from those of Sn–Pb coatings, whose electrical contact resistance is shown in Fig. 9.24. The static contact resistance was only 10 mΩ, but the dynamic contact resistance increased to values as high as 100 mΩ and fluctuated wildly throughout the test. This wild fluctuation is due to the formation of tin and lead oxide particles at the interface.

The examination of the surfaces with SEM and EDX shows that the Sn–Pb coating wore out at 10^5 cycles exposing nickel and some areas were coated with Au–Ag alloy, indicating that metal transfer occurred (Fig. 9.25). Nickel also transferred from the flat surface to the pin. Similar metal transfers were also observed with Au–Co alloy coatings (see Fig. 9.26). It should be noted that the transfer of materials does not necessarily imply that it is due to adhesion, since mechanical interlocking is also possible.

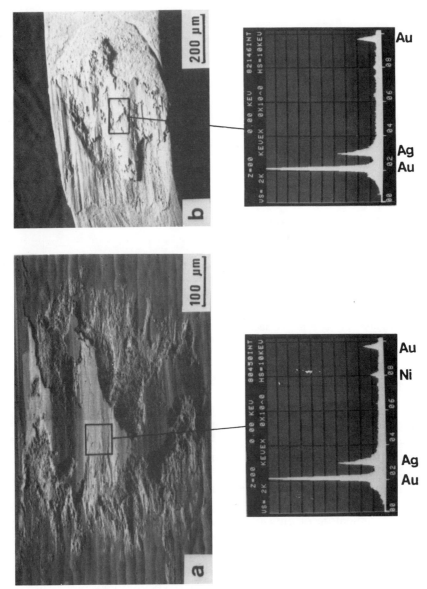

Figure 9.21 Bulk nickel (a) and the Au–30Ag pin (b) after 10^5 cycles.

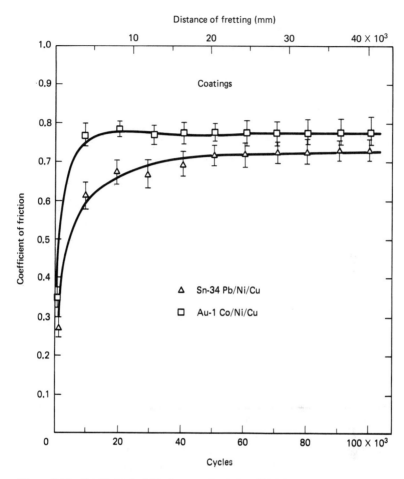

Figure 9.22 Coefficient of friction versus cycles of 1.25-μm Au–1Co and 1.0-μm Sn–34Pb coatings, both on 3.75-μm Ni substrate and Cu laminate.

Sn–Pb alloy coating on the modulated surface of the Ni–Cu substrate. As stated earlier, the Sn–Pb plating on the modulated surface was done to trap wear particles and thus eliminate the wild fluctuation of electric contact resistance without having to use gold alloys. Figure 9.27 shows the friction coefficients of Sn–Pb-coated Ni–Cu specimens with two different thicknesses (i.e., 3.75 and 30 μm) of the nickel layer. The initial coefficient of friction is nearly the same, being around 0.25. However, the steady-state coefficient of friction was different, 0.69 and 0.50 for the 3.75-μm and 30-μm Ni layers, respectively. The increase in the coefficient of friction suggests that plowing took place between the pin and the pad of the modulated surface. The steady values were reached when the number of the newly formed wear particles equals the number of the particles leaving

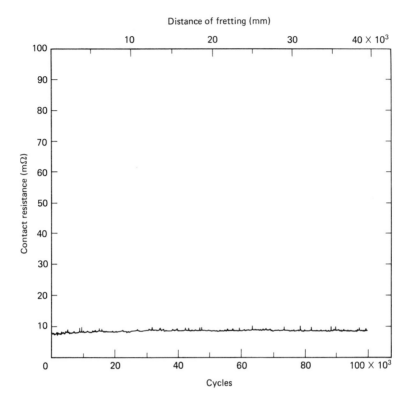

Figure 9.23 Dynamic contacts resistance versus cycles of 1.25-μm Au–1Co coating and 3.75-μm Ni substrate on Cu laminate.

the contact interface. Some of the leaving particles were entrapped in the pocket of the modulated surface. Different steady-state coefficients of friction were reached since the number of entrapped wear particles was less when the nickel layer was present, whereas the 3.75-μm-thick layer was not sufficiently thick to prevent plowing.

The electric contact resistance of the surface with the 3.75-μm-thick nickel layer was 5 mΩ throughout the tests, whereas the specimen with a 30-μm-thick nickel layer showed a large fluctuation. The fluctuation is attributed to the generation and transfer of nickel sheets from the thick nickel plating to the Au–Ag pin, which was observed to occur only when the thick nickel coating is used. When the nickel plates transfer and stick to the Au–Ag surface, the probability of having a metal-to-metal contact is less than when gold–silver slider is rubbing against the nickel layer, due to the formation of very stable nickel oxide layer on nickel. In this case, more contacts will be of NiO–NiO type.

The fact that the nickel sheets were absent when the thickness of the

Sec. 9.3 Control of Electrical Contact Resistance

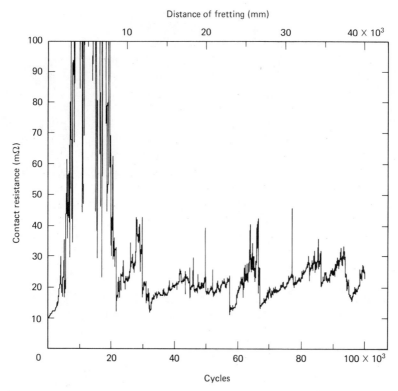

Figure 9.24 Dynamic contact resistance versus cycles of 1.0-μm Sn–34Pb coating and 3.75-μm Ni substrate on Cu laminate.

nickel layer is only 3.75 μm can be explained in terms of the delamination mechanism. When the layer is very thin, the crack nucleation may not be possible due to the large hydrostatic pressure, or alternatively the thickness is so small that it may not lead to work hardening due to the instability of dislocation.

SEM micrographs of the worn surfaces show that the wear particles were pushed into the pockets of the modulated surface. The debris consists of Au–Ag, Sn–Pb, and Ni. Figure 9.28(a) is a micrograph of the Sn–Pb-coated modulated surface with 3.75-μm-thick nickel layer at 10^5 cycles, whereas Fig. 9.28(b) is a micrograph of a similar surface but with 30-μm-thick nickel layer after 10^5 cycles.

The Auger analysis indicates that the surface of Sn–Pb-coated modulated surface and Au–Co alloy surface had an adsorbed layer consisting of oxygen, sulfur, chlorine, and carbon. In these specimens the nickel layer begins to be exposed at the surface after 10^3 cycles of fretting. When the thickness of the nickel layer was 3.75 μm, it took about 10^5 cycles to expose copper.

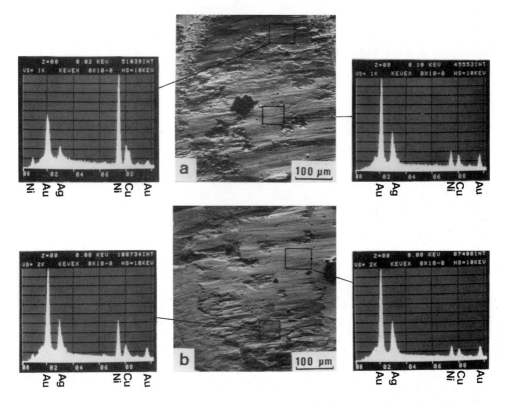

Figure 9.25 1.0-μm Sn–34Pb coating and 3.75-μm Ni substrate on Cu laminate (a), and the Au–30Ag pin (b) after 10^5 cycles. (From Saka, Liou, and Suh, 1984.)

In the case of Au–Co coating, nickel exposure occurred at 10^4 cycles. Even after only 10^2 cycles, the gold–silver alloy begins to transfer from the pin to the plated layer in all cases tested.

Effect of lubrication. Lubricants were applied to the surface of these coated specimens. The coefficients of friction of various specimens are plotted in Fig. 9.29. The coefficients of friction of the Au–Co- and Sn–Pb-coated materials are low, ranging from 0.20 to 0.23 up to 10^5 cycles. The contact resistances are low, less than 10 mΩ, and constant throughout the tests. The reason for this good wear resistance is due to the separation of the surfaces and consequential low adhesion and surface traction. Low wear rates mean a smaller number of trapped wear particles and no separation of electrically conducting surfaces by wear particles.

9.3.3 Discussions and Summary

The coefficients of friction for the flat and modulated surfaces with the same materials have about the same values. The contributions of the

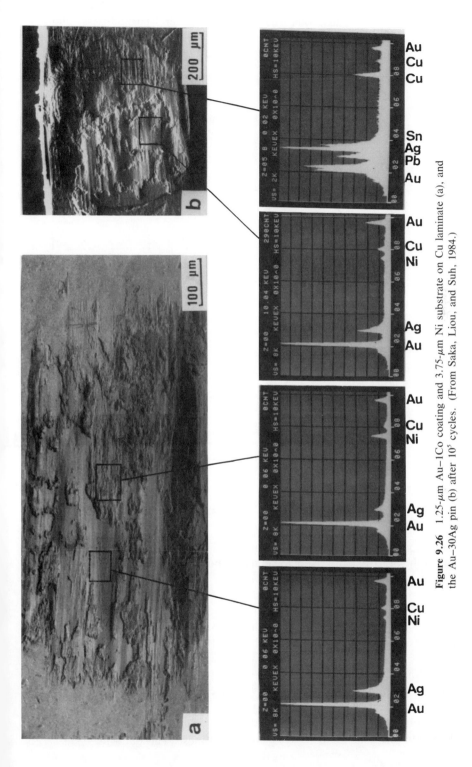

Figure 9.26 1.25-μm Au–1Co coating and 3.75-μm Ni substrate on Cu laminate (a), and the Au–30Ag pin (b) after 10^5 cycles. (From Saka, Liou, and Suh, 1984.)

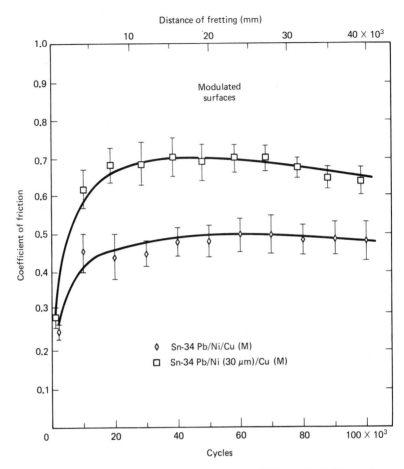

Figure 9.27 Coefficient of friction versus cycles of 1.0-μm Sn–34Pb coatings and Ni substrates (3.75 and 30 μm) on modulated Cu surfaces.

three basic friction generating mechanisms (i.e., asperity deformation, plowing, and adhesion) can be used to analyze the friction coefficients of flat and modulated surfaces.

At the first glance, the fact that the friction coefficients of the modulated surfaces are the same as those of unmodulated surfaces is surprising in view of the results presented in Chapter 3, Section 3.6. The experimental results presented in Section 3.6 showed that a modulated copper surface with a checkerboard pattern had a much lower frictional coefficient than flat surfaces. A careful examination of the worn modulated surfaces obtained in these electric contact experiments revealed that the edges of the modulated surface patterns deformed a great deal during sliding due to the small size of the pin. In running the experiments reported in Chapter 3, a much larger diameter pin was used, that is, 0.25 inches in diameter. Apparently, the

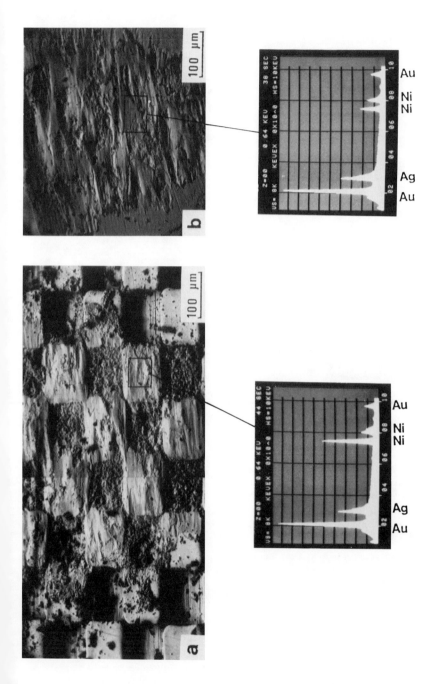

Figure 9.28 (a) 1.0-μm Sn–34Pb coating and 3.75-μm Ni substrate on modulated Cu surface (1), and the Au–30Ag pin (2) after 10^5 cycles; (b) 1.0-μm Sn–34Pb coating and 30-μm Ni substrate on modulated Cu surface after 10^5 cycles. *(continued)*

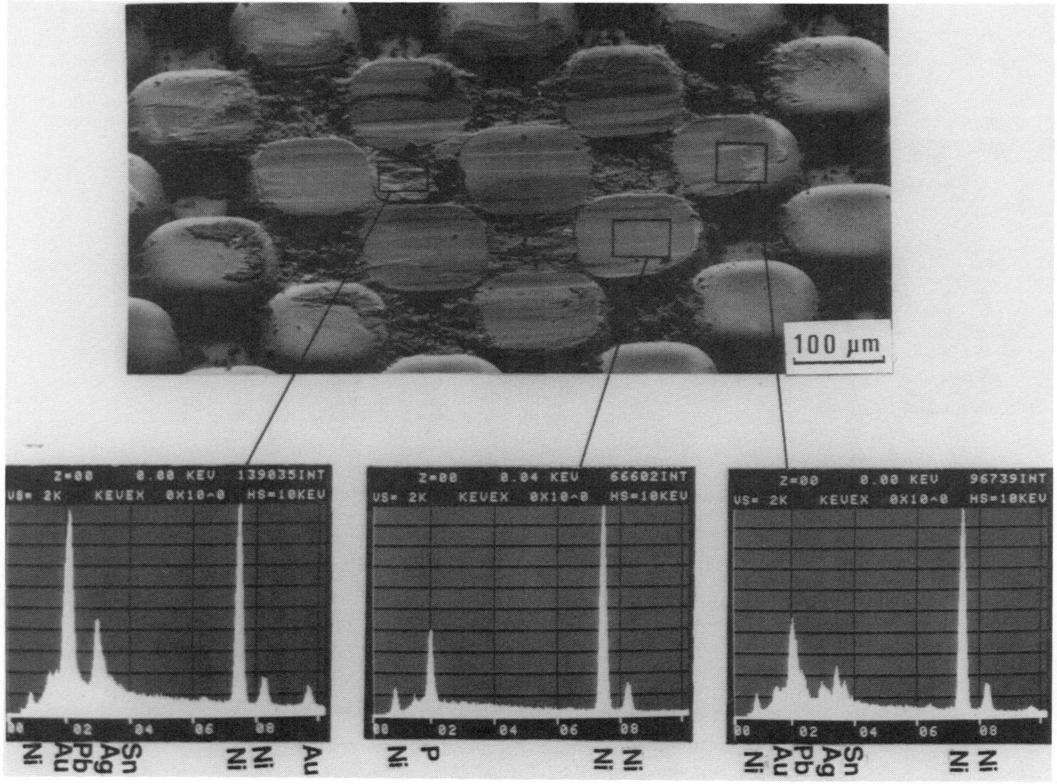

Figure 9.28 (*continued*)

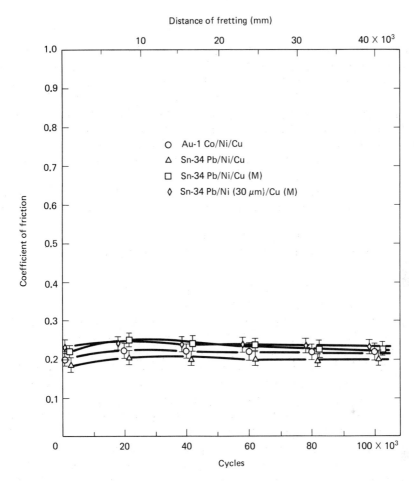

Figure 9.29 Coefficient of friction versus cycles of selected specimens (lubricated tests).

small pins deform the edges of the pattern as they move from the recessed areas to the raised contact surfaces, increasing the frictional force. This conclusion can be verified by performing tests with an AISI 52100 steel slider of 0.25 inches in diameter instead of the small Au–Ag pin. A hemisphered steel slider, which was polished and oxidized at 423 K for 20 hours to generate a coherent oxide film on the surface of the slider and reduce the adhesion, was used to perform fretting tests on both flat and modulated surfaces. The friction coefficients are shown in Fig. 9.30. The friction coefficient for the specimen without modulated surface increases from an initial value of 0.28 to a steady value of 0.63. However, the steady value for the modulated surface with pattern is only 0.36. The reduction of friction

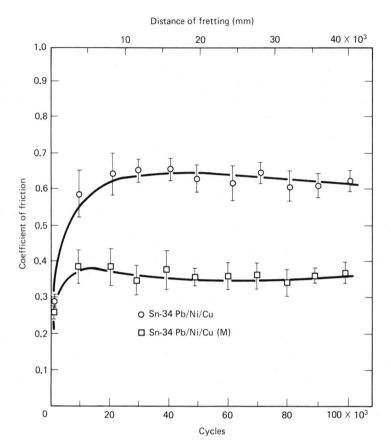

Figure 9.30 Coefficient of friction versus cycles of Sn–34Pb coatings and Ni substrates on flat and modulated surfaces; oxidized AISI 52100 steel sliders.

coefficient is due to a decrease in the plowing component because the wear particles are trapped in the cavities.

The number of entrapped wear particles is expected to be low for the modulated surface. This leads to more metal-to-metal contacts, so that the electrical contact resistance is low. Therefore, the idea that the electrical contact resistance can be kept low even using lead-tin alloy is sound. The experiments reported in this section clearly support the hypothesis that the contact resistance increases when the trapped wear particles oxidize, since whenever the number of trapped particles is reduced such as through lubrication, the contact resistance does not increase. The electrical contact resistance is initially high when the surface consists of metals that can oxidize and form a strong coherent oxide layer. Such is the case with nickel. In general, it is not advisable to use metals as electric contact surface if they are capable of forming strong, coherent oxide layers.

9.4 SOFT-METAL COATINGS ON A HARD SUBSTRATE TO LOWER THE WEAR RATE

As discussed in Chapter 2, the question of whether the outermost surface layer (i.e., about 50 Å from the surface) is soft or hard has not been satisfactorily answered. However, one may assume that when there is no oxide layer blocking the egress of dislocations to the surface, there should always be a zone of low dislocation density due to the image force acting on these dislocations. In this case, the surface layer can deform continuously without much work hardening and fracture. Therefore, one may hypothesize that a thin layer of a soft metal plated on a hard substrate can retard the wear of the harder substrate by preventing the plastic deformation of and crack nucleation in the hard substrate. According to the postulate made regarding the existence of the low-dislocation-density zone, the soft layer must be *less* than a critical value for it to be effective. If the soft layer is thicker than the critical value, delamination can occur within this layer, forming loose wear particles. These wear particles can in turn cause plowing and further plastic deformation. Experiments have been performed to check this hypothesis, which is described in this section following a brief review of the work done in the past.

9.4.1 Review of Past Work

The concept of using a soft metal layer was tried even before the advent of the delamination theory of wear, since the adhesion theory of friction predicts that the coefficient of friction will be low when the flow strength, τ, of the welded function is low, while the hardness, H, is high:

$$\mu = \frac{\tau}{H} \tag{9.1}$$

However, it was shown that the use of the soft metallic layer on a hard substrate does not significantly lower the friction coefficient, although the wear rate is lowered by orders of magnitude (Abrahamson et al., 1974; Jahanmir et al., 1976; Suh, 1976).

Among those who tried to make use of soft metallic layer are Tsuya and Takagi (1964), who electroplated one of the copper sliding surfaces with 6- to 75-μm lead. They found that under a stress of 10 kg/cm^2, in air, the lead film was worn out until a stable layer of 2 μm remained on the surface. Some of this lead was transferred to the unplated copper surface and formed a stable film 2 μm thick. The wear tracks of these samples were very smooth, containing fine parallel lines in the sliding direction. This investigation was extended by Takagi and Liu (1967) to AISI 440C stainless steel and AISI 52100 bearing steel with 0.1 μm of gold electrodeposits upon one of the sliding surfaces. The best results were obtained for the AISI 440C stainless steel sliding on 0.1-μm gold-plated AISI 52100 steel or

440C stainless steel. However, wear was reduced only by a factor of 2 to 3. Takagi and Liu (1968) extended this investigation by considering two gold alloys, silver, copper, and nickel electroplating on AISI 52100 and 440C steels. In these tests only the rotating surface was plated and the normal load used was 60 kg. They found that a 20-μm gold, copper, or silver plate on AISI 52100 steel was needed for wear reduction.

Solomon and Antler (1970) investigated the effect of load and thickness of gold, silver–gold, and nickel–gold alloys electroplated on both sliding copper surfaces. They found that beyond a critical thickness (depending on load and the substrate material) resistance to wear increased dramatically. These results were improved if a very thin nickel was plated on copper before the gold and gold alloy electrodeposition. It may be speculated that the hard nickel layer prevented the subsurface deformation of copper.

Kuczkowski and Buckley (1965) coated nickel and AISI 440C stainless steel with 25-μm plate of various binary and ternary alloys of gallium and indium. A ternary alloy of gallium, indium, and tin reduced the wear rate of AISI 440C stainless steel in vacuum by four orders of magnitude. The wear rate of nickel was also reduced by the same amount when the surface was coated with a binary alloy of gallium and indium.

The process of electroplating the sliding surfaces with precious metals has been applied to ball bearings (Evan and Flatley, 1962, 1963; Flatley, 1964; Brown, 1970) and gears (Rindings, 1965; Lee, 1965) operated in a vacuum. Even though the life was increased in these applications, the components still exhibited a high degree of wear. This is probably due to the thick (30 to 75 μm) coatings used.

Abrahamson et al., (1974) found that thickness of the deposited layer is one of the primary factors if any major improvements in wear characteristics are to be achieved. This phenomenon is discussed further in this section for cadmium and nickel electrodeposited on steel specimens.

9.4.2 Experimental Results with Several Soft-Metal Layers on Hard Substrates

Wear tests were carried out with a cylinder-on-cylinder arrangement (Jahanmir et al., 1976). The specimens were 0.63 cm in diameter and 7.6 cm long with a 0.4-μm (CLA) ground finish prior to plating. In one case the surface was metallographically polished with a 0.25-μm diamond paste. The specimens were rotated at a surface speed of 1.8 μm/min and the stationary pins were pushed against the specimens by a normal load of 2.25 kg. The inert atmosphere tests were carried out dry in a chamber surrounding the mating surfaces under argon flowing at a rate of 10 liters/min.

The substrate materials used had a wide range of hardnesses: AISI 1018 (84 kg/mm^2), AISI 1095 (170 kg/mm^2), and AISI 4140 (270, 370, and 460 kg/mm^2—obtained by different heat treatments). In all cases the sliders

Sec. 9.4 Soft-Metal Coatings on a Hard Substrate to Lower the Wear Rate

were made of the same material as the specimen and were heat-treated and plated in the same manner. The plating thickness varied from 0.05 to 10 μm. The platings tested were gold, gold over a nickel flash, cadmium, nickel, and silver. Some of the gold-plated specimens were plated with a flash of gold first and then heated at 500°C for 2 hours in vacuum to obtain a diffused interface and thus enhanced bonding between the plate and the substrate. The required thickness of gold was then plated on the samples.

The wear data for 1-μm plates of cadmium, silver, gold, and nickel on steel substrates with various hardnesses for 60-minute tests (108 m of sliding) in argon are summarized in Table 9.3. The wear rates of all coatings, with the exception of Ni on steels with a hardness less than 460 kg/mm^2, were lower than the wear rates of unplated materials by at least three orders of magnitude. It should be mentioned that greater wear reductions are possible since the tests were discontinued after 60 minutes while the plates were still effective in reducing wear. To determine the maximum possible wear reduction, the tests should have been continued up to the final failure of the plates.

Effect of plate thickness. The effectiveness of plating in wear reduction is dramatic, as shown in Fig. 9.31 for a nickel plate of initial thickness 1 μm on AISI 4140 steel. The large difference in the size of the wear tracks between the unplated [Fig. 9.31(a)] and the plated sample [Fig. 9.31(b)] should be noted. The unplated sample was tested for 30 minutes, whereas the plated sample was tested for 2 hours. Since no further wear

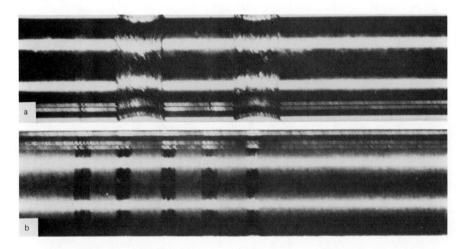

Figure 9.31 Comparison of wear tracks of AISI 4140 steel tested in argon, with a normmal load of 2.25 kg: (a) unplated against unplated (after 30 minutes of testing); (b) 1.0-μm Ni plated against 1.0-μm Ni plated (after 2 hours of testing). (From Jahanmir, S.; Abrahamson, E. P., II; and Suh, N. P., "Sliding Wear Resistance of Metallic Coated Surfaces," *Wear*, Vol. 40, 1976, pp. 75–84.)

could be detected on the plated specimen after 2 hours (216 m of sliding), the test was terminated.

The important effect of the nickel plate thickness on the wear rate of AISI 4140 steel is shown in Fig. 9.32. Wear reduction by three orders of magnitude is observed for an initial nickel plate thickness of 1 μm. This minimum thickness of the plate also decreased the coefficient of friction from 0.63 to 0.45, probably because of less plowing when the plated layer is thin. Figure 9.32 shows that the wear rate increases with the nickel plate thickness for plates thicker than 1 μm. This increase in wear rate is caused by dislocation accumulation leading to strain hardening and delamination within the plate. Cracking and delamination within a thick (25-μm) plate of gold on AISI 1018 steel is shown in Fig. 9.33. The thick layer wears by delamination until the thickness of the layer is reduced to the optimum thickness in which the dislocations presumably are unstable. Afterward,

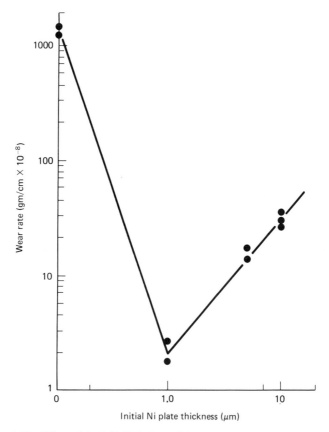

Figure 9.32 Effect of the initial Ni plate thickness of the wear rate of AISI 4140 steel in argon, for 30-minute tests. (From Jahanmir, S.; Abrahamson, E. P., II; and Suh, N. P., "Sliding Wear Resistance of Metallic Coated Surfaces," *Wear*, Vol. 40, 1976, pp. 75–84.)

Sec. 9.4 Soft-Metal Coatings on a Hard Substrate to Lower the Wear Rate

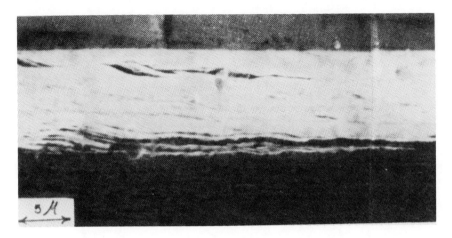

Figure 9.33 Crack formation in a thick Au plate (25 μm initial thickness) after wear testing; AISI 1020 steel substrate. (From Jahanmir, S.; Abrahamson, E. P., II; and Suh, N. P., "Sliding Wear Resistance of Metallic Coated Surfaces," *Wear*, Vol. 40, 1976, pp. 75–84.)

the wear rate is very low, as in the case of thin platings. The transient behavior of thick coatings is shown in Fig. 9.34 for an initial 10-μm nickel plate on AISI 4140 steel. The steady-state wear rate of thick platings may be larger for some sliding conditions where the wear particles cannot be removed from the contact and act as abrasive particles. Similar wear and friction results were obtained with Cd-plated steel, as shown in Fig. 9.35.

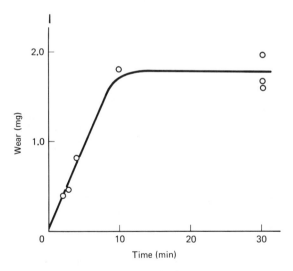

Figure 9.34 Wear of a thick Ni plate (initial thickness 10 μm) versus time. (From Jahanmir, S.; Abrahamson, E. P., II; and Suh, N. P., "Sliding Wear Resistance of Metallic Coated Surfaces," *Wear*, Vol. 40, 1976, pp. 75–84.)

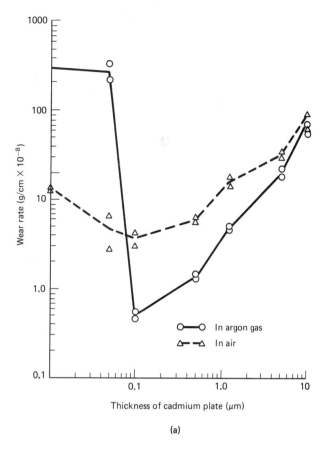
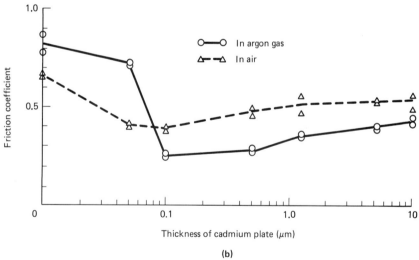

Figure 9.35 (a) Effect of cadmium plate thickness on the wear rate of Cd-plated AISI 1018 steel; (b) friction coefficient of Cd-plated steel as a function of plate thickness. (From Jahanmir, S.; Suh, N. P.; and Abrahamson, E. P., II, "The Delamination Theory of Wear and the Wear of a Composite Surface," *Wear*, Vol. 32, 1975, pp. 33–49.)

Effect of surface roughness and bond strength. The steel samples that were plated with a thin layer of gold without any special treatments failed immediately at the beginning of sliding. This failure was found to have been caused by the weakness of the bond between the plated layer and the substrate, which was accentuated by the roughness of the substrate surface at the time of plating. In these tests, since the optimum plate thickness is very small, the substrate surface roughness may play a major role in the life of the coatings. The bond strength between electrodeposited gold and steel has been reported to be very low and to become even weaker with increasing substrate surface roughness (Safranck, 1974). Under sliding conditions, the roughness can cause further deterioration of the bond strength by the deformation and final fracture (Fig. 9.36) of the original substrate asperities. Therefore, the gold-plated steel samples were specially treated to achieve a good bond strength. By diffusion of a very thin layer of gold or by plating a flash of nickel over the substrate before gold plating, it was possible to increase the life of the thin gold plates from immediate failure to more than 108 m of sliding (60 minutes).

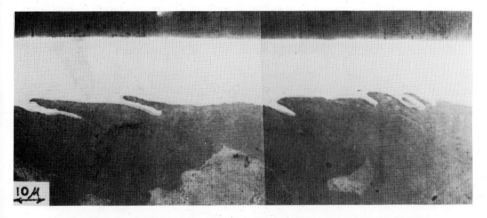

Figure 9.36 Deformation of substrate asperities due to the sliding action on a thick plate of Au on AISI 1020 steel substrate. (From Jahanmir, S.; Abrahamson, E. P., II; and Suh, N. P., "Sliding Wear Resistance of Metallic Coated Surfaces," *Wear*, Vol. 40, 1976, pp. 75–84.)

The influence of the substrate surface finish on cadmium-plated specimens was checked with a 0.05-μm cadmium coating on AISI 1018 steel with both a fine-ground finish and a metallographically polished finish. The surface roughness had only a moderate effect on the life of the cadmium plates, and polishing only increased the life of the coatings from 25 m to 31 m of sliding. Since the steel–cadmium adhesion strength is much greater than the adhesion strength of steel–gold (Safranck, 1974), this result suggests that moderate changes of surface roughness may only influence plating–substrate combinations which have a low adhesive strength.

Hardness matching between the coating and the substrate. According to the delamination theory, the plated material must be softer than the substrate to minimize the wear rate. The results on the wear rate of cadmium- and nickel-plated steels in Table 9.3 support this hypothesis. The wear-rate data indicate that nickel plate on steel specimens which are softer than 460 kg/mm^2 fails immediately at the start of sliding. However, the wear rate of a 1-μm nickel-plated 4140 steel specimen with a hardness of 460 kg/mm^2 is very low. Table 9.3 also indicates that a 1-μm cadmium plate with a hardness of 30 to 50 kg/mm^2 was successful on all substrates with hardness ranging from 84 to 370 kg/mm^2.

TABLE 9.3 Experimental Results on the Wear Resistance of 1-μm Plates on Steel[a]

Coating	Substrate and Hardness (kg/mm^2)	Wear Rate ($\times 10^{-8}$ mg/cm)	Coefficient of Friction
Cd	AISI 1018 (84)	3.6	0.35
Cd	AISI 1095 (170)	1.8	0.25
Cd	AISI 4140 (270)	3.6	0.35
Cd	AISI 4140 (370)	3.6	0.25
Ag	AISI 1095 (170)	1.8	0.33
Au (diffused)	AISI 1095 (170)	1.8	0.85
Au (Ni underlayer)	AISI 1095 (170)	-1.8[b]	0.9
Ni	AISI 1018 (84)	Immediate failure	—
Ni	AISI 1095 (170)	Immediate failure	—
Ni	AISI 4140 (270)	Immediate failure	—
Ni	AISI 4140 (370)	Immediate failure	—
Ni	AISI 4140 (460)	1.8	0.45

[a] Experimental conditions: wear-tested in argon, normal load 2.25 kg, sliding distance 108 m.
[b] These specimens gained weight.

Effect of environment. The wear tests conducted in air on a 1-μm plate of various materials on steel indicated that only gold platings (with special treatments) were effective in wear reduction. Other platings, such as cadmium, silver, and nickel, were not successful in air, because they oxidized and were removed early in the tests. The oxide particles of the plating materials caused plowing when they became loose and were entrapped at the sliding interface. Plowing increases the surface traction and thus accelerates the delamination wear.

The wear track of 0.1-μm Cd-plated steel tested in argon [Fig. 9.37(a)] is compared with the wear track in air [Fig. 9.37(b)]. The micrograph indicates that delamination has occurred quite extensively in air.

Since the poor performance of Cd-plated steel in air was believed to be caused by oxidation, a series of tests were conducted under boundary

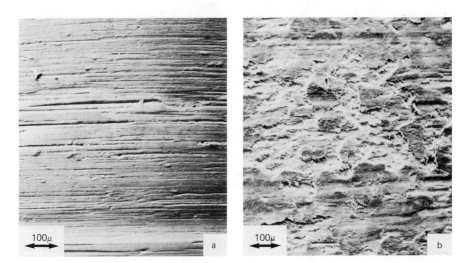

Figure 9.37 Wear track of 0.1-μm Cd-plated steel: (a) tested in argon; (b) tested in air. (From Jahanmir, S.; Suh, N. P.; and Abrahamson, E. P., II, "The Delamination Theory of Wear and Wear of a Composite Surface," *Wear*, Vol. 32, 1975, pp. 33–49.)

lubrication to minimize oxidation. The first set of tests was run with Gulftex 39 under a load of 2.25 kg for a 216-m sliding distance. The unplated steel and 0.1-μm Cd-plated steel exhibited a similar wear rate of 4.6×10^{-9} g/cm. The wear track of both specimens became dark, indicating that the Gulftex oil had reacted with the specimens and caused the high wear rate of Cd-plated steel. Therefore, for the next set of tests a pure mineral oil (Nujol) was used. Under the same conditions as the tests with Gulftex 39, the unplated steel had a wear rate of 3.2×10^{-9} g/cm when lubricated with mineral oil, whereas the 0.1-μm Cd-plated steel had experienced such low wear that no weight change could be detected. Therefore, the test duration was tripled (648 m). After the tests the unplated steel had a wear rate of 7×10^{-9} g/cm, while the wear rate of 0.1-μm Cd-plated steel was only 0.5×10^{-9} g/cm. It is believed that longer tests will result in larger wear-rate reductions, similar to the tests performed in argon.

It should be noted that in all these tests the wear tracks of the unplated samples appeared to be very rough compared with plated steel surfaces tested in argon. As shown in Fig. 9.38, the cratered appearance of the unplated steel indicates that many wear sheets have been separated from the wear track. The wear track of the plated sample, on the other hand, is much smoother, with parallel furrows in the sliding direction. Only a few indications of delamination were observed on the 0.1-μm Cd-plated steel specimens.

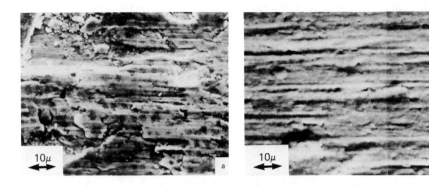

Figure 9.38 Wear tracks in argon gas under a load of 2.25 kg after a 45-m sliding distance: (a) unplated steel; (b) 0.1-μm Cd-plated steel. (From Jahanmir, S.; Suh, N. P.; and Abrahamson, E. P., II, "The Delamination Theory of Wear and the Wear of a Composite Surface," *Wear,* Vol. 32, 1975, pp. 33–49.)

To make sure that these results were not limited to the experimental geometry employed in these tests, a series of wear tests were conducted in argon on annular specimens of similar materials. An annular specimen of 2.5 cm outside diameter, having a 5.3-cm^2 contact area, was pressed under a load of 22.5 kg on another annular sample rotating at 60 rpm for 15 minutes. Similar results were obtained both for the 0.1-μm Cd-plated steel and for the thicker coatings.

The results presented are consistent with the predictions made by the delamination theory of wear on the wear resistance of soft metal coatings. For best wear resistance the plated material must be softer than the substrate, thinner than a critical thickness, and bonded strongly to the substrate. The condition of the surface of the substrate is an important factor since it was shown that smoother plated surfaces last longer. It has been shown (Jahanmir and Suh, 1977) that subsurface damage caused by machining operation greatly influences initial wear behavior. Therefore, components for sliding applications must be prepared carefully so as to minimize the damage to the substrate during machining. Otherwise, the damaged substrate may cause premature failure and offset the beneficial effects of plating. If the delaminated particles are not removed from the contact, they may oxidize and serve as abrasive particles, thus degrading the coating and enhancing delamination.

Soft-metal coatings have been tested by many investigators previously for increasing wear resistance, as discussed earlier. Some investigators assumed that the metal layer softens during sliding and acts as a lubricant, which is contradicted by the fact that even a high-melting-point metal such as nickel can be effective in wear reduction under low-speed sliding wear, and that the thickness of the plate has a pronounced influence. It is not

Sec. 9.4 Soft-Metal Coatings on a Hard Substrate to Lower the Wear Rate

the absolute softness that is important; it is the ratio of flow stresses of the substrate and the coating which is the determining factor.

The relative lattice spacings, elastic moduli, and crystal structures of the coating and the substrate may also influence the effectiveness of the coating material. When the shear modulus of the coating is less than that of the substrate, the dislocations generated in the coating during sliding will be repelled by the interface, while the dislocations generated in the substrate will be attracted toward the interface. If the coating is sufficiently thin, some of the mobile dislocations in the plated layer may be eliminated by the image force, and also by the stress field established by other dislocations, when the surface is unloaded after the slider asperity has passed. In this case the coating will remain soft and function as a protective layer.

Another important influence of these relative material properties is the dislocation mobility across the plate–substrate interface. Ideally, when the substrate is hard and resists plastic deformation, the dislocations in the substrate should not cross the interface and move into the coating; rather, they should tangle up near the interface. This phenomenon will arise if the lattice spacing and the crystal structure and orientation of the coating and the substrate are sufficiently different to prevent the dislocations generated in the substrate form penetrating across the interface. In this respect the choice of f.c.c. metals (such as nickel) or h.c.p. metals (such as cadmium) for the coating material and b.c.c. metals (such as steel) for the substrate is quite appropriate.

The above discussion presumes that there is a strong bond between the substrate and the coating material. It is well known that metals which readily form an alloy also form a strong adhesive bond. However, the requirements of solubility of one metal in another are precisely the qualities that tend to promote penetration of dislocations across the interface. Therefore, the material choice is rather limited. What may help the situation, however, is the texturing of the surface layer during deformation. The surface layer orients during deformation in such a manner that the primary slip planes become nearly parallel to the surface. If the substrate metal has a different crystallographic structure from the plating materials, the orientation of crystallographic planes in the coating will be different from that in the substrate and the interface will retard the transport of dislocations across the interface.

The experimental results on the role of coating in sliding wear presented here are not direct and explicit proof of the assumed dislocation instability caused by the image force, but they are in accordance with the original predictions made by the delamination theory of wear. It appears, however, that original postulates predicting the effects of thickness and hardness of the metallic coatings are justified by the indirect evidence provided by the wear tests.

In the preceding paragraphs the role of thin soft coatings was explained in terms of dislocation mechanics. However, other plausible explanations were also attempted without much success. One may attempt to explain the observed phenomenon in terms of the state of stress during sliding wear and the stress relaxation of the material. To explain the nonhardening nature of the surface layer, one may speculate that stress relaxation counters work hardening. However, these arguments fail since they cannot explain why work hardening occurs when the plated layer is thick. Another explanation may be that crack nucleation cannot occur in thin plated layers due to the large compressive stress near the surface. Further research to clarify these points is needed, however.

9.4.3 Conclusions on the Effectiveness of Soft-Metal Layers

Based on the foregoing results, it may be concluded that:

1. Major reduction in wear can be achieved by coating a hard substrate with a material that is much softer than the substrate.
2. There exists an optimum thickness of soft coating for wear resistance. This thickness is, in general, less than 1 μm for steel plated with Cd, Ag, Au, or Ni.
3. Cd, Ag, and Ni plates are effective only in an inert atmosphere, but Au is effective both in air and in an inert atmosphere.
4. The surface roughness of the substrate and the coating–substrate bond strength are two important factors for the wear resistance of soft metallic coatings.

9.5 HARD COATING OF SINGLE-PHASE MATERIALS ON THE SURFACE

The delamination theory of wear of Chapter 5, the genesis of friction discussed in Chapter 3, and the solution wear due to chemical instability analyzed in Chapter 8 suggest that an ideal wear-resistant material is a tough and hard material which is chemically stable. However, in most commercially available materials hardness is obtained at the expense of decreased toughness. A means of obtaining both hardness and toughness is to coat the surface of a tough material with a hard surface layer. A hard surface layer reduces the frictional force and the wear rate when sliding against a relatively soft material if the pair is chemically stable and if the coating is well bonded and mechanically compatible with the substrate. The role of the hard layer is to prevent plowing, whereas the chemical insolubility criterion is to assure that there be minimal dissolution. Hard coating can be particularly useful in an abrasive environment. The fact that the relative hardness of materials affects the wear rate is illustrated in Fig. 9.39.

Sec. 9.5 Hard Coating of Single-Phase Materials on the Surface

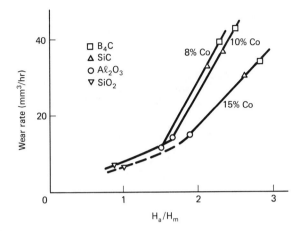

Figure 9.39 Wear rate as a function of the relative hardness of the abrasives to that of the workpiece.

Typical hard materials used in sliding applications are carbides, oxides, nitrides, borides, and amorphous glasses. Normally, carbides have higher melting points and higher hardness than their corresponding oxides and nitrides. Figure 9.40 gives a plot of the hardness of several carbides as a function of temperature. The hardness of oxides is given in Table 9.4. It is generally more difficult to create bonding with oxides than with carbides because of the extreme chemical stability of oxides, as discussed in Chapter 8.

One of the basic considerations in choosing a coating technique is the quality of bonding between the substrate and the coating. The bond is established chemically or mechanically or both. To create a primary chemical bonding, there must be either a chemical reaction between the substrate material and the coating or diffusion of elements between them to form a solid solution at the interface.

There are several different means of creating a chemical bond in the case of carbides. In a process of metal deposition developed by Cook and Kramer (1978), there is a coupling reaction and diffusion, as in

$$WC + Hf \longrightarrow HfC + W \qquad (9.2)$$

The carbon is supplied solely from the substrate and there is no external supply from the gas phase. The bond created by this type of reaction is so strong that when the bonded interface is indented with a diamond indenter, the crack does not run parallel to the interface.

Another less successful means of creating a chemical bonding in the presence of carbon-containing gases is to supply less than the stoichiometrically required amount of carbon. This may be illustrated for the case of TiC coating as follows:

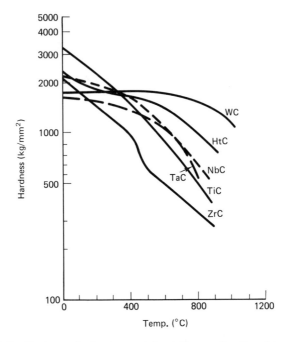

Figure 9.40 Hardness of refractory metal carbides as a function of temperature. (Reprinted with permission from Toth, L. E., *Transition Metal Carbides and Nitrides*. Copyright 1971 by Academic Press, New York.)

TABLE 9.4 Hardness at Specified Temperatures of the Phases That May Be Present in a Steel

Phase	Hardness, H_v (kgf/mm²)		
	400°C	600°C	800°C
Iron	45	27	10
Iron and interstitials	90	27	10
TiO	1300	1000	650
FeO	350	210	50
MgO	320	220	130
NiO	200	140	100
MnO	120	60	45
Al_2O_3	1300	1000	650
SiO_2	700	2100	300
ZrO_2	650	400	350
TiO_2	380	250	160
$MgAl_2O_4$	1250	1200	1050
$ZrSiO_4$	400	290	140

Sources: Data from Ramalingam and Watson (1973) and Westbrook (1966).

$$a\text{TiCl}_4 + b\text{CH}_4 + \text{H}_2 \longrightarrow b\text{TiC} + (a-b)\text{Ti} + 4a\text{HCl}$$
$$+ (1 + 2b - 2a)\text{H}_2 \quad (9.3)$$
$$(a-b)\text{Ti} + (a-b)\text{WC} \longrightarrow (a-b)\text{TiC} + (a-b)\text{W}$$

If the molar concentration of methane in Eq. (9.3) is less than the molar concentration of titanium chloride (i.e., $a > b$), free titanium metal forms, which then reacts with WC to form additional titanium carbide.

The solubility of tungsten in TiC is high enough at the deposition temperature that a solid solution may form between TiC and W. Upon cooling, tungsten precipitates out in the TiC matrix. On the other hand, the reaction products of Eq. (9.2) do not readily form a solid solution. Therefore, tungsten always remains on the substrate side of the substrate–coating interface and the reaction occurs by diffusion of carbon.

The chemical bonding between the substrate and Al_2O_3 is a difficult problem to solve. A patent by the General Electric Company claims that the formation of cobalt aluminate between the substrate and the Al_2O_3 coating provides a strong bonding, but the apparent lack of good bonding is still a problem with alumina-coated tools.

When the service temperature of the coating is different from the deposition temperature, the coefficient of thermal expansion of the coated layer should not be too different from that of the substrate, so as to minimize the interfacial stress. Since a compressive state of residual stress in the surface layer is desirable to prevent crack propagation, the coefficient of expansion of the coated layer should be slightly less than that of the substrate if the coating is done at a temperature higher than the service temperature, provided that the coating can be prevented from delaminating. The coefficient of expansion of several carbides and nitrides are given in Table 9.5.

Coating techniques used to deposit hard layers on substrates may be classified into two categories: chemical vapor deposition (CVD) and physical vapor deposition (PVD). In the CVD process for carbides, metal salt and carbon-containing gases such as methane are introduced to a vacuum chamber and reacted on a heated tool surface. This leaves metal carbides on the tool surface. Pure metal vapor may also be reacted with carbon-containing gases, nitrogen gas, or oxygen gas (or water vapor) to form metal carbide, metal nitride, and metal oxide, respectively, which is sometimes known as *activated reactive evaporation* (Bunshah and Raghuram, 1972). In the PVD process the coating material is evaporated by heating or chemical reactions (Galski, 1972). When gases are well mixed in a gaseous phase and deposited on a substrate, amorphous, glassy materials can be created and deposited on the surface. Commonly used commercial processes are CVD, sputtering, and plasma spraying techniques. Many of these processes are proprietary. Table 9.6 gives a list of coating techniques and references.

Many of the CVD techniques require heating of the substrate to induce

TABLE 9.5 Values of the Coefficient of Thermal Expansion

Materials and Composition (wt %)			Coefficient of Expansion (10^{-6} °C^{-1})	Reference
WC		Co		
100		0	5.7–7.2	Schwarzkopf
94–94.5		5.5–6.0	5	and Kieffer
85		15	6	(1960)
Wc	TiC	Co		
94	1	5	5	Schwarzkopf
84.5	2.5	13	5.5	and Kieffer
86	5	9	5.5	(1960)
78	14	8	6.2	
78	16	6	6	
76	16	8	6	
69	25	6	7	
34	60	6	7.5	
TiC$_{0.97}$			7.4	Toth (1971)
ZrC$_{0.99}$			—	
HfC$_{0.99}$			6.6	
VC$_{0.37}$			—	
NbC$_{0.99}$			6.6	
TaC$_{0.99}$			6.3	
WC				
Parallel to a			5.0	
Parallel to b			4.2	
TiN			9.35	Toth (1971)
ZrN			7.24	
HfN			6.9	
NbN			10.1	
TaN			3.6	

TABLE 9.6 Coating Techniques and Materials

Type of Coating	Coating Technique	References
TiC	CVD	Ekmar (1970), Richman and Lee (1974), Schintlmeister et al. (1975)
TiC	Activated reactive evaporation	Nakamura and Inagawa (1975), Bunshah and Raghuram (1972)
TiN	CVD	Schintlmeister et al. (1975), Kieffer et al. (1970)
TiO$_x$C$_y$	Sputtering	Carson et al. (1976)
Al$_2$O$_3$	CVD	Funk et al. (1975), Hale (1971), Lindstromand and Johannesson (1975)
TiC–TiN double coating	CVD	Hauser (1975)

chemical reaction on the surface, which limits their applicability to cemented carbide tools. Normal sputtering and plasma arc sputtering can be used to deposit hard layers on cold substrates, but the quality of bonding may not be as good as those deposited on hot substrates.

Hard coatings have been used in many different applications, ranging from parts of watches to industrial cutting tools. Hard coating of TiN has been used to prolong the life of high-speed tools (Su and Cook, 1977) and drills; TiC and TiN are routinely coated on carbide tools to increase the chemical stability of the tool; Hf can be deposited on WC to create HfC layer for improved chemical stability (Kramer and Cook, 1980); and glass can be deposited to improve life at low speeds.

9.6 ION IMPLANTATION

A novel method of hardening and changing the chemistry of the surface layer is ion implantation. Ion implantation modifies the microstructure and chemistry of a surface layer by driving ionized elements into the surface layer in a strong electric field (see Fig. 9.41). This technique has been originally developed in recent years to dope semiconductors, but has also been used to enhance the surface properties of bearings and sliding surfaces (Poate, 1979; Singer et al., 1980; Shepard and Suh, 1982). The depth of the penetration is only about 0.5 μm, and therefore its application in tribology is limited to reasonably well-lubricated cases.

9.6.1 Experimental Technique

Pin-on-disk friction and wear tests were conducted by Shepard (1981) using 3.8- to 4.5-cm-diameter wear disks, and 0.635-cm hemispherically tipped pins made from 99.9999% pure iron, OFHC copper, and commercially pure titanium. Both sides of each disk were mechanically polished with 1.0- and 0.3-μm alumina powder. The ends of the pins were also polished with 1.0- and 0.3-μm alumina. Both pins and disks were Freon-vapor degreased, rinsed in isopropyl alcohol, dried, annealed in a vacuum furnace at 1073 K for 1 hour, and stored in a vacuum desiccator prior to implantation.

The wear tracks of the implanted and unimplanted disks were sectioned and observed for subsurface deformation. Specimens were nickel plated, and a pie-shape section was cut from the worn disk. This section was then mounted in Bakelite, giving a tapered section (Rabinowicz, 1950) intersecting the wear track at approximately 60°. This mounting configuration aids in preventing surface and subsurface damage which may occur during sample preparation. The specimens were then polished using the same procedure described for the friction and wear specimens. They were subsequently Freon-vapor degreased, rinsed in isopropyl alcohol, dried, etched, and then stored in isopropyl alcohol prior to observation in the SEM.

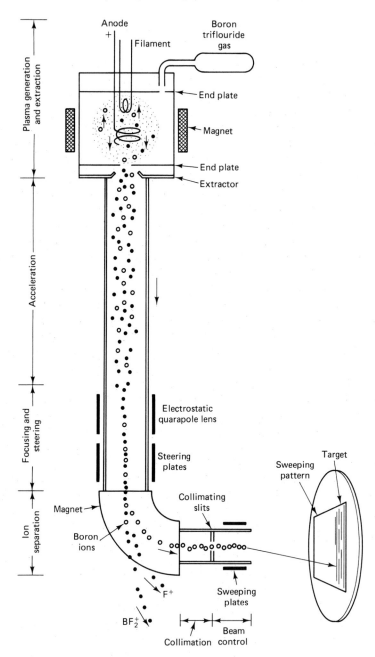

Figure 9.41 Schematic of the implantation process. (From Morehead, F. F., Jr., and Crowder, B. L., "Ion Implantation," *Scientific American,* Vol. 228, No. 4, 1973, pp. 65–71.)

Sec. 9.6   Ion Implantation   461

Prior to implantation, the samples were degreased and cleaned with isopropyl alcohol. The disks and pins were implanted in a Model 200-CF5 Varian/Extrion ion implanter. During implantation the disks and pins were clamped against a Freon- or water-cooled heat sink that kept the samples below 373 K. Both the pins and disks were implanted to a fluence of 10^{17} ions/cm^2 in a target chamber which was held at about 5×10^{-7} torr. The ion beam was scanned over the specimen surface to give a uniform current density between 4.3 to 6.2 μA/cm^2.

The ion–substrate testing combinations are as follows: N$^+$ into iron, titanium, and copper; Zn$^+$ into copper; and Al$^+$ into iron. Although precautions were taken to prevent oxidation of the samples before and after implantation, it was inevitable that some oxidation took place. It has not been clarified if the ions were implanted into an oxide layer or if the oxide layer was sputtered off during implantation.

The implanted and unimplanted specimens were tested using a pin-on-disk geometry. The specimen (disk) was rotated and the slider (pin) was held stationary in a holder attached to a strain ring. The tangential force was continuously measured by the strain gages and recorded.

All tests were conducted at approximately 293 K in a controlled humidity environment. Samples were lubricated with mineral oil, and tested for a duration of 5 hours at a sliding velocity ranging from 0.5 to 2 m/min (40 rev/min) with a normal load of 400 g unless otherwise specified. Every test combination was conducted a minimum of three times and the treatment of the pin surface was the same as the disk surface for all cases.

At the end of the tests, the lubricant and any loose particles were rinsed off with isopropyl alcohol. The pins and disks were vapor degreased and wear volume was determined. Because of very low wear rates, specimen weight loss could not be used as a measure of wear volume. Therefore, wear volume was estimated from a Talysurf profilometer trace taken perpendicular to the sliding direction for each test. Wear scars on both the pins and disks were examined using optical and scanning electron microscopy.

9.6.2 Experimental Results

The friction coefficient of the iron, copper, and titanium pairs versus sliding distance are shown in Figs. 9.42, 9.43, and 9.44, respectively. In the case of N$^+$ implanted iron, the friction coefficient, $\mu_{N,Fe} = 0.065$, is almost half that of the unimplanted surface. Similarly, the Al$^+$ implanted iron reduces the friction coefficient from 0.128 to 0.035. After a sliding distance of approximately 200 m, the friction coefficient of the implanted surface decreased, while that of the unimplanted surface increased. In the titanium system, the N$^+$ implanted surface again had a substantially lower friction coefficient, ($\mu_{N,Ti} = 0.1$) than that of the unimplanted surface ($\mu_{Ti} = 0.47$). The reduction in this case is more substantial than in the Al$^+$ implanted iron system.

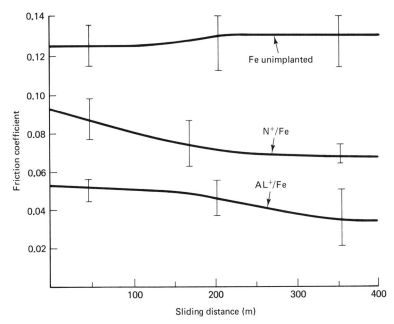

Figure 9.42 Change in friction coefficient with sliding distance of unimplanted iron, Al$^+$ implanted iron, and N$^+$ implanted iron systems. (From Shepard, 1981.)

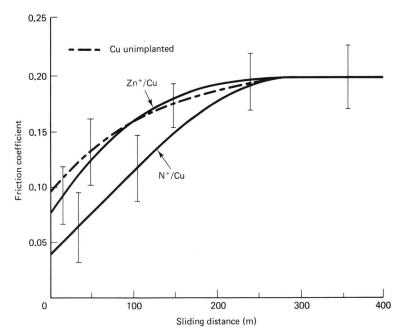

Figure 9.43 Change in friction coefficient with sliding distance of unimplanted Cu, Zn$^+$ implanted Cu, and N$^+$ implanted Cu systems. (From Shepard, 1981.)

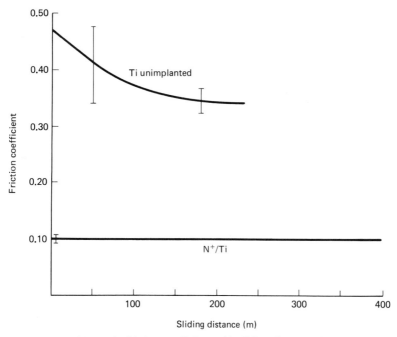

Figure 9.44 Change in friction coefficient with sliding distance of unimplanted Ti and N^+ implanted Ti systems. (From Shepard, 1981.)

Unlike the iron and titanium systems, nitrogen implanted into copper had little effect on the friction coefficient. An initial friction coefficient of 0.09 increased to that of the unimplanted material after a sliding distance of roughly 100 m. The effect of the Zn^+ implantation into the copper was negligible and μ_{Cu} and $\mu_{Zn,Cu}$ ranged from 0.06 to 0.2.

Selected surface profiles of the unimplanted and implanted specimens are shown in Figs. 9.45 through 9.47. In the case of iron implanted with N^+ and Al^+, the wear of the disk was not measurable. However, on the Al^+ implanted disk, wide, shallow plowing grooves were observed (Fig. 9.48). Similarly, the N^+ implanted iron occasionally exhibited deep plowing marks (Fig. 9.49), although "smearing" of material on the surface was more common (Fig. 9.50). The Al^+ implanted iron pin surface wore at the same rate as the unimplanted pin. Only the N^+ implanted iron pin showed a significant decrease in the wear rate. Note that the pin was more severely deformed than the disk since the pin is continuously loaded, whereas any location on the disk undergoes cyclic loading.

In contrast to the implanted iron specimens, the unimplanted disk wore at a rate of 7.92×10^{-17} m^3/sec ($K = 1.04 \times 10^{-10}$).[1] Even after

---

[1] The wear coefficient reported is normalized with respect to hardness; that is, the bulk hardness is used in the calculation of K. Present methods cannot accurately measure the actual hardness of such a thin layer; only the relative hardness can be obtained.

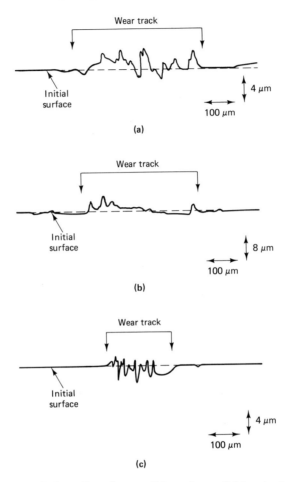

Figure 9.45 Typical profiles of worn disk surfaces of (a) unimplanted Fe, (b) N^+ implanted Fe, and (c) Al^+ implanted Fe. (From Shepard, 1981.)

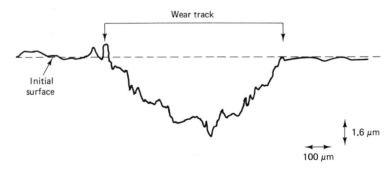

Figure 9.46 Typical profiles of a worn unimplanted Ti disk surface. (From Shepard, 1981.)

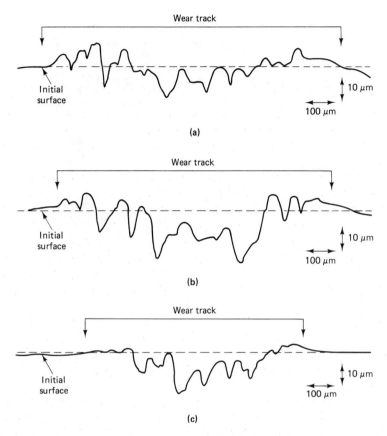

Figure 9.47 Typical profile of a worn disk surface of unimplanted Cu, Zn^+ implanted Cu, and N^+ implanted Cu. (From Shepard, 1981.)

one revolution of sliding, the pin began to wear and numerous plowing grooves were formed on the disk surface.

Titanium implanted with nitrogen showed the greatest improvement in wear resistance. Plowing was undetectable on the surface of the disk using a profilometer. However, when foreign particles or embedded wear particles were present, occasional plowing grooves were visible with the aid of the SEM (Fig. 9.51). The pins did exhibit small wear scars, but the scarred area on these pins was almost 15 times smaller than that observed on any other system. Also, the scarred surface of the pin was worn smooth with the exception of the surface around the perimeter of the scarred area. In contrast, the unimplanted titanium wore at a rate of 2.8×10^{-14} m^3/sec ($K = 2.2 \times 10^{-7}$). Figure 9.52 shows a typical wear track after 12,000 rev of sliding.

Like the friction coefficient, the wear rate of Zn^+ implanted copper was not significantly reduced. The implanted disk wore at a rate of 1.64 ×

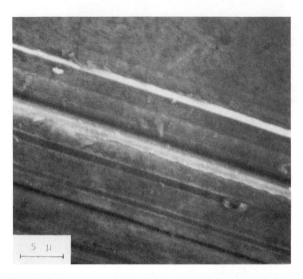

Figure 9.48 Typical wear track of Al$^+$ implanted Fe disk; shallow plowing grooves. (From Shepard, S. R., and Suh, N. P., "The Effect of Ion Implantation on Friction and Wear of Metals," *Journal of Laboratory Technology,* Transactions of the ASME, Vol. 104, 1982, pp. 29–38. Reprinted by permission of the American Society of Mechanical Engineers.)

10^{-14} m^3/sec ($K = 8 \times 10^{-9}$), whereas the unimplanted disk wore at a rate of 8.62×10^{-15} m^3/sec ($K = 4 \times 10^{-9}$). The N$^+$ implanted copper showed reduced wear during the first 200 m of sliding. However, at greater sliding distances, the wear rate approached that of the unimplanted copper.

There was severe plastic deformation below the wear track of both implanted and unimplanted copper specimens in a region very close to the surface (i.e., within 1 μm of the surface). Figure 9.53 shows the subsurface for a typical wear groove on the N$^+$ implanted copper specimen. Similar grooves were observed for the Zn$^+$ implanted and unimplanted copper wear tracks. The grains along the worn grooves are elongated in the direction of the sliding. At only 1 μm below the surface, the grains meet at approximately a 120° angle, as do recrystallized metals.

The unimplanted iron and titanium specimens showed subsurface deformation similar to that of the copper samples except that the grains did not align themselves along the wear grooves. Figure 9.54 shows typical wear grooves for unimplanted iron. There is a significant change in the size of the grains close to the surface. Slightly below the surface, and often above the depth of the wear groove, the size of the grains returns to the grain size in the undisturbed material.

The subsurface of an unimplanted titanium specimen after only 2400 passes is shown in Fig. 9.55. The wear grooves are deepest and most severe in this sample and, accordingly, subsurface deformation was severe.

Figure 9.49 Wear track of N^+ implanted Fe disk; deep plowing grooves. (From Shepard, S. R., and Suh, N. P., "The Effect of Ion Implantation on Friction and Wear of Metals," *Journal of Laboratory Technology*, Transactions of the ASME, Vol. 104, 1982, pp. 29–38. Reprinted by permission of the American Society of Mechanical Engineers.)

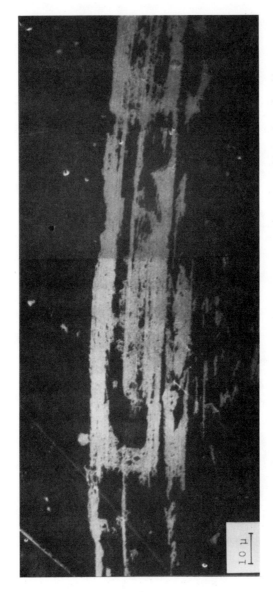

Figure 9.50 Wear track of N^+ implanted Fe disk; severe plastic deformation of a thin layer. (From Shepard, S. R., and Suh, N. P., "The Effect of Ion Implantation on Friction and Wear of Metals," *Journal of Laboratory Technology*, Transactions of the ASME, Vol. 104, 1982, pp. 29–38. Reprinted by permission of the American Society of Mechanical Engineers.)

Sec. 9.6   Ion Implantation   469

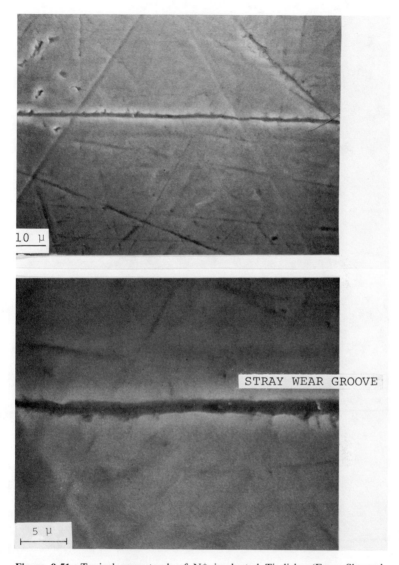

Figure 9.51 Typical wear track of N^+ implanted Ti disk. (From Shepard, S. R., and Suh, N. P., "The Effect of Ion Implantation on Friction and Wear of Metals," *Journal of Laboratory Technology,* Transactions of the ASME, Vol. 104, 1982, pp. 29–38. Reprinted by permission of the American Society of Mechanical Engineers.)

In contrast, there was no measurable distortion in the grains of the N^+ implanted titanium and iron specimens, nor in the Al^+ implanted iron specimens.

A chemical analysis of the Al^+ implanted iron disk was performed using SIMS and ISS. The unimplanted disk surface had a corrected Fe/Al

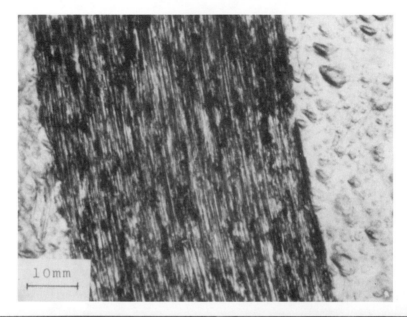

Figure 9.52 Typical wear track of an unimplanted Ti disk. (From Shepard, S. R., and Suh, N. P., "The Effect of Ion Implantation on Friction and Wear of Metals," *Journal of Laboratory Technology,* Transactions of the ASME, Vol. 104, 1982, pp. 29–38. Reprinted by permission of the American Society of Mechanical Engineers.)

Figure 9.53 Subsurface deformation of a typical worn N$^+$ implanted Cu specimen. (From Shepard, S. R., and Suh, N. P., "The Effect of Ion Implantation on Friction and Wear of Metals," *Journal of Laboratory Technology*, Transactions of the ASME, Vol. 104, 1982, pp. 29–38. Reprinted by permission of the American Society of Mechanical Engineers.)

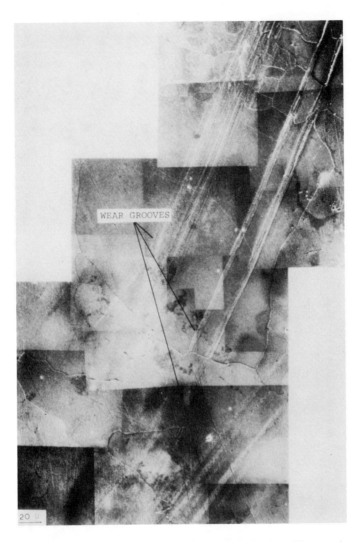

Figure 9.54 Subsurface deformation of a typical worn unimplanted iron specimen. (From Shepard, S. R., and Suh, N. P., "The Effect of Ion Implantation on Friction and Wear of Metals," *Journal of Laboratory Technology,* Transactions of the ASME, Vol. 104, 1982, pp. 29–38. Reprinted by permission of the American Society of Mechanical Engineers.)

ratio of 33.86 using ISS, while the Al^+ implanted iron disk surface had a ratio of 2.88. In the wear groove of the Al^+ implanted iron disk, after 12,000 rev of sliding and under a load of 400 g in lubricated conditions, the Fe/Al ratio was 6.47. Note that this ratio gives only the relative concentration of aluminum. Their absolute values are meaningless unless compared with the unimplanted sample. Similarly, using SIMS, the Fe/Al ratios

Sec. 9.6   Ion Implantation   473

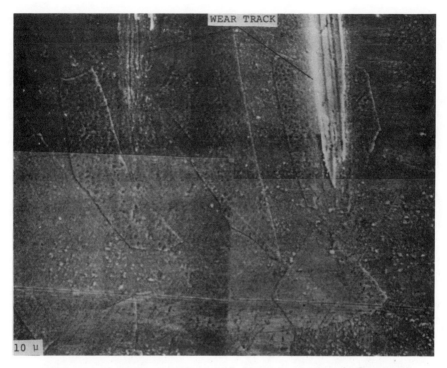

Figure 9.55 Subsurface deformation of a typical worn unimplanted Ti specimen. (From Shepard, S. R., and Suh, N. P., "The Effect of Ion Implantation on Friction and Wear of Metals," *Journal of Laboratory Technology*, Transactions of the ASME, Vol. 104, 1982, pp. 29–38. Reprinted by permission of the American Society of Mechanical Engineers.)

were 1.52, 0.866, and 1.21 for the unimplanted iron, the Al^+ implanted iron disk, and the wear track of the implanted disk, respectively.

Aluminum was present on the unimplanted surface because the samples were polished with Al_2O_3. These results still show, however, that in the wear groove, which is at a depth several orders of magnitude greater than the depth of the implanted layer, over one-third of the aluminum implanted ions still remained.

9.6.3 Mechanisms of Friction and Wear

For a variety of ion–substrate combinations, it has been shown that implantation to fluences of 10^{17} ions/cm^2 increases the hardness of the material in the near-surface region. Bolster and Singer (1980) determined the relative hardness of implanted layers as thin as 25 nm using an abrasive wear technique similar to that developed by Rabinowicz (1977). They found that for N^+ implanted steels, the hardness increased substantially in the implanted region, which, in turn, increased wear resistance by a factor of

100. This formation of a hard thin layer appears to be related to surface alloying. The existence of these alloys in implanted layers has been reported by Poate (1979). He observed the formation of metastable solid solutions and amorphous alloys in iron, nickel, and copper of high ion concentrations.

Since specific alloyed surfaces show a substantial increase in hardness over the bulk material, it is clear that the formation of an alloyed surface minimizes the friction coefficient. This, in turn, decreases surface and subsurface plastic deformation and thus inhibits the delamination wear process (i.e., crack nucleation caused by large subsurface plastic deformation, crack propagation, and eventual wear particle formation; see Chapter 4).

As discussed in Chapter 3, the friction force is caused by three mechanisms: plowing of the surface, adhesion at the asperity contacts, and deformation of the surface asperities. The initial dynamic friction for an initially smooth surface is primarily the result of plowing on either the pin or disk surface. Experiments show that the plowing of the specimen surface results when "embedded wear particles" on the surface of the pin dig into the specimen surface.

The embedded wear particles that cause plowing form by delamination wear, as modeled in Fig. 9.56. Surface and subsurface deformation occurs as soon as the counterface slides over the surface. When the subsurface is sufficiently deformed, cracks nucleate below the surface and will eventually extend and propagate. When these cracked surface layers finally shear to the surface, they deform the surrounding material and pile up against an obstacle, which may be a surface layer about to delaminate. The resulting wear sheet is the embedded wear particle. Figure 9.57 illustrates the formation and piling of several wear sheets on the surface of an Al^+ implanted slider after a sliding distance of several centimeters. These attached wear particles

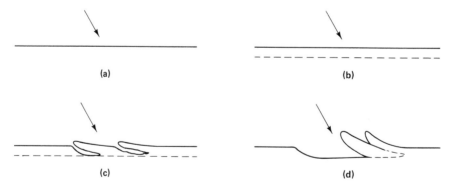

Figure 9.56 Schematic representation of wear particle formation during sliding on an initially smooth surface. (a) plastically deforming zone; (b) crack nucleation; (c) crack propagation and wear sheet formation; (d) final topography of surface. (From Shepard, S. R., and Suh, N. P., "The Effect of Ion Implantation on Friction and Wear of Metals," *Journal of Laboratory Technology,* Transactions of the ASME, Vol. 104, 1982, pp. 29–38. Reprinted by permission of the American Society of Mechanical Engineers.)

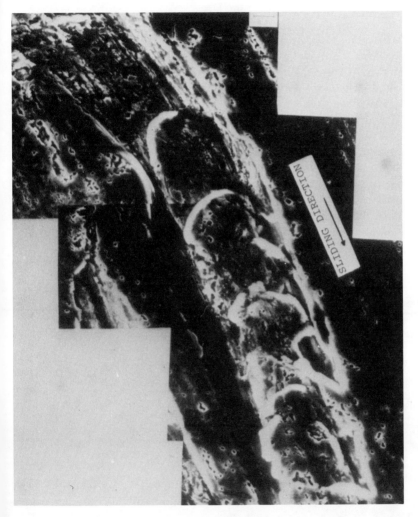

Figure 9.57 Wear particle formation on the surface of an Al^+ implanted Fe pin after 5 cm of sliding on a lubricated Al^+ implanted disk. (From Shepard, S. R., and Suh, N. P., "The Effect of Ion Implantation on Friction and Wear of Metals," Journal of Laboratory Technology, Transactions of the ASME, Vol. 104, 1982, pp. 29–38. Reprinted by permission of the American Society of Mechanical Engineers.)

will serve as plowing tools. Note that the wear particle formed may break off from the bulk material and either become embedded elsewhere in the slider or specimen, or remain loose and act as an abrasive wear particle. In either case it may continue to plow the surface.

When the specimen surface is hard enough to resist the forces exerted by the wear particle without undergoing large plastic deformation, plowing of the sliding surface by this wear particle is minimized. Also, if a smooth, hard, and thick layer can be created on the pin surface, the delamination process by which the plowing particles are created will be slowed down if the deformation of the pin surface can be reduced by a hard layer. When plowing is reduced on either surface, the friction coefficient will be correspondingly reduced.

To minimize plowing and consequently friction, the number of embedded wear particles (either from the slider–specimen material or from the environment) must be minimized and/or a smooth hard surface used. The implanted specimens that showed a reduction in the friction coefficient (and wear) had a hard thin layer on the surface. Figure 9.58 shows evidence of a hard layer on the N^+ implanted titanium pin surface. Under the applied load, the thin layer appears to have been "crushed in." The N^+ implanted iron pin, shown in Fig. 9.59, also has this hard layer, signified by the sharp fracture lines along the edge of the plowing scars. A more ductile deformation

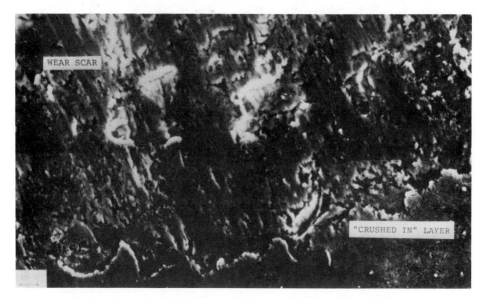

Figure 9.58 Typical N^+ implanted titanium pin surface. (From Shepard, S. R., and Suh, N. P., "The Effect of Ion Implantation on Friction and Wear of Metals," *Journal of Laboratory Technology*, Transactions of the ASME, Vol. 104, 1982, pp. 29–38. Reprinted by permission of the American Society of Mechanical Engineers.)

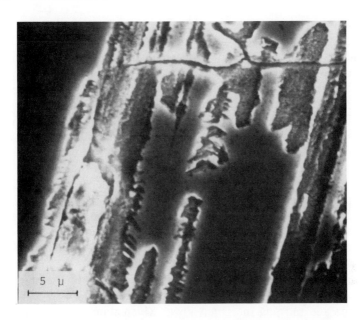

Figure 9.59 Typical N^+ implanted Fe pin surface after 25 cm of sliding. (From Shepard, S. R., and Suh, N. P., "The Effect of Ion Implantation on Friction and Wear of Metals," *Journal of Laboratory Technology,* Transactions of the ASME, Vol. 104, 1982, pp. 29–38. Reprinted by permission of the American Society of Mechanical Engineers.)

occurs on the unimplanted surface. This relative increase in hardness in both specimens was observed using a scratch test. However, the magnitude of the increased hardness in the thin surface layer could not be provided by such means.

The hard layer found on the N^+ implanted titanium and iron disks could be a result of one or both of the following phenomena: (1) lattice distortion and/or (2) formation of hard compounds. In the former case, the increase in the internal stress may impede dislocation motion and hence produce a hardened layer. In the latter case, it is speculated that $Ti_xN_yO_z$ and $Fe_xN_yO_z$ are formed during the implantation of N^+ into titanium and iron, respectively. Oxygen has been included since it was present on the surface prior to implantation. The thickness of these layers are estimated to be 2000 Å (2×10^{-7} m) and 1375 Å (1.375×10^{-7} m), respectively. The scarring of the N^+ implanted iron specimen varies from an unscarred surface to one with occasional plowing grooves. Only when wear particles or foreign matter are embedded into the pin or disk are plowing grooves evident. Both implanted pin surfaces were worn, but not to the extent of the unimplanted pins.

Due to a reduction in the number of plowing grooves on the specimen surfaces and the amount of material plastically deformed on the pin surface, the friction coefficient is correspondingly reduced. With the reduction of friction coefficient, the wear rate is also reduced.

Unlike the N^+–Ti or N^+–Fe specimens, the Al^+–Fe specimen is deformed and plowed in numerous areas of the wear track. These plowing grooves, however, are much shallower than those on the unimplanted specimen. This explains the reduction in the friction coefficient. As expected, after a sliding distance of several inches, the Al^+ implanted iron pin forms embedded wear particles, which cause the observed plowing. Although plowing occurs, loose wear particles were sparsely distributed in the lubricant and wear was not measurable even after 56 hours of continuous sliding. The presence of aluminum on the surface and in the wear groove appears to increase the hardness of the near-surface region, which explains the increased wear resistance observed. It is speculated that the aluminum present was in the form of Al_2O_3.

The persistent wear resistance and presence of aluminum at depths greater than that of the implanted layer in the Al^+–Fe specimens can be explained as follows. Plowing of the surface can occur by the plowing tool actually digging into the surface or by the asperity sliding on the top of the surface and pushing the layer down, which causes plastic flow of the material along the edges of the wear groove. In the latter case, as modeled in Fig. 9.60, the pin is still sliding on a hardened layer even though plowing occurs. The presence of aluminum at depths much greater than the implanted

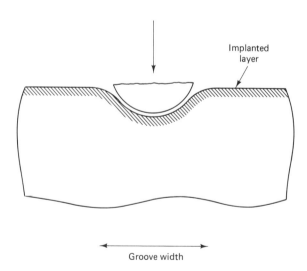

Figure 9.60 Model of contact between a rigid asperity and an Al^+ implanted Fe surface. (From Shepard, S. R., and Suh, N. P., "The Effect of Ion Implantation on Friction and Wear of Metals," *Journal of Laboratory Technology,* Transactions of the ASME, Vol. 104, 1982, pp. 29–38. Reprinted by permission of the American Society of Mechanical Engineers.)

layer, as shown by SIMS and ISS, cannot be explained in terms of the diffusion process, since predicted diffusion rates of aluminum into iron are only on the order of several atomic diameters for the testing time.

It is interesting to note that the friction coefficient of the Al^+ implanted iron system is half that of the N^+ implanted iron system even though the specimen surface of Al^+–Fe specimen has a larger number of wear grooves. This may be attributed to the difference in adhesion, since the aluminum implanted surface may have become Al_2O_3.

Plowing was not reduced on the Zn^+ implanted copper pin or specimen, and hence the friction coefficient was the same as that for the unimplanted copper. Zinc atoms usually substitute for copper atoms. Thus the material hardness does not increase substantially since substitutional solid solution hardening has a relatively mild effect on hardness compared with interstitial solid solution hardening. The absence of a hard layer to reduce plowing explains the observed severe subsurface plastic deformation and the unchanged tribological properties.

Although N^+ implanted into copper had no long-term effects on the friction and wear behavior, there appeared to be a change in the chemical state of the near-surface region, as observed by the formation of a blue film found only on the implanted surface. The implanted layer was either too thin or not hard enough to minimize plowing, probably the latter.

9.6.4 Conclusions

Experimental and theoretical investigations summarized in this section explain the effects of ion implantation on friction and wear of metals. Improved tribological properties of implanted metals are a result of the formation of a hardened layer in the near-surface region. It is speculated that hardening is created by changes in mechanical properties and surface chemistry caused by alloying effects. The thin hard layer does not serve to support the load, but rather to decrease the plowing component of friction under relatively light loads. A reduction in the friction coefficient, in turn, reduces wear.

Of the ion–substrate combinations investigated, only the N^+ implanted titanium and iron, and the Al^+ implanted iron, showed an improvement in the friction and wear properties. For the test conditions, the Zn^+ and N^+ implanted copper had little effect on the tribological properties.

9.7 MINIMIZATION OF WEAR OF POLYMERS BY PLASMA TREATMENT

In Chapter 6 it was shown that the highly crystalline polymers without large, unsymmetric pendant groups wear due to plastic deformation of the surface layer, which orients molecules parallel to the surface, eventually leading to the transfer of thin films to the counterface during the asperity interactions. Formation of thin films occurs because the intermolecular force between the layers of oriented molecules is less than the force applied externally. It follows, therefore, that if the molecules near the surface can be prevented from plastically deforming (without making the entire surface brittle), this mode of deformation and wear can be minimized. One such a method has been tried at MIT as part of the Master's thesis of Youn (1980). The technique consists of exposing the surface of the polymers to a helium plasma to cross-link the molecules near the surface to prevent molecular orientation and thin-film transfer.

9.7.1 Experimental Procedure

Two crystalline polymers, HDPE and POM, and two amorphous polymers, polymethyl methacrylate (PMMA) and polycarbonate (PC), were selected for the cross-linking and wear experiments. They were injection molded in the form of 0.3-cm-thick disks. All the specimens were cut into square plates (0.043 m × 0.043 m) for the surface treatment and friction and wear tests. An AISI 52100 bearing steel pin of 0.3 cm outside diameter was used as the counter face.

The polymer specimens were surface treated by the CASING (Cross-linking by Activated Species of Inert Gases) technique (Hansen and Schonhorn, 1966). The apparatus used for CASING is similar to those described elsewhere (Hansen et al., 1966; Collins et al., 1973).

Helium plasma was selected since it does not affect the wetability or surface roughness (Schonhorn and Hansen, 1967). The specimens were placed in the discharge tube and the reactor was evacuated with a mechanical pump, allowing a continuous flow of the gas through a flow meter. They were treated for different time intervals (500 and 1000 seconds) under the following conditions: pressure of 1 torr, frequency of 13.56 MHz, radio-frequency power of 100 W, and room temperature. For high-temperature treatment, a band heater was used on the reactor tube while the reactor was evacuated. The treatment time was 1000 seconds for the high temperature treatment.

Friction and wear tests were conducted with the pin-on-disk geometry (shown in the appendix of Chapter 3). The polymer specimen was rotated on an aluminum disk which was connected to a speed-controlled dc motor. The steel pin was held stationary in a holder which was attached to a strain gage transducer. The weight of the specimen was measured before and after the experiment to an accuracy of 10^{-5} g with a microbalance. Friction forces were recorded continuously using a chart recorder which was connected to the strain gages. All experiments were conducted in air and the temperature (22°C) and relative humidity (65%) were controlled. The normal load was 4.4 N and the sliding speed was 3.3 cm/sec.

The thickness of the cross-linked layer on the helium-plasma-treated polymers was determined by dissolving the uncross-linked bulk using appropriate solvents. HDPE was extracted by *p*-xylene, PMMA by toluene, POM by aniline, and PC by ethylene dichloride with a small amount of antioxidants. The remaining gel was dried in a vacuum oven at 80°C for 2 hours and weighed in a microbalance.

The surfaces of the helium plasma treated and untreated polymers were characterized by ESCA (electron spectroscopy for chemical analysis). Since the thickness of the cross-linked layer was less than 1 μm, ESCA is the most suitable technique for the analysis of the plasma-treated polymers.

9.7.2 Experimental Results

Figure 9.61 shows the measured thickness of a cross-linked layer as a function of the plasma treatment time. HDPE, PMMA, and POM had a cross-linked layer of 0.1 to 0.5 μm, but PC did not have any cross-linking. The photoelectron spectra were obtained from the ESCA analysis for the room-temperature and high-temperature treated specimens. The elemental peaks were identified and the atom fraction (Table 9.7) was calculated based

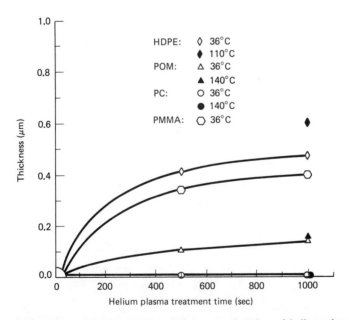

Figure 9.61 Thickness of cross-linked layer as a function of helium plasma treatment time and temperature. Plasma conditions: radio frequency, 13.56 MHz; RF power, 100 W; vacuum pressure, 1.0 mmHg. For HDPE at 36°C, there is a temperature rise of 16°C from the room temperature during the treatment. (From Youn and Suh, 1981.)

TABLE 9.7 Atom Fraction Calculated from the ESCA Peak

Specimen	C_{1s} (285 eV)	C_{1s} (289 eV)	O_{1s} (533 eV)
HDPE control	93.9	—	6.1
He plasma (room temperature)	81.6	—	18.4
He plasma (110°C)	83.3	—	16.7
POM control	12.3	45.1	42.6
He plasma (room temperature)	6.2	49.0	44.8
He plasma (140°C)	22.2	40.6	37.2

on the peak area measurement and the ASF (atomic sensitivity factor). All the specimens had a higher oxidation level after the helium plasma treatment at room temeprature. The ratio of oxygen was smaller when HDPE was treated at 110°C than when it was treated at room temperature, but still larger than when it was untreated. The fraction of oxygen at the POM surface layer increased after the room temperature treatment and became smaller than that of untreated POM when it was treated at 140°C.

Figure 9.62 shows the steady-state friction coefficient of surface-treated polymers as a function of exposure time to helium plasma. The wear rates of the treated polymers are shown in Fig. 9.63 with respect to the helium plasma exposure time. All the friction coefficients of polymers are almost constant. The wear rates of HDPE and PMMA decreased dramatically with the surface treatment time, by a factor of more than 10. The wear

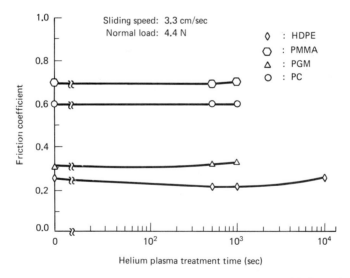

Figure 9.62 Friction coefficient of polymers as a function of helium plasma treatment time. (From Youn and Suh, 1981.)

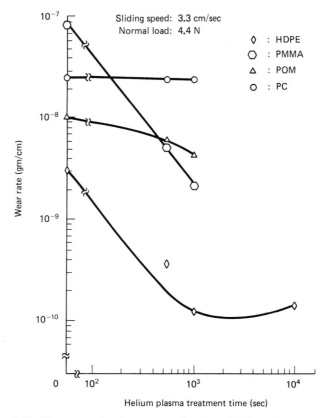

Figure 9.63 Wear rate of polymers as a function of helium plasma treatment time. (From Youn and Suh, 1981.)

rate of POM is decreased only by a factor of 2 to 3 and there is no change in the wear rate of PC. The friction coefficients and wear rates of the surface-treated polymers are plotted with respect to the surface treatment temperature in Figs. 9.64 and 9.65, respectively. While the friction coefficients of HDPE and POM are slightly increased, that of PC is slightly decreased. The wear rate of POM is reduced by a factor of 7 to 8, but the wear rate of PC has not changed.

9.7.3 Discussion

Many gases can be excited by electrodeless RF discharge to yield an activated plasma which consists of ionized species, free electrons, free radicals, and excited molecules or atoms, as well as unchanged gas. The possible mechanisms of polymers on contact with the activated gases are formation of free radicals, production of vinylene unsaturation, formation of intermolecular cross-links, chain degradation, oxidation, and decay of

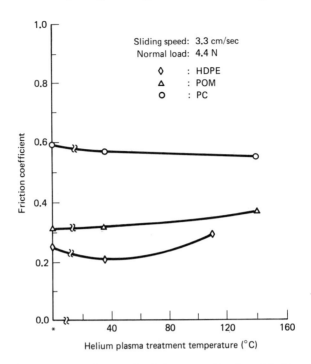

Figure 9.64 Friction coefficient of polymers as a function of helium plasma treatment temperature. (From Youn and Suh, 1981.)

vinyl, vinylene, and vinylidene unsaturation (Hansen and Schonhorn, 1966; Collins et al., 1973). Free radicals are formed as intermediate products. Cross-linking occurs when two radicals are formed in close proximity or radicals can migrate until they find the partners. The cross-link density depends on the chemical structure of the polymers and the treatment conditions. The cross-linking density of HDPE and POM increases when they are cross-linked at high temperatures because the formation and reaction rates of the radicals would increase at high temperature and the trapped radicals would react further to yield cross-linking when they are stored in the helium environment at the high temperature (Collins et al., 1973). The ESCA results (Table 9.7) indicate the fraction of carbon and oxygen atoms as well as the ratio of carbons which are oxidized and not oxidized due to the shift of binding energy levels. The amount of oxygen on HDPE and POM was reduced after surface treatment at high temperature, which implies the possible increase of cross-links.

The friction coefficient is important because it determines the contact stress distribution. It is affected by both the shear strength and the hardness. Although both properties must have changed when the surface was cross-linked, they could not be measured, due to their very small thickness, using conventional hardness testers. The shear strength of helium-plasma-treated

Sec. 9.7 Minimization of Wear of Polymers by Plasma Treatment

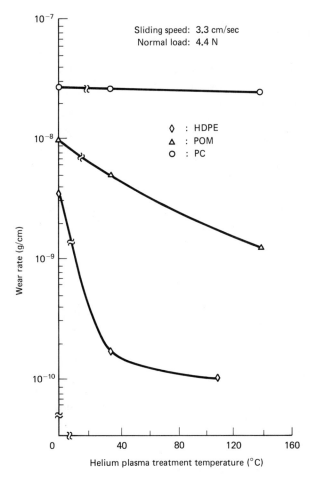

Figure 9.65 Wear rate of polymers as a function of helium plasma treatment temperature. (From Youn and Suh, 1981.)

polymers was measured by Schonhorn and Hansen (1966), who conducted a lap shear test using a sandwich specimen of aluminum plate–epoxy adhesive–polymer specimen–epoxy adhesive–aluminum plate. They found that the helium plasma treatment increases the shear strength of polyethylene and PTFE by a factor of 5.

The friction coefficients of the treated polymers remain almost unchanged, although the shear strength of the surface cross-linked polymers may undergo large increases. This indicates that the three components of the frictional force (i.e., plowing of the surface layer, adhesion, and asperity deformation) were not significantly affected by the formation of their cross-linked layer.

Wear rates are calculated based on the weight loss and the sliding distance. The surface cross-linked polymers show the greatest reduction in the wear rate, while PC, which had no cross-linking, does not show any reduction of wear rate. This indicates that cross-linking prevented plastic deformation of the surface layer and consequently, thin-film transfer. Even if a limited amount of plastic deformation occurs, cross-linking will not allow the alignment of molecules parallel to the surface, thus reducing the wear rate. Therefore, it may be reasonable to assume that the cross-linking made the applied shear stress at the surface be less than the surface shear strength of the material (see Chapter 6).

It should be noted that cross-linking the bulk plastic will not necessarily improve the wear resistance since it will behave like a glassy polymer. Therefore, for the cross-linking action to be effective in reducing wear, it must confine the cross-linking process to a very thin surface layer on the order of 1 to 2 μm. The optimum thickness of the cross-linked layer will obviously depend on the loading conditions.

9.7.4 Conclusions

Based on the experimental results presented in this section, the following conclusions may be made:

1. The wear resistance of polymers can be improved significantly when crystalline polymers are cross-linked by the activated plasma treatment. The experimental results show that the wear rates of surface-cross-linked polymers decrease by orders of magnitude.
2. The wear mechanism of polymers can be explained by extending the delamination theory of wear. Based on the theory, the criteria for minimum wear can be expressed in terms of applied shear stresses and the material strength.
3. The wear rate decreases with an increase in the thickness of the cross-linked layer. However, if the cross-linked region is too deep, the wear rate will approach that of glassy polymers.
4. The most effective way to surface treat helium plasma for wear application is to treat the specimens at a high temperature and store them more than an hour in a helium environment. This gives the maximum cross-link density under the given plasma treatment conditions.

9.8 CONCLUDING REMARKS

Chapter 9 presented several examples of how the theories discussed in earlier chapters can be applied to solving practical tribological problems. Many of the theories presented in this chapter have been developed over a period of about 10 years beginning in the early 1970s, because insight to tribological problems could not be gained from the widely accepted "old" theories. Although there remains much to be done, the examples cited in this chapter illustrate the possibility that the friction and wear theories presented in this book can be useful in devising practical solutions to tribological problems. Theories are useless unless they can predict the plausible outcome of an event that may occur in the future.

REFERENCES

ABRAHAMSON, E. P., II, JAHANMIR, S., SUH, N. P., and COLLING, D. A., "Application of the Delamination Theory of Wear to a Composite Metal Surface," *Proceedings of the International Conference on Production,* Tokyo, 1974.

BOLSTER, R. N., and SINGER, I. L., "Surface Hardness and Abrasive Wear Resistance of Ion-Implanted Steels," ASME/ASLE Lubrication Conference, San Francisco, August 1980.

BROWN, P. E., "Bearing Retainer Material for Modern Jet Engines," *ASLE Transactions,* Vol. 13, 1970, pp. 225–239.

BUNSHAH, R. F., and RAGHURAM, A. C., "Activated Reactive Evaporation Process for High Rate Deposition of Compounds," *Journal of Vacuum Science Technology,* Vol. 9, 1972, p. 1385.

BURGESS, S., "Friction and Wear of Composites," S.M. thesis, MIT, 1983.

CARSON, W. W., LEUNG, C. L., and SUH, N. P., "Metal Oxycarbides as Cutting Tool Materials," *Journal of Engineering for Industry,* Transactions of the ASME, Vol. 98, 1976, pp. 279–286.

COLLINS, G. C. S., LOWE, A. C., and NICHOLAS, D., "An Analysis of PTFE Surfaces Modified by Exposure to Glow Discharge," *European Polymer Journal,* Vol. 9, 1973, pp. 1173–1185.

COOK, N. H., and KRAMER, B. M., "Tungsten Carbide Tools Treated with Group IVB and VB Metals," U.S. Patent 4,066,821, 1978.

EKMAR, C. S., "Carbide-Coated Cemented Metal Carbide," German Patent 2,007,427, issued to Sardvikens Jernverks Aktiebolag, September 10, 1970.

EVAN, H. E., and FLATLEY, T. W., "Bearings for Vacuum Operations, Retainer Material and Design," NASA Technical Note D-1339, 1962.

EVAN, H. E., and FLATLEY, T. W., "High Speed Vacuum Performance of Gold-Plated Miniature Ball Bearings with Various Retainer Materials and Configurations," NASA Technical Note D-2101, 1963.

FLATLEY, T. W., "High Speed Vacuum Performance of Miniature Ball Bearings Lubricated with a Combination of Barium, Gold, and Silver Films," NASA Technical Note D-2304, 1964.

GALSKI, F. A., ed., *Proceedings of the 3rd International Conference on Chemical Vapor Deposition,* American Nuclear Society, 1972.

HALE, T. E., "Coated Cemented Carbide Product," U.S. Patent 3,736,107, 1971.

HANSEN, R. H., and SCHONHORN, H., "A New Technique for Preparing Low Surface

Energy Polymers for Adhesive Bonding," *Journal of Polymer Science*, Vol. 84, 1966, pp. 203-209.

HAUSER, C., private communcation, Stellram, Nyon, Switzerland, 1975 (see also "Recouvrements protecteurs adaptés aux outils de coupe," *Revue Polytech*, March 1976).

JAHANMIR, S., and SUH, N. P., "Surface Topography and Integrity Effects on Sliding Wear," *Wear*, Vol. 44, 1977, pp. 87-99.

JAHANMIR, S., ABRAHAMSON, E. P., II, and SUH, N. P., "Sliding Wear Resistance of Metallic Coated Surfaces," *Wear*, Vol. 40, 1976, pp. 75-84.

KIEFFER, R., REITER, N., and FISTER, D., "New Developments in the Field of Cemented-Carbide and Ceramic Cutting Tools," *Materials for Metal Cutting*, The Iron and Steel Institute, 1970, p. 157.

KUCZKOWSKI, T. J., and BUCKLEY, D. H., "Friction and Wear of Low Melting Binary and Ternary Gallium Alloy Films in Argon and in Vacuum," NASA Technical Note D-2721, 1965.

LEE, R. D., JR., "Lubrication of Heavily Loaded, Low Velocity Bearings and Gears Operating in Aerospace Environmental Facilities," AEDC-TR-65-19, 1965.

LINDSTROMAND, J. N., and JOHANNESSON, R. T., "Nucleation of Al_2O_3 Layers on Cemented Carbide Tools," *Proceedings of Conference on Chemical Vapor Deposition*, Electrochemical Society, Inc., 1975, p. 453.

LIOU, M. J., "Fretting Corrosion and Contact Resistance of Edge Card Connectors," S.M. thesis, MIT, 1983.

MOREHEAD, F. F., JR., and CROWDER, B. L., "Ion Implantation," *Scientific American*, Vol. 228, No. 4, 1973, pp. 65-71.

NAKAMURA, K., and INAGAWA, K., "Superhard Cutting Tips Coated with TiC," *Japanese Journal (Kinzoku) of Metals*, August 1975, p. 35.

POATE, J. M., "Metastable Alloy Formation by Ion Implantation," *Thin Solid Films*, Vol. 58, 1979, pp. 133-143.

RABINOWICZ, E., "Taper Sectioning, a Method for the Examination of Metal Surfaces," *Journal of the Metal Industry*, Vol. 76, 1950, pp. 813-886.

RABINOWICZ, E., "Abrasive Wear Resistance as a Materials Test," *Lubrication Engineering*, Vol. 33, 1977, pp. 378-381.

RAMALINGAM, S., and WATSON, J. D., "Tool Life Distribution, Part 4: Minor Phases in Work Material and Multiple-Injury Failure," *Journal of Engineering for Industry*, Transactions of the ASME, Vol. 100, 1973, pp. 201-209.

RICHMAN, M. H., and LEE, M., "Some Properties of TiC-Coated Cemented Tungsten Carbide," *Metals Technology*, Vol. 1, Part 12, 1974.

RINDINGS, T. L., "Operational Evaluation of Dry Thin Film Lubricated Bearings and Gears for Use in Aerospace Environment Chambers," AEDE-TR-65-1, 1965.

SAFRANCK, W. H., *The Properties of Electrodeposited Metals and Alloys*, Elsevier, New York, 1974.

SCHINTLMEISTER, W., PACHER, K., and PFAFFINGER, K., "Structure and Strength Effects in CVD Titanium Carbide and Titanium Nitride Coatings," *Proceedings of the 5th International Conference on CVD*, 1975, pp. 523-539.

SCHONHORN, H., and HANSEN, R. H., "Surface Treatment of Polymers for Adhesive Bonding," *Journal of Applied Polymer Science,* Vol. 11, 1967, pp. 1461–1474.

SCHWARZKOPF, P., and KIEFFER, R., *Cemented Carbides,* Macmillan, New York, 1960.

SHEPARD, S. R., "The Effect of Ion Implantation on Friction and Wear of Metals," S.M. thesis, MIT, 1981.

SHEPARD, S. R., and SUH, N. P., "The Effect of Ion Implantation on Friction and Wear of Metals," *Journal of Laboratory Technology,* Transactions of the ASME, Vol. 104, 1982, pp. 29–38.

SINGER, I. L., CAROSELLA, C. A., AND REED, J. R., "Friction Behavior of 52100 Steel Modified by Ion Implanted TI," *Proceedings of the Ion Beam Modification of Materials Conference,* Albany, New York, 1980.

SOLOMON, A. J., and ANTLER, M., "Wear Mechanisms of Gold Electrodeposits," Plating, East Orange, N.J., 1970, pp. 812–816.

SU, K. Y., and COOK, N. H., "Enhancement of High Speed Steel Tool Life by Titanium Nitride Sputter Coating," *Proceedings of the 5th North American Metalworking Research Conference,* Society of Manufacturing Engineers, 1977, pp. 297–302.

SUH, N. P., "Microstructural Effects in Sliding Wear of Metals," in *Fundamental Aspects of Structural Alloy Design,* Plenum Press, New York, 1976.

SUH, N. P., and BURGESS, S., U.S. Patent Pending, 1984.

SUH, N. P., LIOU, M. J., and SAKA, N., U.S. Patent Pending, 1984.

TAKAGI, R., and LIU, T., "The Lubrication of Steel by Electroplated Gold," *ASLE Transactions,* American Society of Lubrication Engineers, Vol. 10, 1967, pp. 115–123.

TAKAGI, R., and LIU, T., "Lubrication of Bearing Steels with Electroplated Gold under Heavy Loads," *ASLE Transactions,* American Society of Lubrication Engineers, Vol. 11, 1968, pp. 64–71.

TOTH, L. E., *Transition Metal Carbides and Nitrides,* Academic Press, New York, 1971.

TSUYA, Y., and TAKAGI, R., "Lubricating Properties of Lead Films on Coppers," *Wear,* Vol. 7, 1964, pp. 131–143.

WESTBROOK, J. H., "Temperature Dependence of Hardness of Some Oxides," *Revue des Hautes Températures et des Réfractaires,* Vol. 3, 1966, p. 47.

YOUN, J. R., "An Investigation of Friction and Wear Behavior of Polymers," S.M. thesis, MIT, 1980.

YOUN, J. R., and SUH, N. P., "Tribological Characteristics of Surface Treated Polymers," *Proceedings of the Society of Plastics Engineers,* 39th ANTEC, Boston, May 1981.

INDEX

Abebe, M., 285, 299, 343, 346, 350
Abrahamson II, E. P., 20, 22, 65, 90, 197–203, 221, 222, 415, 443–49, 451, 452, 487, 488
Abrasive wear, 264
  classical model, 23, 304
  experimental investigation, 274
  grit size effect, 279–82
  mechanisms, 282
  phenomenological aspects, 266
  slip-line field solution, 285
  test equipment and methods, 266
  wear rates, 268–71, 279–82, 455
Adachi, M., 161, 173
Adelsberg, L. M., 377–79
Adhesion theory of friction, 4, 64, 95
  effect of surface energy, 7, 64
Adhesion theory of wear, 11
Adsorption, 31, 33
Allan, A. J. G., 5, 22
Amonton, G., 63, 91
Anand, S. C., 138, 172
Antler, M., 444, 489
Appl, F. C., 285, 299, 343, 346, 350
Application of theories, 413–89
Archard, J. F., 13, 22, 226, 257
Argon, A. S., 47, 48, 142, 143, 172, 174

Ashelby, D. W., 157, 173
Asperity deformation, 76, 92, 196
Ast, D. G., 231, 259
Atomic radius, 33
Augustsson, G., 38, 46, 139, 143, 172
Avient, B. W. E., 268, 271, 343

Bacon, R., 250, 259
Baer, E., 231, 258
Bahadur, S., 230, 258
Baronel, C. N., 117, 173
Barrow, G., 381, 412
Basu, D., 366, 380
Bathe, K.-J., 161, 172
Bellman, Jr., R., 308, 312, 343
Bely, V. A., 230, 259
Bhushan, B., 380, 395, 405–10, 486
Billinghurst, P. R., 231, 258
Bitter, J. G. A., 332, 343
Blok, H., 379, 395, 396
Bolster, R. N., 473, 487
Boothroyd, G., 363, 379
Bose, A. C., 35, 47
Boundary lubrication, 2, 34, 35, 413
Boussinesq, J., 105, 172, 189
Boussinesq solution, 186
Bowden, F. P., 64, 91, 96
Bowers, R. C., 231, 258

Briscoe, B. J., 230, 258
Brooks, C. A., 231, 258
Brown, P. E., 444, 486
Bryant, M. D., 157, 174
Buckley, D. H., 9, 22, 32, 33, 35, 46, 64, 91, 161, 175, 444, 488
BUE (Built-up edge), 357
Bunshah, R. F., 457, 458, 487
Burgess, S., 248, 249, 258, 415–23, 487

Cadoff, L. H., 377–79
Cahn, J. W., 42, 44, 46
Carosella, C. A., 459, 489
Carson, W. W., 373, 379, 458, 487
Ceramics, 30
Challen, J. M., 78, 91, 95
Characteristics of materials, 27
Chemical instability, 356
Chemical state of surfaces, 26, 34
Chemical vapor deposition (CVD), 455, 457, 458
Chemisorption, 31
Chip formation, 357
Chung, C. I., 244, 258
Clerico, M., 171, 173, 199, 221, 231, 232, 234–38, 258
Clinton, W. C., 231, 258
Close, L. J., 174, 187
Coating of the surface, 443, 454–59
Coefficient of friction, 5, 428 (*See also* Friction)
Coefficient of thermal expansion, 458
Coefficient of wear, 14, 16, 18, 24 (*See also* Wear)
Cohesive energy density, 33
Colling, D. A., 444, 487
Collins, G. C. S., 480, 482, 483, 487
Compatibility chart of friction, 8
Composites, 223, 232, 239, 248, 249, 416
Compton, W. A., 306, 345
Conrad, J., 313, 338, 342, 345
Constitutive relations:
  Hooke's law, 50
  power law, 51
  Prandtl-Reuss equation, 51
Continuum mechanics, 48
Cook, N. H., 22, 24, 209, 222, 358, 360, 365, 366, 379, 380, 395, 397, 398, 405–10, 415, 455, 459, 487, 489

Coulomb, C. A., 63, 91
Coy, R. C., 35, 46
Crack:
  formation in elastic solids, 122
  growth in sliding wear, 186
  nucleation, 141
  propagation, 153–71
  propagation mechanism, 161
  tip sliding displacement (CTSD), 161, 166, 186
  trajectory, 159
Cross-linking by activated species of inert gases (CASING), 480
Crowder, B. L., 460, 488
Cutting tool materials, 364

Date, S. W., 271, 343
Dautzenberg, J. H., 38, 46, 52
Davis, G. de V., 381, 396
Delamination theory of wear, 74, 129, 196, 199, 209
Demer, L. J., 36, 47
Demirici, A. H., 161, 174
Dench, W., 381, 389, 392
Dewhurst, P., 360, 380
Diffendorf, R. J., 251, 252, 258
Diffusion controlled wear, 376
Dislocations in metals, 27
  cells, 27
  friction stress, 37
  image force, 38
  low-density dislocation zone, 38
Double layer, 31
Douthwaite, R. M., 271, 344
Doyle, E. D., 357, 358, 360, 382
Drucker, D. C., 351
Ductile fracture, 176
Duke, C. B., 31, 42, 46
Dunn, L. A., 271, 345
Duquette, D. J., 151, 175, 206, 222

Edwards, R., 374, 380, 392, 394
Eggum, G. E., 333, 343
Eiss, Jr., N. S., 231, 259
Ekmar, C. S., 458
Elastic deformation of the surface, 105–22
Elastoplastic deformation, 130
Electrical contact resistance, 424
Electrical double layer, 32
Elliot, J., 374, 380, 388, 393, 394
Elliot, R., 380, 389, 391, 393, 394

Index

Erdogan, F., 159, 173
Erosive wear, 264, 305
  data, 317–29
  empirical equations, 332
  mechanisms, 318, 332
  microstructural observations, 307
  models, 321, 333
  phenomenological observations, 306, 313
  test apparatus and methods, 306
Equilibrium equation, 48
Equivalent strain, 50
Equivalent stress, 49
Ernst, H., 64, 91
Eshelby, J. D., 153, 173
Espenschade, P. W., 346
Euler, L., 64, 91
Evan, H. E., 444, 487
Evans, A. G., 126–28, 173, 333, 334, 336–39, 340, 343
Evans, E., 381, 390, 392
Excess free energy, 371, 387
Experimental techniques, 65, 100, 266, 274, 306, 417, 424, 444, 454, 461

Fatigue fracture, 176
Fatigue under cyclic loading, 185
Ferry, J. D., 263
Field, J. E., 330, 344
Fillion, P., 364, 381
Finnie, I., 105, 174, 306, 321, 329–33, 337–39, 343, 345, 360, 381
Fister, D., 458, 488
Flatley, T. W., 444, 487
Fleming, J. F., 154–57, 172, 198, 200, 221
Flom, D. G., 359, 360, 380
Forsyth, P. I. E., 160, 173
Fourie, J. T., 36, 46
Fracture, 176
Fracture mechanics, 178
Free energies:
  estimation of, 388–95
  excess, 371, 387
  Gibbs, 383
  Helmholtz, 384
Fretting wear, 13, 206
Friction:
  abrasion (due to), 264
  adhesion theory of, 3, 64, 78
  analysis of friction generating mechanisms, 75, 83
  coefficient of, 5, 6, 67, 212, 217, 225, 226, 234, 240–44, 246, 277, 429, 448, 450, 462, 463, 482, 484
  Early theories, 3, 11
  Effect of:
    asperity deformation, 76, 83
    hard particles, 264
    ion implantation, 473
    normal load, 4
    plasma treatment, 479
    plowing, 80, 84, 96
  experimental results, 65–72
  genesis of, 63, 73
  measurement of, 100
  mechanisms, 73, 237, 283, 473
  modulated surfaces, 87, 433
  phenomenological aspects, 3, 65, 227, 231
  polymers, 223
  relative contribution of three causes, 83, 84
  six stages of, 73–75
  sliding distance (as a function of), 72
  sliding surfaces, 64
Friction space, 64, 85–87, 237
Frictional behavior of polymers, 223, 237
Frictional behavior of steel, 65–72
Fuchs, S., 110, 173, 186
Funk, 458

Galski, F. A., 457, 487
Gane, N., 41, 46
Gaul, D. J., 153, 194, 206, 222
Georges, J. M., 35, 47
Ghate, B., 366, 380
Gilchrist, A., 307, 332, 344
Gill, S., 132, 173
Gleiser, M., 374, 380, 388, 393, 394
Goddard, J., 91, 96, 271, 343
Godfrey, D., 35, 47
Goodier, J. N., 106, 113, 187, 335, 345
Goodman, L. E., 222
Graphite fiber, 249–52
Green, A. P., 9, 78, 91
Griffith's criterion for brittle fracture, 176
Grosch, K. A., 245, 246, 258

Gulden, M. E., 306, 313, 316, 333, 343, 345
Gupta, P. K., 209, 222

Hale, T. E., 458, 487
Hamilton, G. M., 222
Hansen, J. S., 317–29, 344
Hansen, M., 380, 389, 391, 393, 394
Hansen, R. H., 480, 484, 487
Hard coatings, 454
Hard particles, 272
Hardness:
  carbides, 456
  cyclic loading (due to), 41
  experimental results, 36, 37, 39, 40
  metals, 212, 216, 275, 376, 427, 450
  oxides, 273, 376, 456
  polymers, 226, 233
  surfaces, 36–41
Hardy, C., 117, 173
Haritos, G. K., 157, 174
Harker, H. J., 271, 343
Hartstein, F., 306, 346
Hartung, P., 366, 376–78, 380
Hasselman, D. P., 366, 380
Hauser, C., 458, 488
Hazra, J., 359, 361, 380
Hearle, A. D., 156, 173
Herring, C., 31, 47
Hertz, H., 110, 173, 186
Hertzian stress, 109, 186
Hildebrand, 387
Hill, R., 47, 48, 59
Hilliard, J. E., 42, 44, 46, 380, 388
Hills, D. A., 156, 173
Hinduja, S., 381, 412
Hironaka, S., 35, 48
Hirst, W., 226, 257
Hirth, J. P., 141, 173
Hisakado, T., 91, 96
Hoersch, V. A., 175, 187
Huber, M. T., 110, 173, 186, 187, 191
Hultgren, R., 374, 380, 394
Humenik, M., 364, 380
Hutchings, I. M., 310, 329, 344
Hydrostatic stress, 49

Inagawa, K., 458, 488
Ion implantation, 459
Ives, L. K., 308, 309, 344

Jacquet, M., 35, 47
Jaeger, J. C., 380, 395, 399, 400, 401, 403, 404
Jahanmir, S., 22, 35, 47, 90, 130, 131, 133, 135–40, 144, 145–52, 173, 192–94, 197–203, 221, 222, 306, 331, 344, 415, 443–49, 451, 452, 488
Jain, V. K., 230, 258
James, D. I., 231, 258
Jefferis, J. A., 130, 174, 192
Johanesson, R. T., 458, 488
Johnson, K. L., 112, 117, 130, 157, 173–74, 192
Johnson, R. W., 271, 344
Johnson, W., 47, 48
Jones, W. J. D., 271, 344
Jono, M., 166, 174

Kabil, Y., 306, 343
Kane, G. E., 366, 380
Kar, M. K., 230, 258
Kato, S., 77, 91
Kaufman, L., 380, 388
Keer, L. M., 156
Khruschov, M. M., 266–68, 344
Kieffer, R., 364, 375, 381, 458, 488
Kikukawa, M., 166, 174
Kim, C. H., 366, 380
Kirk, J. A., 36, 40, 47
Klaus, E. E., 35, 47
Kobayashi, A., 77, 91
Komanduri, R., 359, 360, 380
Komvopoulos, K., 91, 96–99, 299–305, 344, 346, 350
Kragelski, I. B., 10, 22
Kramer, B. M., 13, 22, 366–68, 370, 372, 373–77, 380, 388–92, 455, 459, 487
Kramer, E. J., 259
Kramer, I. R., 36, 45, 47
Krause, H., 161, 174
Kuczkowski, T. J., 444, 488
Kuwahara, K., 69, 91

Lamy, B., 35, 47
Larsen-Basse, J., 268, 270, 271, 344
Latanision, R., 31, 35, 37, 47
Lawn, B. R., 107, 108, 124–27, 174
Lee, M., 458, 488
Lee, Jr., R. D., 444, 487

Leung, C. L., 373, 379, 458, 487
Levy, A. V., 308, 312, 315, 343, 344
Lewis, R. B., 223, 227, 258
Limit analysis, 351
  lower bound theorem, 351
  upper bound theorem, 353
Lindstromand, J. N., 458, 488
Liou, M. J., 415, 424, 425, 430, 436, 438, 488
Liu, C. K., 113, 115, 116, 118–23, 174
Liu, T., 443, 488, 489
Loading due to impact of spheres, 334
Loladze, T. N., 366, 381
Low dislocation zone, 36
Lowe, A. C., 480, 484, 488
Lowen, E. G., 381, 395, 410, 412
Ludema, K., 83, 91, 245, 247, 258
Lundberg, G., 174, 186
Lv-limit, 223

Macks, E. F., 10, 23, 64, 91
Makinson, R. K., 230, 231
Malkin, S., 271, 343
Marui, E., 77, 91
Masumoto, H., 69, 91
Mathia, T., 35, 47
Matsubayushi, T., 77, 91
Matsushige, K., 231, 258
McClintock, F. A., 47, 48, 142, 144, 160, 166, 174
McLaren, K. G., 6, 230, 258
Mechanical properties of solid surfaces, 35, 103, 225, 233, 239
Mehrotra, P. K., 313, 338, 342, 345
Meille, G., 35, 47
Mellor, P. B., 47, 48
Melting points of metals, 375
Merchant, M. E., 64, 91, 357, 358, 360, 381
Merwin, J. E., 130, 174, 192
Merwin-Johnson method, 192
Metal cutting, 357
Metals (*See also* Friction; Wear):
  copper solutions, 210
  desirable characteristics, 210
  introduction to, 27
  sliding wear, 195
Method of characteristics, 57
Michell, J. H., 113, 174

Microstructural effects on sliding wear, 209
Mikosza, A. G., 124, 174
Miller, L. E., 231, 259
Modulated surfaces, 87, 433
Moore, A. J. W., 64, 91
Moore, M. A., 271, 344
Morehead, Jr., F. F., 460, 488
Morrow, J., 41, 47
Morton, W. B., 174, 186
Mulhearn, T. O., 271, 344
Muraka, P. D., 381, 412
Mutis, A., 271, 345

Nabarro, F. R. N., 37, 47
Nakamura, K., 458, 488
Nathan, G. K., 271, 344
Nayak, R. N., 366, 380
Nezbeda, C. W., 250, 259
Nguyen, L. T., 33, 48
Nicholas, D., 480, 484, 487
Nielson, J. H., 307, 332, 344

O'Connor, J. J., 112, 174
Oh, H. L., 105, 174, 175, 333, 337–39, 344
Oh, K. D. L., 332, 337–39, 344
Oliver, W. C., 39, 47
O'Rourke, J. T., 227, 258
Oxley, P. L. B., 78, 91, 95, 360, 381

Pacher, O., 366, 381, 458, 488
Pamies-Teixeira, J. P., 146, 198, 199, 210–19, 222
Parikh, N. M., 364, 380
Peng, Y. I., 366, 381
Peterson, T. L., 231, 259
Pethica, J. B., 39, 41, 47
Pfaffinger, K., 366, 381, 458, 488
Physical state of materials, 26, 30
Physical vapor deposition, 457
Physisorption, 31
Pinchibeck, P. H., 5, 22
Plasma treatment of polymers, 479
Plastic deformation, 128
Plowing, 10, 73, 80, 96, 196, 433
Poate, J. M., 459, 474
Polymers:
  friction of, 223, 237
  hygroscopic nature, 29
  introduction to, 28, 227
  mechanisms for friction, 237

Polymers (cont.)
  molecular structure, 228, 229
  plasma treatment, 479
  time-temperature superposition, 260
  tribological application, 30
  wear mechanisms, 255
  wear of, 223, 479
Pooley, C. M., 230, 231, 258
Prandtl-Reuss equation, 51, 131
Prevention of wear, 416, 443, 454
Properties of materials, 27
Prophet, H., 374, 381, 394
Puttick, K. E., 231, 259
Pv-limit, 223

Quinn, T. F. J., 35, 46

Rabinowicz, E., 7, 8, 16, 18, 22, 23, 64, 65, 87, 91, 207, 222, 271, 344, 459, 473, 488
Raghuram, A. C., 457, 458, 487
Raine, T., 366, 374, 380, 381, 393, 394
Ramakrishna, V., 374, 380, 388, 393, 394
Ramalingam, S., 271–73, 345, 366, 375, 381, 456
Rebinder, P. A., 35, 48
Reconstruction of surface atoms, 31
Reed, J. R., 459, 489
Reiter, N., 458, 487
Residual stress in elastoplastic solids, 133, 134, 138, 139
Response of materials to surface traction, 103
Rhee, S. S., 144, 175
Richardson, R. D. C., 267, 269, 270, 345
Richman, M. H., 458, 488
Rigney, D. A., 141, 173
Rindings, T. L., 444, 488
Ritchie, R. D., 155, 175
Rollarson, E. C., 358, 382
Rosenblatt, M., 333, 343
Rosenfield, A. R., 156, 175
Ruff, A. W., 41, 48, 308, 309, 344, 346
Ruland, W., 250, 259
Russel, P. G., 271, 345

Safranck, W. H., 449, 488
Sage, W., 306
Saka, N., 23, 91, 92, 146, 198, 199, 210–19, 222, 271, 274, 275, 277, 279–98, 299–305, 344–46, 350, 415, 424, 425, 430, 436, 438, 489
Sakamoto, T., 92, 96
Sakurai, T., 35, 48
Samuel, L. E., 271, 344
Sander, B. I., 40, 41, 48
Sargent, G. A., 313, 338, 342, 345
Savkin, V. G., 230, 259
Savitskii, K. B., 36, 40, 48
Schick, H., 374, 381
Schintlmeister, W., 366, 381, 458, 488
Schonhorn, H., 29, 48, 480, 484, 487, 489
Schroeder, T., 359, 360, 380
Schroter, K., 364, 381
Schwarzkopf, P., 364, 375, 381, 458, 488
Sclar, C. R., 368, 380
Senior, J. M., 231, 258
Shaw, M. C., 10, 22, 23, 64, 91, 360, 381, 395, 410, 412
Sheldon, G. L., 306, 332, 345
Shepard, S., 415, 459, 462–78
Shirakashi, T., 360, 382
Shunk, F., 381, 389, 391, 393–95
Sin, H.-C., 17, 23, 65, 80, 92, 96, 130, 136, 140, 141, 157, 158–61, 162–70, 174, 200, 208, 271, 274, 275, 277, 279–98, 345
Singer, I. L., 459, 473, 487, 489
Sliding wear:
  of composites, 223
  of metals, 195
  of polymers, 223
  of tool materials, 356
Slip-line field method, 59, 92–96, 346
Smeltzer, C. E., 306, 345
Smith, J. O., 113, 175
Smith, L. S. A., 231, 259
Smith, W. C., 366, 380
Smurigov, V. A., 230, 259
Soft metal coatings, 443
Solomon, A. J., 444, 489
Solubilities of tool materials, 373, 374
Solution wear, 367, 375
Somerjai, G. A., 31, 48
Soule, D. E., 250, 259

Index

Soven, P., 31, 48
Sproles, E. S., 153, 175, 206, 222
Steffens, D. A., 250, 259
Steijn, R. P., 225, 230, 243, 259
Stevenson, M. G., 381, 396
Stowers, I. F., 14, 23
Strain gradient at the surface, 143
Strain invariants, 50
Stress:
  deviator, 50
  hydrostatic, 49
  invariants, 49
Stress distribution:
  at the crack tip, 160, 180
  in elastoplastic solids, 130
  in perfectly elastic solid, 105
Stress intensity factor, 153–56, 179
Stull, D., 374, 381, 394
Su, K. Y., 147, 153, 175, 415, 459, 489
Subsurface plastic deformation, 149–51, 192
Suh, N. P., 13, 21–23, 33, 48, 65, 69, 70, 74, 85, 87, 90–92, 130, 131, 133, 135–57, 161, 162–67, 173, 175–86, 192–94, 197–203, 210–19, 221, 222, 233, 240, 241, 248, 256, 259, 260, 271, 274, 275, 277–306, 310, 314, 344–46, 350, 351, 357, 364, 366–68, 370, 373, 374, 379–81, 388–92, 415, 416, 424, 425, 430, 436, 438, 443–49, 451, 452, 458, 459, 462–78, 481, 485, 487–89
Sung, N.-H., 233, 240, 241, 259
Surface ridge formation, 129
Surfaces:
  adsorption, 31
  characteristics, 30
  chemisorption, 31
  deformation of, 52, 105–39
  double layer, 31
  hardness, 36–41
  mechanical properties, 35
  physisorption, 31
  reconstruction, 31
  residual stress, 133, 134, 138, 139
  stress distribution, 105
  surface ridges, 129
Surface roughness, effect on wear, 209

Svivodyonok, A. I., 230, 259
Swain, M. V., 107, 108, 171, 174, 175, 374, 381
Swanson, T. D., 36, 40, 47
Swedlow, J. L., 159, 175

Tabor, D., 6, 22, 41, 47, 64, 91, 230, 231, 245, 247, 258
Takagi, R., 443, 444, 488, 489
Tanaka, K., 230, 231, 259
Tanouye, P. A., 270, 271
Tay, A. O., 381, 396
Taylor, F. W., 364, 382
Temperature distribution at the sliding interface, 395
  cutting tool interface, 363, 407
  energy partition, 396
  governing equations, 395
  moving heat source, 396
  multiple asperity contacts, 404
  single asperity contait, 397
Test apparatus and methods (*See* Experimental techniques)
Tewksbury, E. J., 35, 47
Thermal expansion, 458
Thermodynamics:
  interface, 41–45
  introduction to, 382
  properties, 388
Thomas, H. R., 175, 187
Tilly, G. P., 306, 345
Time-temperature superposition, 260
Timoshenko, S. P., 106, 113, 175, 187, 335, 345
Tipnis, V., 360, 362, 382
Tohkai, M., 92, 207, 222
Tokarsky, E., 251, 252, 258
Tomaru, M., 35, 48
Tool materials, 364
Tool wear, 364
Tool wear mechanisms, 365
Tordion, G. V., 117, 173
Toth, L. E., 456, 458, 489
Tresca yield criterion, 50
Tribology:
  choice of materials, 210
  definition, 1
  metals, 195
  novel methods, 413
  polymers, 223
  phenomenological observations, 4

Tribology (cont.)
  surface topography, 19
  three fundamental aspects, 1
Tsuya, Y., 443, 489
Turley, D. M., 357, 359, 360, 382
Turner, A. P. L., 48, 176–86, 201, 222, 260, 345, 351
Tzukizoe, T., 92, 96

Uchiyama, Y., 230, 231, 259
Usui, E., 360, 382

Vaidyanathan, S., 333, 337–39, 345
Van Den Boogaart, A., 231, 259
Void nucleation, 141
von Mises flow rule, 51, 131
von Turkovich, B., 359, 360, 380

Wagman, et al., 382, 390
Wahl, H., 306, 346
Warren, J. H., 231, 259
Watson, J. D., 271–73, 345, 366, 375, 381, 456
Wear (*See also* specific wear topics, e.g., Erosive wear)
  abrasive wear, 264, 295
  carbide, wear rate of, 15, 373
  chemical stability (due to), 356
  classification, 12
  coefficient, definition of, 13, 205
  coefficients (Experimentally measured), 14, 15, 18, 68, 205, 208, 212, 217, 218, 225–27, 234, 240–43, 248, 268, 271, 274, 279–82, 302, 307, 373, 447, 448, 450, 455, 483, 485
  composites, wear of, 223, 232, 248, 416
  cutting tool, 357, 376
  delamination theory of wear, 74, 129, 196, 199, 209
  diffusion wear, 366, 376
  early theories, 11
  effect:
    of cold work, 18
    of ductility on abrasion, 17
    of ion implantation, 459
    of microstructure, 209
    of plasma treatment, 479
    of soft coating, 443
    of surface roughness, 20
  erosive wear, 264, 305
  factor, 224
  fretting, 206
  hard particles, due to, 264
  mechanisms:
    of abrasive wear, 282
    of erosive wear, 319, 332
    of sliding wear, 196, 473
    of tool wear, 365
  models, 200, 248, 255, 283, 318, 332, 367
  polymers, 223, 255, 479
  prediction of wear coefficients and rates, 208, 248, 302, 373
  sliding wear of metals, 195
  sliding wear of polymers, 223
  solution wear, 367
West, C. H., 231, 259
Westbrook, J. H., 273, 346, 382, 456, 489
Wheeler, D. R., 161, 175
Whitehouse, D. J., 19, 23, 209, 222
Williams, J. E., 358, 382
Williams, W. S., 250, 259
Wilman, H., 91, 96, 271, 343
Wilshaw, T. R., 125–28, 173, 333, 334, 336–38, 340, 343
Winter, R. E., 330, 344
Wolak, J., 306, 343
Wood, C. D., 346
Woodward, A. J., 112, 174

Yield criteria:
  Tresca, 50
  von Mises, 51
Youn, J. R., 256, 259, 415, 480, 481–85
Young, J. P., 308, 346

Zaat, J. H., 38, 46, 52
Zisman, W. H., 231, 258

FEB

HETERICK MEMORIAL LIBRARY
621.89 S947t   onuu
Suh, Nam P./Tribophysics

3 5111 00148 5022